Gene Regulation and Epigenetics

Carsten Carlberg

Gene Regulation and Epigenetics

How Science Works

 Springer

Carsten Carlberg
Institute of Animal Reproduction and Food
Research
Polish Academy of Sciences in Olsztyn
Olsztyn, Poland

Institute of Biomedicine
School of Medicine
University of Eastern Finland
Kuopio, Finland

ISBN 978-3-031-68729-7   ISBN 978-3-031-68730-3   (eBook)
https://doi.org/10.1007/978-3-031-68730-3

This Springer imprint is published by the registered company Springer Nature Switzerland AG
The registered company address is: Gewerbestrasse 11, 6330 Cham, Switzerland

If disposing of this product, please recycle the paper.

Preface

This book describes the fascinating area of eukaryotic gene regulation and epigenetics. Specific expression of genes is shaping the phenotype of our cells and tissues. Accordingly, the regulation of gene expression, *i.e.*, their up- and downregulation, an essential fundamental aspect of nearly all processes in physiology, both in health and in disease. These processes are in part very dynamic and respond to divergent daily challenges, such as incoming diet or infections. Therefore, not only biologists and biochemists should be aware of this topic, but **all students of biomedical disciplines will benefit from being introduced into the concepts of gene regulation**. This will provide them with a good basis for their specialized disciplines.

Epigenetics describes the packaging and accessibility of the genome that we carry in each of the trillions of cells forming our body. The prefix "epi" (meaning "upon", "above", or "beyond") indicates that, in contrast to genetics, epigenetic processes do not affect the DNA sequence of the human genome. This means that there exists a layer of information beyond that encoded in the human genome. Genomic DNA is wrapped around complexes of histone proteins that help to fit the genome into a cell nucleus with a diameter of less than 10 μm. This protein-DNA complex is referred to as **chromatin**.

The differentiation of embryonic stem (ES) cells into specialized cell types happens during embryogenesis within the first weeks of the life of a fetus. Moreover, in every moment of an adult's life stem cells in the bone marrow, the skin and the intestine are growing and differentiating into specialized cells that replace cell loss, such as of our immune system or of our body's outer and inner surface. The underlying mechanism of all these differentiation processes is epigenetic programming of chromatin structure, *i.e.,* a change of the so-called **epigenetic landscape**. Thus, the most important function of chromatin is to keep in a cell- and tissue-specific manner some 90% of the human genome inaccessible to transcription factors and polymerases. In other words, chromatin acts as a gatekeeper for undesired gene activation.

Each of the 400 tissues and cell types forming our body uses a different subset of the 20,000 protein-coding genes of the human genome. Accordingly, **each of us has**

only one genome but at least 400 different epigenomes. Epigenetics provides a molecular explanation of how the human genome is connected with environmental signals. It associates with our lifestyle and environment during intra-uterine as well as post-natal conditions. The dynamic component of epigenetics implies that **some epigenetic programming events are reversible**. For example, if an unhealthy lifestyle paired with food excess and physical inactivity over many years causes epigenetic programming of metabolic tissues that results in insulin resistance, this process may be reversed by significant lifestyle changes leading to a reprogramming of the dynamic component of the epigenome. As long as no irreversible tissue damage has happened, this reprogramming can have significant consequences for our health, *i.e.*, to a certain extent **we may have it in our own hands to reverse a disease condition**. Thus, not only our mental memory, such as memories of our childhood, is a learning process that is based on epigenetic programming of neurons, but each cell of our body has an epigenetic memory recording the perturbations that the cell had been exposed to.

The activation of intracellular signal transduction cascades *via* extracellular signals, such as peptide hormones, cytokines, or growth factors, results in the activation of transcription factors and chromatin-modifying enzymes. The actions of these nuclear proteins cause local changes in the epigenome, which enable and modulate the transcription of specific target genes of the different signals affecting a cell. There are 1600 human genes encoding for transcription factors, the most important of which we will discuss in this book, such as **p53** in the context of cellular stress, **NFκB** together with inflammatory response, **OCT4** at the example of cellular differentiation or **nuclear receptors** in sensing steroid hormones as well as macro- and micronutrients. Thus, a good understanding of transcription factors and the processes that alter their activity is a fundamental aspect of modern life science research.

The content of the book is linked to the course in "Molecular Medicine and Genetics", which I lecture in different forms since 2002 at the University of Eastern Finland in Kuopio. The book represents an updated version of the textbooks *Mechanisms of Gene Regulation* (ISBN 978-94-017-7741-4) and *Human Epigenomics* (ISBN 978-981-10-7614-8). I thank my coauthor on both books, Prof. Ferdinand Molnár, for allowing me to use figures, which he created, as the basis for the figures of this book. The present book is subdivided into 16 chapters: following two introductory chapters, four chapters take a view on gene regulation from the perspective of transcription factors, and four further look on it from the angle of chromatin and non-coding RNA (ncRNA). Then three chapters explain the impact of epigenetics from the standpoint of health and the three last chapters from the view of diseases. A glossary in the appendix will explain the major specialist's terms.

I hope the readers will enjoy this rather visual book and get as enthusiastic about gene regulation and epigenetics as the author is since many decades.

Olsztyn, Poland/Kuopio, Finland   Carsten Carlberg
June 2024

Contents

Abbreviations

1,25(OH)$_2$D$_3$	1,25-dihydroxyvitamin D$_3$
25(OH)D$_3$	25-hydroxyvitamin D$_3$
3C	Chromosome conformation capture
3D	3-dimensional
4C	Circularized chromosome conformation capture
5C	Chromosome conformation capture carbon copy
5caC	5-carboxylcytosine
5fC	5-formylcytosine
5hmC	5-hydroxymethylcytosine
5hmU	5-hydroxyuracil
5mC	5-methylcytosine
A	Adenine
ABC	ATP-binding cassette
ABL1	ABL proto-oncogene 1, non-receptor tyrosine kinase
ACTL6A	Actin like 6A
ACTR	Actin-related protein
AGO2	Argonaute RISC catalytic component 2
ALL	Acute lymphoblastic leukemia
AML	Acute myeloid leukemia
AMPK	Adenosine monophosphate-activated protein kinase
AP1	Activator protein 1 (JUN-FOS heterodimer)
APO	Apolipoprotein
AR	Androgen receptor
ARID	AT-rich interaction domain
ARL4C	ADP-ribosylation factor-like
ASH1L	ASH1 like histone lysine methyltransferase
ASIP	Agouti-signaling protein
ATAC-seq	Assay for transposase-accessible chromatin using sequencing
ATF	Activating transcription factor
ATM	ATM serine/threonine kinase
ATP	Adenosine triphosphate

ATR	ATR serine/threonine kinase
A^{vy}	Agouti viable yellow
BCL3	BCL3 transcription coactivator
BDNF	Brain-derived neurotrophic factor
BER	Base excision repair
BET	Bromodomain and extraterminal
BHLH	Basic helix-loop-helix
BMAL1	Basic helix-loop-helix ARNT like 1
BMI1	BMI1 proto-oncogene, Polycomb ring finger
BMP	Bone morphogenetic protein
BPTF	Bromodomain PHD finger transcription factor
BRD	Bromodomain containing
C	Cytosine
CAGE	Cap analysis of gene expression
cAMP	Cyclic adenosine monophosphate
CAMP	Cathelicidin
CAR	Constitutive androstane receptor, also called NR1I3
Cas	CRISPR-associated
CBFB	Corebinding factor subunit β
CBX	Chromobox
CCL	Chemokine C-C motif ligand
CCN	Cyclin
CCR	C-C chemokine receptor
CD	Cluster of differentiation
CD40LG	CD40 ligand
CDK	Cyclin-dependent kinase
CDKN	Cyclin-dependent kinase inhibitor
CEBP	CCAAT-enhancer binding protein
CETP	Cholesteryl ester transfer protein
CHD	Chromodomain helicase DNA-binding
ChIP	Chromatin immunoprecipitation
ChIP-seq	Chromatin immunoprecipitation sequencing
CIMP	CpG island methylator phenotype
CLOCK	Clock circadian regulator
CLP	Common lymphoid progenitor
CNS	Central nervous system
CNV	Copy number variation
CpG	CpG dinucleotide
CREB1	cAMP-response element binding protein
CREBBP	CREB-binding protein, also called KAT3A
CRISPR	Clustered regularly interspaced short palindromic repeats
CRY1	Cryptochrome circadian clock 1
CSNK2A1	Casein kinase 2α 1
CTCF	CCCTC binding factor
CTLA4	Cytotoxic T-lymphocyte associated protein 4

CTNNB1	Catenin β1
CVD	Cardiovascular disease
CXCL	Chemokine C-X-C motif ligand
CXCR	C-X-C motif chemokine receptor
CXXC1	CXXC finger protein 1
CYP	Cytochrome P450
Da	Dalton
DACH1	Dachshund family transcription factor 1
DAG	Diacylglycerol
DAMP	Damage-associated molecular pattern
DANT1	DXZ4 associated non-coding transcript 1, proximal
DBD	DNA-binding domain
DC	Dendritic cell
DDR	DNA damage response
DEF	Defensin
DGCR8	DiGeorge syndrome critical region gene 8
DNase-seq	DNase I hypersensitivity followed by sequencing
DNMT	DNA methyltransferase
DOHaD	Developmental origins of health and disease
DOT1L	DOT1 like histone lysine methyltransferase
DPE	Downstream promoter element
DR	Direct repeat
DVL	Dishevelled segment polarity protein
EED	Embryonic ectoderm development
EGF	Epidermal growth factor
EGFR	Epidermal growth factor receptor
EHMT	Euchromatic histone-lysine N-methyltransferase
ELK1	ETS transcription factor ELK1
EMT	Epithelial-mesenchymal transition
ENCODE	Encyclopedia of DNA elements
EP300	E1A-binding protein p300, also called KAT3B
ER	Endoplasmic reticulum
ERCC	ERCC excision repair, TFIIH core complex helicase subunit
eRNA	Enhancer RNA
ES	Embryonic stem
EZH	Enhancer of zeste homolog
FAD	Flavin adenine dinucleotide
FAIRE	Formaldehyde-assisted isolation of regulatory elements
FANTOM	Functional annotation of the mammalian genome
FASN	Fatty acid synthase
FGF	Fibroblast growth factor
FGFR4	FGF receptor 4
FISH	Fluorescence in situ hybridization
FMR1	Fragile X mental retardation 1
FOS	FOS proto-oncogene, AP1 transcription factor subunit

FOX	Forkhead box
FTO	Fat mass and obesity associated
FXN	Frataxin
FXR	Farnesoid X receptor, also called NR1H5
G	Guanine
G6PC	Glucose-6-phosphatase
GABA	Gamma-aminobutyric acid
GAS5	Growth arrest-specific 5
GATA	GATA-binding protein
GH1	Growth hormone 1
GLI	GLI family zinc finger
GLUT	Glucose transporter
GMP	Granulocyte-monocyte progenitor
GPR	G protein-coupled receptor
GR	Glucocorticoid receptor
GSK3	Glycogen synthase kinase 3
GTEx	Genotype tissue expression
GTF	General transcription factor
GWAS	Genome-wide association study
HAT	Histone acetyltransferase
HDAC	Histone deacetylase
HDL	High-density lipoprotein
Hi-C	High-throughput chromosome capture
HIV	Human immunodeficiency virus
HMGA1	High mobility group AT-hook 1
HNF	Hepatocyte nuclear factor
HNRNPU	Heterogeneous nuclear ribonucleoprotein U
HOTAIR	HOX transcript antisense RNA
HOTTIP	HOXA transcript at the distal tip
HOX	Homeobox
HP1	Heterochromatin protein 1, official name CBX5
HSC	Hematopoietic stem cell
HSF1	Heat shock transcription factor 1
HSP	Heat shock protein
HTT	Huntingtin
IAP	Intracisternal A particle
ICR	Imprinting control region
IDH	Isocitrate dehydrogenase
IDOL	Inducible degrader of LDLR
IGF	Insulin-like growth factor
IHEC	International human epigenome consortium
IKBKG	Inhibitor of NFκB kinase regulatory subunit gamma, also called NEMO
IKK	IκB kinase
IL	Interleukin

ILC	Innate lymphoid cell
indel	Insertion-deletion
INF	Interferon
INO80	INO80 complex subunit
Inr	Initiator
iPS	Induced pluripotent stem
IRF	Interferon regulatory factor
IRS	Insulin receptor substrate
ISWI	Imitation SWI
JAK	Janus kinase
JUN	JUN proto-oncogene, AP1 transcription factor subunit
KAT	Lysine acetyltransferase
K^{ATP}	ATP-sensitive K^+
kb	Kilobase pairs (1000 base pairs)
KCNQ1	Potassium voltage-gated channel subfamily Q, member 1
kD	Kilo Dalton
KDM	Lysine demethylase
KIF11	Kinesin family member 11
KLF	Krüppel-like factor
KLRK1	Killer cell lectin like receptor K1
KMT	Lysine methyltransferase
KRAB	Krüppel associated box
LAD	Lamin-associated domain
LBD	Ligand-binding domain
LBR	Lamin B receptor
LCK	LCK proto-oncogene, SRC family tyrosine kinase
LDL	Low-density lipoprotein
LDLR	LDL receptor
LIN28A	LIN-28 homolog A
LINE	Long interspersed element
LOCK	Large organized chromatin K9-modification
LPCAT3	Lysophospholipid acyltransferase 3
LPL	Lipoprotein lipase
LPS	Lipopolysaccharide
LRH-1	Liver receptor homolog-1, also called NR5A2
LSD1	Lysine-specific demethylase 1, also called KDM1A
LT	Lymphotoxin
LTA4H	Leukotriene A4 hydrolase
LXR	Liver X receptor, also called NR1H3 and NR1H2
MAF	Minor allele frequency
MAGEL2	MAGE family member L2
MALAT1	Metastasis-associated lung adenocarcinoma transcript 1
MAPK	Mitogen-activated protein kinase
Mb	Mega base pairs (1,000,000 base pairs)
MBD	Methyl-DNA-binding domain

MC4R	Melanocortin 4 receptor
mCH	Non-CpG methylation
MDM2	MDM2 proto-oncogene, E3 ubiquitin protein ligase
MECP2	Methyl-CpG-binding protein 2
MED	Mediator
MeDIP-seq	Methylated DNA immunoprecipitation sequencing
MEIS1	Meis homeobox 1
MEN1	Menin 1
MEP	Megakaryocyte-erythrocyte progenitor
MGMT	O-6-methylguanine-DNA methyltransferase
MHC	Major histocompatibility complex
MIC	MHC class I polypeptide-related sequence
miRNA	MicroRNA
MLH1	MutL homolog 1
MMR	Mismatch repair
MNAT1	MNAT1 component of CDK activating kinase
MPP	Multipotent progenitor
MR	Mineralocorticoid receptor
mRNA	Messenger RNA
MS	Multiple sclerosis
MTHFR	Methylenetetrahydrofolate reductase
MYC	MYC proto-oncogene, BHLH transcription factor
MYOD1	Myoblast determination protein 1
NAD	Nicotinamide adenine dinucleotide
NAFLD	Non-alcoholic fatty liver disease
NAMPT	Nicotinamide mononucleotide phosphoribosyltransferase, also called visfatin
NANOG	Nanog homeobox
NCOA	Nuclear receptor coactivator
NCOR	Nuclear receptor corepressor
ncRNA	Non-coding RNA
NER	Nucleotide excision repair
NFAT	Nuclear factor of activated T cells
NFκB	Nuclear factor kappa B
NGS	Next-generation sequencing
NICD	NOTCH intracellular domain
NIK	NFκB inducing kinase
NK	Natural killer
NLS	Nuclear localization sequence
NOR1	Neuron-derived orphan receptor 1, also called NR4A3
NOTCH	NOTCH receptor
OCT4	Octamer-binding transcription factor 4, encoded by the *POU5F1* gene
PABPC1	Poly(A)-binding protein cytoplasmic 1
PAMP	Pathogen-associated molecular pattern

PARP1	Poly(ADP-ribose) polymerase 1
PBMC	Peripheral blood mononuclear cell
PCK	Phosphoenolpyruvate carboxykinase
PCR	Polymerase chain reaction
PDCD1	Programmed cell death 1, also called PD1
PDGFRA	Platelet-derived growth factor receptor α
PER1	Period circadian clock 1
PGC	Primordial germ cell
PGR	Progesterone receptor
PHD	Plant homeodomain
PLTP	Phospholipid transfer protein
Pol II	RNA polymerase II
PPAR	Peroxisome proliferator-activated receptor
PPARGC1	PPARγ, coactivator 1
PRC	Polycomb repressive complex
pre-miRNA	Precursor miRNA
pri-miRNA	Primary miRNA
PRK	Protein kinase
PRKDC	Protein kinase, DNA-activated, catalytic subunit
PRR	Pattern-recognition receptor
PTCH	Patched receptor
P-TEFb	Positive transcription elongation factor
PUFA	Polyunsaturated fatty acid
PXR	Pregnane X receptor, also called NR1I2
RA	Retinoic acid
RAR	Retinoic acid receptor
RB1	RB transcriptional corepressor 1
RBBP4	RB-binding protein 4, chromatin remodeling factor
RBPJ	Recombination signal-binding protein for immunoglobulin kappa J
RCOR	REST corepressor
RE	Response element
REL	REL proto-oncogene, NFκB subunit
REST	RE1-silencing transcription factor
REV-ERB	Reverse-Erb, also called NR1D1
RISC	RNA-induced silencing complex
RNAi	RNA interference
RNA-seq	RNA sequencing
ROR	RAR-related orphan receptor
rRNA	Ribosomal RNA
RTK	Receptor tyrosine kinase
RUNX1	Runt-related transcription factor 1
RUVBL	RuvB like AAA ATPase
RXR	Retinoid X receptor
SAH	S-adenosylhomocysteine
SAM	S-adenosyl-L-methionine

SCN	Suprachiasmatic nucleus
SETD2	SET domain containing 2
SETDB1	SET domain bifurcated histone lysine methyltransferase 1
SF1	Steroidogenic factor 1
SHARP	SMRT/HDAC1-associated repressor protein
SIN3A	SIN3 transcription regulator family member A
SINE	Short interspersed element
siRNA	Small interfering RNA
SIRT	Sirtuin
SLC	Solute carrier family
SLCO	Organic anion transporter
SMAD	SMA- and MAD-related protein
SMARC	SWI/SNF-related matrix-associated actin-dependent regulators of chromatin
snoRNA	Small nucleolar RNA
SNP	Single nucleotide polymorphism
snRNA	Small nuclear RNA
SNV	Single nucleotide variant
SOX2	SRY-box 2
SP1	Specificity protein 1
SPI1	Spleen focus forming virus proviral integration oncogene, also called PU.1
SRC	SRC proto-oncogene, non-receptor tyrosine kinase
SREBF1	Sterol regulatory element-binding transcription factor 1
SRF	Serum response factor
STAT	Signal transducer and activator of transcription
SULT2A1	Sulfotransferase family 2A, member 1
SUV39H	Suppressor of variegation 3-9 homolog
SWI/SNF	Switching/sucrose non-fermenting
T	Thymine
T1D	Type 1 diabetes
T2D	Type 2 diabetes
T_3	Triiodothyronine
TAD	Topologically associated domain
TAF	TBP-associated factor
TAL1	TAL BHLH transcription factor 1, erythroid differentiation factor
TBL1X	Transducin β-like 1 X-linked
TBL1XR1	TBL1X receptor 1
TBP	TATA box-binding protein
TCF7L2	Transcription factor 7-like 2
TCGA	The Cancer Genome Atlas
TCR	T cell receptor
TDG	Thymine-DNA glycosylase
TERC	Telomerase RNA component
TET	Ten-eleven translocation

TF	Transcription factor
TFBS	Transcription factor binding site
TGFβ	Transforming growth factor β
T_H	T helper
THR	Thyroid hormone receptor
THRSP	Thyroid hormone responsive
TLF	TBP-like factor
TLR	Toll-like receptor
TNF	Tumor necrosis factor
TNFSF	TNF superfamily member
TOR(C)	Target of rapamycin (complex)
TP53	Tumor protein p53
TRBP	Transactivation-response RNA-binding protein
T_{REG}	Regulatory T
TRIM	Tripartite motif-containing
tRNA	Transfer RNA
TSS	Transcription start site
U	Uracil
UBE3A	Ubiquitin protein ligase E3A
UGT2B4	UDP glucuronosyltransferase 2 family, polypeptide B4
UHRF1	Ubiquitin-like plant homeodomain and RING finger domain 1
UTR	Untranslated region
UV	Ultraviolet
VDR	Vitamin D receptor
VLDL	Very low-density lipoprotein
VNN	Vanin 1
WNT	Wingless-type MMTV integration site family member
XCI	X chromosome inactivation
Xist	X-inactive specific transcript
YY1	YY1 transcription factor
ZBTB33	Zinc finger and BTB domain containing 33
ZFP	Zinc finger protein
β-OHB	β-hydroxybutyrate

Chapter 1
The Human Genome and Its Variations

Abstract In this chapter, we will briefly describe the genetic adaption of anatomically modern humans due to migration from Africa to new geographic and climatic environments in Asia and Europe. This led not only to obvious differences in skin color and other andromorphic features among the populations, but also in different resistance to diseases and diversity in dietary intake. The genetic basis of the variation of human populations and individuals has recently been studied and catalogued by large consortia, such as the *1000 Genomes* project. Genome-wide genotyping and whole genome sequencing allows the study and analysis of complex diseases, such as T2D (type 2 diabetes) and cardiovascular disease (CVD), on the basis of dozens to hundreds of genetic variants, such as single nucleotide variants (SNVs) and copy number variations (CNVs).

Keywords *Homo sapiens* · Evolution · Human migrations · Human populations · SNVs · CNVs · Haplotype blocks · *1000 Genomes* project

1.1 Diversity of Human Populations

Approximately 200–300,000 years ago anatomically modern humans developed in East Africa. **The main characteristic of *Homo sapiens* is a superior locomotive ability that is essential for encountering predators and food procurement**. Some 50–75,000 years ago, reasonable numbers of these modern humans started to migrate to Asia and Europe and replaced there through interbreeding already prevalent archaic (i.e., nowadays extinct) human species, such as the Neanderthals (Fig. 1.1). Due to their new environments our ancestors were exposed to a number of divergent selective pressures, such as thermoregulation in colder climates, tolerance to hypoxia (i.e., oxygen supply deprivation) at high altitude and light skin pigmentation in regions with lower levels of ultraviolet (UV)-B radiation.

After spreading to different continents, many human populations became geographically isolated, so that new gene variations could not be spread anymore to all members of our species. Since in the past distant human populations were less likely

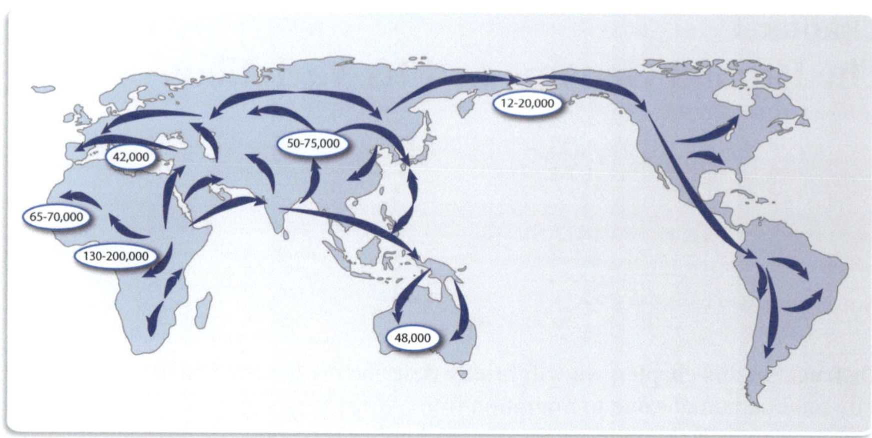

Fig. 1.1 **Migrations of *Homo sapiens*.** The spread of anatomically modern humans from East Africa over the rest of the continent was followed by an expansion from the same area to Asia, probably by both a southern and northern route some 50–75,000 years ago. Oceania, Europe and the Americas were settled from Asia in that order. The migration patterns are primarily based on analyses of changes in mitochondrial DNA

to exchange migrants, they cluster genetically in relation to their geographic distance from each other. Thus, human genetic variation diverted geographically, when individuals accumulated further mutations during the past 50,000 years. Since anatomically modern humans lived in Africa already for some 300,000 years, populations on this continent are more diverse than in the rest of the world.

There are obvious phenotypic differences of human populations concerning skin color, body height and facial features, but there are no absolute genetic differences between them. For example, there is no single nucleotide difference that can distinguish Africans from Eurasians. In contrast, population differences are based on thousands of gene loci. This implies that a property (often referred to as a "trait"), such as skin color, can change rather rapidly, when the allele (gene variant) frequencies shift at the respective loci contributing to the trait (Box 1.1). Some 500 years ago, when navigation over the oceans became possible, voluntary and involuntary (slaves) migration started, which caused significant population admixtures in particular in the Americas but also in other continents. Before that time there had been at least two major events of admixture in Europe: some 9000 years ago early farmers from Anatolia interfered with indigenous European hunters and gatherers and then some 5000 years ago Yamnaya pastoralists from the Eurasian steppe migrated to Europe. Thus, **the phenotype of present-day Europeans as well as European decent individuals is largely the product of this Bronze Age collision of these three ancestral tribes**.

Box 1.1: Natural Selection in Evolution Positive natural selection, i.e., the force that drives the increase in prevalence of advantageous traits, has played a central role in human evolution. Individuals with advantages (referred to as "adaptive traits") tend to be more successful in reproduction, i.e., they contribute with more offspring to the following generation than others. Due to inheritance from one generation to the other, the selection process increases the prevalence of the respective adaptive trait. Under persistent selection pressure such adaptive traits, step by step, may become universal to the population. **Factors fostering selection (i.e., evolutionary pressures) include limits on resources, such as food, or threats, such as pathogens**. However, adaptive traits not always become prevalent within a population. Gene frequency alteration in a population can also happen via a genetic drift of genes that are not under selection (Sect. 12.1). In this context, even a deleterious allele may become universal to the members of a population, e.g., under the influence of a weak selection or in small populations.

Hundreds of complex phenotypic traits determine how an individual looks and behaves as well as his/her risks to develop non-communicable diseases. Furthermore, each complex trait is based on dozens to hundreds of gene variants and environmental influences, i.e., for understanding the molecular basis of a trait its genetic architecture needs to be uncovered, which is based on (i) the number of variants that influence a heritable phenotype, (ii) their relative magnitude concerning different traits, (iii) the population frequency of the respective variants and (iv) their interactions with each other and the environment.

The genome of today's human populations were investigated by whole genome sequencing and it has been found that every individual carries, on average, 5 million SNVs, 600,000 indels (insertion-deletions) below 50 base pairs in length, 15,000 larger insertions, 10,000 larger deletions and 140 large inversion (Sect. 1.3) covering in total about 36 million base pairs (Mb) of sequence (1.2% of the haploid genome). Most of these genetic variants are neutral, i.e., they do not contribute to phenotypic differences or disease risk and achieved simply by chance significant frequencies within respective populations.

During evolution of our species up to 10% of all protein-coding genes, i.e., some 2000 genes, may have been affected by positive natural selection. In particular, the immune system, the digestive tract and the skin (including hair, sweat glands and sensory organs) had been susceptible to positive selection. This is due to the fact that these organ systems are more likely in contact with the environment than other parts of our body. For example, variants of genes with key importance in the innate and adaptive immune system (Chap. 13), such as those encoding for membrane immune receptors, had been under special positive selection by pathogenic microbes. An interesting example is the CCR5 (C–C chemokine receptor 5) receptor on T cells, which is essential for the entry of the human immunodeficiency virus (HIV) 1 specifically to this cell type. Importantly, a 32 base pair deletion in the *CCR5* gene, which significantly affects the functionality of the encoded protein, protects its carriers from HIV-1 infection, i.e., from virus entry into their T cells. This mutation is currently under **positive selection** in populations, such as in South Africa, where HIV-1 infections occur on larger scale. Other examples of positive selections are alleles that were introduced into modern humans already longer times ago through interbreeding with archaic human species. A variant of the *SLC16A11* (solute carrier family 16 member 11) gene, which encodes for a lipid transporter in the endoplasmatic reticulum (ER), reached high frequencies, e.g., in native Americans. Initially, this allele allowed better survival at conditions of low food availability but nowadays, at conditions of high fat Western diet (Box 1.2), it is associated with increased T2D risk.

> **Box 1.2: Western Diet** A dietary pattern that characterized by high intakes of pre-packaged foods, refined grains, red meat, processed meat, high-sugar drinks, sweets, fried foods, butter and other high-fat dairy products is referred to as "Western diet". It was distributed by US American fast-food and supermarket chains to Europe and has arrived in nearly all countries and human populations. It also involves the higher consumption of eggs, potatoes, corn (and high-fructose corn syrup) and low intakes of fruits, vegetables, whole grains, pasture-raised animal products, fish, nuts and seeds. In contrast to Mediterranean and Nordic diet, Western diet significantly increases the risk of obesity, T2D and CVD.

1.2   Genetic Variants of the Human Genome

The reference haploid sequence of the human genome (Box 1.3) was released in 2001 by the first "Big Biology" project (Sect. 6.1), the *Human Genome* project. Interestingly, it took another 20 years to fill all gaps in the sequence from telomere to telomere. In contrast, the annotation of the human genome is still not finished, i.e., there are still protein-coding and in particular non-coding genes, for which we do not know the function. A prime focus of today's genetics is the characterization of the associations between genetic variation and phenotype on the genome-wide level. Moreover, the causal mechanisms by which genetic variation influences human biology are intensively investigated.

The variation of the genome from individual to individual is a challenge. These variations are primarily SNVs, which are variants of the sequence of the reference genome where exactly one nucleotide (A, T, G or C) is altered (Fig. 1.2). In contrast, structural variants of the genome mostly affect more than one nucleotide. These can be indels, where in most cases only a few bases are added or removed, respectively. Indels that are not multiples of 3 base pairs in length and are located within protein-coding regions result in frameshift mutations, i.e., from the position of the mutation onwards the whole amino acid sequence of the encoded protein is changed. Furthermore, CNVs consist of deletions or insertions of DNA stretches in one genome compared to another. These variants can be heterozygous or homozygous. A predominant class of insertions is that of ancient transposons. These DNA stretches persist in the genome as short interspersed elements (SINEs, e.g., *Alu* elements) and long interspersed nuclear elements (LINEs).

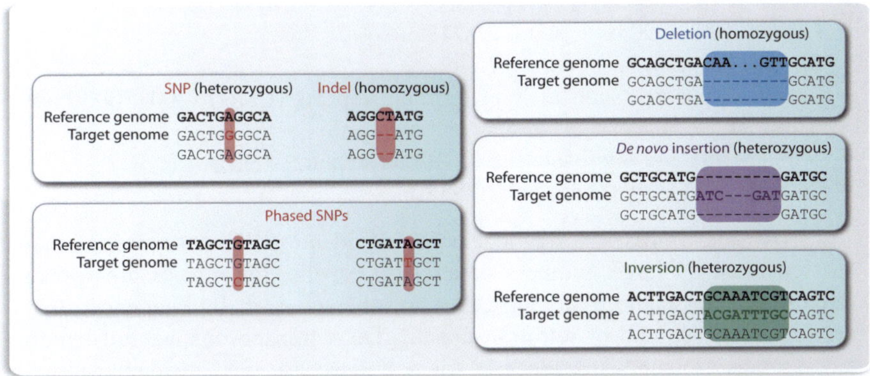

Fig. 1.2 Types of variations present in human genome sequences. The haploid reference genome is indicated at the top of each variant example, while the individual's diploid genome is shown below. The genetic variants can be either heterozygous or homozygous. The process of statistical estimation of haplotypes is referred to as "phasing", i.e., in phased SNVs their order within a haplotype is determined

Box 1.3: The Human Genome The human genome is the complete sequence of the anatomically modern human and was obtained by the *Human Genome* project (www.genome.gov/10001772) via whole genome sequencing. This reference sequence represents the assembly of the genomes of a few young healthy male donors. With the exception of germ cells, i.e., female oocytes and male sperm, each human cell contains a diploid genome formed by 2 × 3.05 billion base pairs (Gb) that is distributed on 2 × 22 autosomal chromosomes and two X chromosomes for females and a XY chromosome set for males. In addition, every mitochondrion contains 16.6 kb (kilo base pairs) mitochondrial DNA. The haploid human genome contains some 20,000 protein-coding genes and even more of ncRNA genes (Chap. 10). The protein-coding sequence covers less than 2% of the human genome, i.e., **the 98% of the genome is non-coding and primarily has regulatory function**. Furthermore, some 54% of the sequence of the human genome is formed by repetitive DNA (often also referred to as "junk DNA"), which is sorted into the following categories (by order of frequency):

- LINEs (500–8000 base pairs) 20.71%
- SINEs (100–300 base pairs) 12.79%
- Retrotransposons, such as long terminal repeats (LTRs) (200–5000 base pairs) 8.85%
- Minisatellite and microsatellite (2–100 base pairs) 4.93%
- Satellites (200–2000 base pairs) 2.54%.

LINEs and SINEs are identical or nearly identical DNA sequences that are separated by large numbers of nucleotides, i.e., the repeats are spread throughout the whole genome. LTRs are characterized by sequences that are found at each end of retrotransposons. DNA transposons are full-length autonomous elements that encode for a transposase, i.e., an enzyme that transposes DNA from one to another position in the genome (also known as "jumping DNA").

The different types of human genetic variants are referred to as common (or polymorphisms), when they have a minor allele frequency (MAF) of at least 1% within the studied population, or as rare, when they have a MAF of less than 1%. Thus, high frequency SNVs are also referred to single nucleotide polymorphisms (SNPs). SNVs represent the most frequent class of genetic variations among individuals and approximately 7 million of them show a MAF of more than 5% (www.ncbi.nlm. nih.gov/SNP). The *1000 Genomes* project (Sect. 1.3) and other large whole genome sequencing projects indicated that in addition there is a huge number of rare and novel SNVs. Nevertheless, **the majority of variants of any given individual are common in the whole population**. Therefore, the average difference in nucleotide sequence of a pair of unrelated humans lies in the order of only 1 in 1000, i.e., SNVs contribute to some 0.1% to the variation of human genomes. This proportion is low compared with other species and **confirms the recent origin of our species from a small founding population**.

The impact of SNVs on the coding sequence of the human genome is well established. Synonymous mutations do not alter the encoded protein, while non-synonymous mutations cause a change in the amino acid sequence (missense) or introduce a premature stop codon (nonsense). In average, a typical human genome contains some 150 SNVs resulting in protein truncation, 10,000 SNVs result in change amino acids and 500,000 SNVs affect transcription factor binding sites (TFBSs) (Sect. 1.3). Interestingly, **each individual is heterozygous for 50–100 genetic variants that can cause inherited disorders in homozygous offspring**, including different types of cancer or cancer predisposition syndromes (Chap. 15). This provides a large demand and challenge for genetic counseling based on whole genome sequencing. Moreover, gene-environment interactions provided by lifestyle factors, such as the personal choice of food, will create an additional level of complexity (Chap. 16).

Indels as well as CNVs in exonic sequences can result in either non-frameshift or frameshift mutations. Moreover, CNVs in intronic sequences may lead to alternative splicing. Some 60,000 unique CNVs are known and some of them are quite common in human populations (Sect. 1.3). However, the vast majority of SNVs and CNVs are located in regulatory and not in coding regions of genes, i.e., **the phenotypic consequences of most genetic variants are rather based on an gene regulatory or epigenetic processes than on a change in protein function**. However, the vast majority of these variants are not yet mechanistically characterized.

1.3 Measuring Human Genetic Variations

Mendelian disorders, such as cystic fibrosis or Huntington's disease, are mono-genic, i.e., in these cases a single SNV can explain the occurrence of the rare disease. However, common diseases like T2D, CVDs or cancer have a multigenic basis and were investigated by genome-wide association studies (GWASs). These studies employ an agnostic approach in the search for unknown disease variants, i.e., hundreds of thousands of SNPs were tested for association with a disease in large cohorts of patients *versus* healthy controls. With an average SNP density of 1 in 1000 nucleotides these studies require the testing of millions of SNPs per individual and hundreds to thousands of individuals.

GWASs with 2000 to 5000 individuals confidently identified common variants with effect sizes, referred to as **odds ratios** (ORs), of 1.5 or greater, i.e., a 50% increased risk for the tested trait. Larger sample sizes were achieved by pooling several GWASs through meta-analyses. For example, sample sizes of at least 60,000 individuals provide sufficient power to identify the majority of variants with ORs of 1.1, i.e., a 10% increased risk. Disease- and trait-associated genomic loci can be found in the database GWAS Catalog (www.ebi.ac.uk/gwas). Monogenic forms of diseases or traits have high ORs of 50 or above (Fig. 1.3, left). In contrast, there is a large number of common variants with a small to modest OR that have a dominant role in common complex diseases and traits (Fig. 1.3, right). For example, the trait "body height" is dependent on at least 180 gene loci, i.e., it is a paradigm of a complex/polygenic trait. In Europe, this trait has changed significantly (in average by 10 additional centimeters) within the last few generations under the environmental trigger of improved quality and quantity of nutrition.

The genetic approach of linkage analysis was used in the past, in order to identify genes responsible for monogenic disorders. However, these rare diseases represent only a relatively small fraction of the population. In contrast, **basically all common diseases have a complex origin**, i.e., they involve many gene loci. Moreover, for most diseases and traits a larger number of SNVs are described, which show high linkage disequilibrium to variants on stretches of 10–100 kb genomic DNA. These so-called **haplotype blocks** are inherited from generation to generation, i.e., they are not interrupted by meiotic recombination events (Fig. 1.4). Thus, the borders of haplotype blocks represent recombination events that had happened many generations ago in our ancestors.

Interestingly, **gene conversion during meiosis is some 100-times more frequent than point mutations and therefore a more efficient mechanism of evolution**. This is one of the major consequences of sexual reproduction. Since African populations

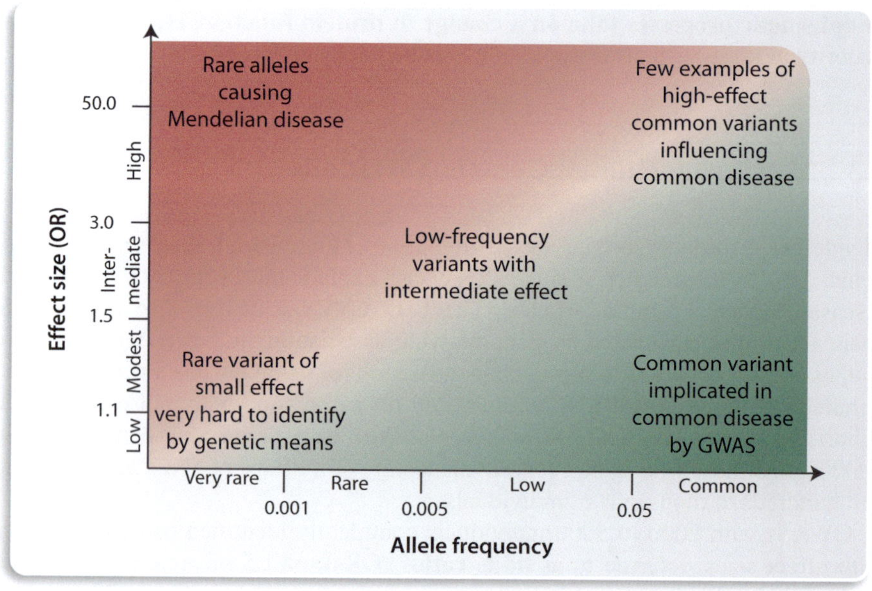

Fig. 1.3 Identifying genetic variants by risk allele frequency. The strength of a genetic effect is indicated by ORs. Most emphasis and interest lies in identifying associations with characteristics shown within the diagonal box. Whole genome sequencing of large numbers of individuals identifies far more low frequency SNPs with intermediate ORs (**center**)

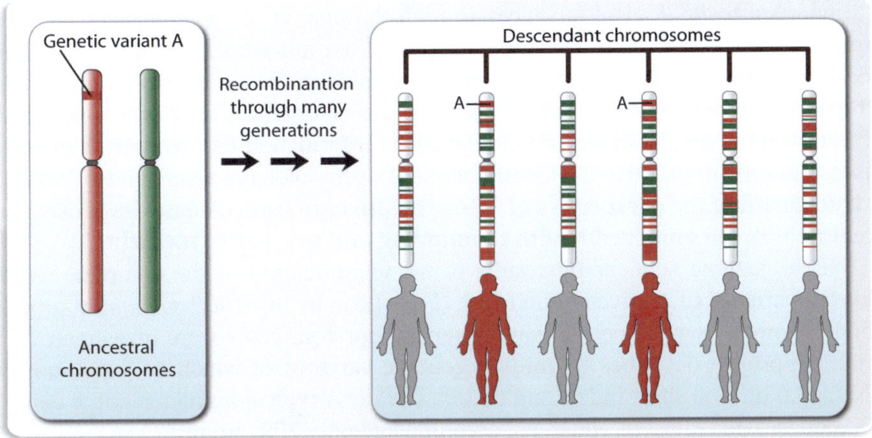

Fig. 1.4 The origin of haplotypes. Two plain ancestral example chromosomes get scrambled through meiotic recombination over many generations, in order to yield different descendant chromosomes. For example, after 30,000 years a typical chromosome will have undergone more than one crossover per 100 kb. In the case of a genetic variant (marked by an A) on one ancestral chromosome the risk of a particular disease increases. Thus, the two individuals (red) in the current generation who inherited that region of the ancestral chromosome will be at increased risk. Within the haplotype block that carries the disease-causing variant there are many SNPs that can be used to identify the location of the variant

have existed far longer than European and Asian populations, their haplotype blocks are shorter, i.e., the blocks had more time to decay because of the accumulation of recombination events in a higher number of generations. In contrast, all non-African individuals derived from a small population of eastern African origin (Sect. 1.1), i.e., they went through a demographical bottleneck that is clearly visible in the limited diversity of their genomes.

Despite some notable successes in revealing numerous novel SNPs and genomic loci associated with complex phenotypes, in average **not much more than 10% of the heritability of most complex, polygenic traits and diseases have been explained by common variants**. When exclusively SNV analysis is performed, the missing or unsolved heritability does not allow assigning an individual with any reliable estimation about his/her risk for a particular disease. The only well-known exceptions are age-related macular degeneration and type 1 diabetes (T1D), for which the combinations of common and rare variants can provide a quantifiable risk profile. For comparison, the heritability of only 20% of coronary heart disease cases is explained by in total 80 genetic loci, 20% of T2D by some 100 loci, 20% of inherited breast cancer by some 150 loci, 33% of inherited prostate cancer by some 100 loci and 30% of Alzheimer's disease by some 20 loci. Some of the missing heritability may be explained by rare variants with high ORs, which are poorly captured by standard GWASs. In addition, **environmental exposures, including those experienced as fetus, affect the epigenome and may explain large parts of the missing heritability** (Sect. 12.2).

Important technological advances in high-throughput sequencing have led to a rapid decrease in the costs of DNA sequencing. As a result, whole genome sequencing became an affordable tool in understanding the genomic basis of health and disease. At present, already huge amounts of data have been obtained from whole genomes of both healthy and diseased individuals. This information has not only helped in disease stratification and in the identification of their molecular mechanisms, but also is **transforming the perspective of future health care from disease diagnosis and treatment to personalized health monitoring and preventive medicine**.

Whole genome sequencing results in the identification of the complete set of genetic variants of a given individual (Fig. 1.5). In the *1000 Genomes* project 2504 genomes from 26 populations covering four continents were investigated. In total, the project describes 88 million genome variants, of which 84.7 million are SNVs, 3.6 million short indels and 60,000 CNVs. A typical human genome carries 200,000 variants, most of which are common and only 20% are rare (MAF < 0.5%). Individuals from African ancestry populations show most variant sites confirming the human origin from Africa (Sect. 1.1).

Data of the *1000 Genomes* project are publicly available and are central to human genetics, in order to understand population genetic variation. Moreover, the data serve as a global reference for ancestry and admixture of individuals and populations. For example, they define control allele frequencies for both human genetics studies and

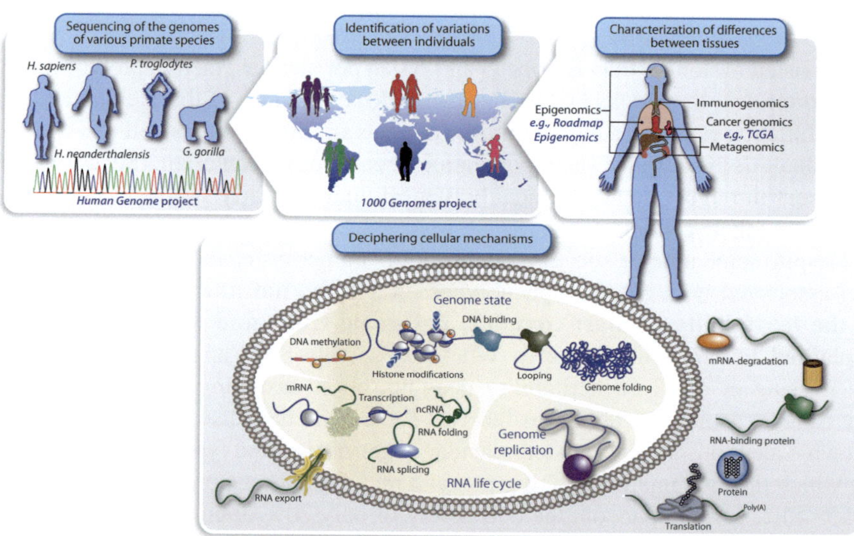

Fig. 1.5 Roadmap of sequencing science. The *Human Genome* project (Box 1.3) created a reference genome. Nowadays, also the genomes of all other primate species are known including some extinct human species (**top left**). Whole genome sequencing of several thousand individuals was performed in large consortia, such as the *1000 Genomes* project or the *UK Biobank* (**top center**). Moreover, the genetic and epigenetic differences between tissues and cell types of the same individual were collected in cancer genomics and epigenomics projects, such as *The Cancer Genome Atlas (TCGA)* and *Roadmap Epigenomics* (**top right**). The application of different NGS methods allows investigating and integrating many different processes within the cell (**bottom**)

medical genetics, where the identification of rare and novel variants is central for clinical diagnoses. A more recent set of *1000 Genome* project data include another benchmark, which is whole genome sequences of more than 600 family trios, i.e., father, mother and child. This further increases the accuracy of SNV calling, in particular in non-coding regions of the genome. This drastically increased the number of indel variants.

> **(Clinical) conclusion**: Fundamental to nearly all aspects of molecular medicine is some knowledge about the human genome and its variations. This is basis for understanding our origin as well as individual-specific disease risks. However, the genetic contribution to the traits describing the human body and its risk for diseases is average less than 20%.

Additional Reading

Amaral, P., Carbonell-Sala, S., De La Vega, F.M., Faial, T., Frankish, A., Gingeras, T., Guigo, R., Harrow, J.L., Hatzigeorgiou, A.G., Johnson, R., *et al.* (2023). The status of the human gene catalogue. Nature *622*, 41–47.

Claussnitzer, M., Cho, J.H., Collins, R., Cox, N.J., Dermitzakis, E.T., Hurles, M.E., Kathiresan, S., Kenny, E.E., Lindgren, C.M., MacArthur, D.G., *et al.* (2020). A brief history of human disease genetics. Nature *577*, 179–189.

Lappalainen, T. and MacArthur, D.G. (2021). From variant to function in human disease genetics. Science *373*, 1464–1468.

Lovell, J.T. and Grimwood, J. (2022). The first complete human genome. Nature *606*, 468–469.

Porubsky, D. and Eichler, E.E. (2024). A 25-year odyssey of genomic technology advances and structural variant discovery. Cell *187*, 1024–1037.

Tam, V., Patel, N., Turcotte, M., Bosse, Y., Pare, G. and Meyre, D. (2019). Benefits and limitations of genome-wide association studies. Nat Rev Genet *20*, 467–484.

Chapter 2
Gene Expression and Chromatin

Abstract In this chapter, principles of gene expression are discussed. An essential condition that a gene can be expressed, i.e., transcribed into RNA, is that its regulatory regions, such as transcription start sites (TSSs) and enhancers, are located within euchromatin. Chromatin is the physical representation of epigenetics. In a tissue- and cell type-specific fashion the majority of the genome is in heterochromatin and only approximately half of all genes are expressed, i.e., transcribed. The wrapping of genomic DNA around nucleosomes and the posttranslational modification of histones determine the density of chromatin packing. Tissue- and signal-specific gene expression is the central mechanism to control the general properties of a cell and its response to environmental perturbations. Large protein complexes that are formed by transcription factors, polymerases and other nuclear non-histone proteins organize the 3-dimensional (3D) architecture of the chromatin into functional units being used for most efficiently coordinated gene expression.

Keywords Central dogma of molecular biology · Euchromatin · Heterochromatin · Epigenetics · Histone proteins · Nucleosome · Chromatin architecture · Promoter · Enhancer · Gene expression · Chromatin architecture

2.1 Central Dogma of Molecular Biology

The "central dogma of molecular biology" indicates a clear direction in the flow of information from DNA to RNA to protein (Fig. 2.1). This means that besides a few exceptions, such as reverse transcription of the RNA genome of retroviruses, genomic DNA stores the building plan of all organisms. Accordingly, **genes are defined as those regions of genomic DNA that can be transcribed into RNA.**

C. Carlberg, *Gene Regulation and Epigenetics*,
https://doi.org/10.1007/978-3-031-68730-3_2

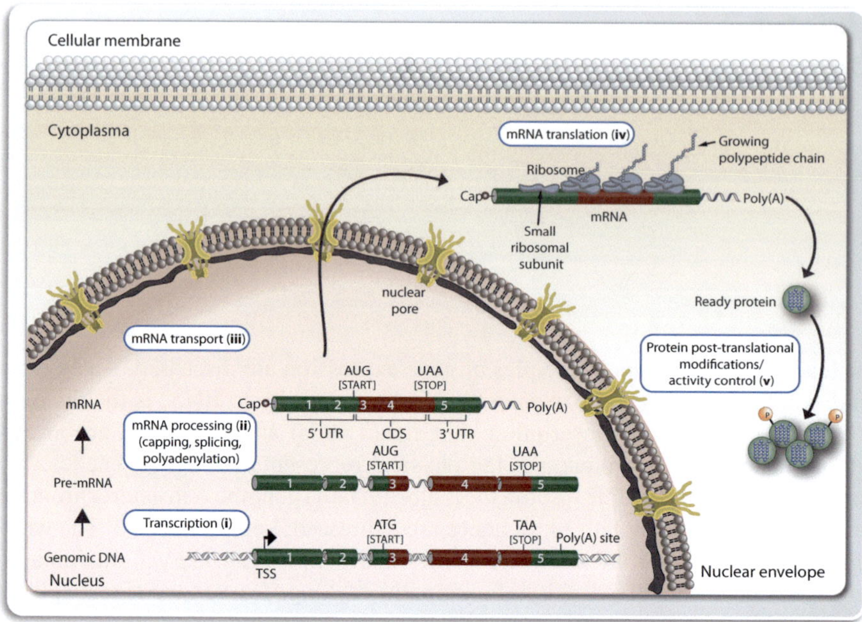

Fig. 2.1 Flow of information from DNA to RNA. The TSS of a gene is the first nucleotide that is transcribed into mRNA, i.e., the TSS defines the "start" of a gene body but has no defined sequence. In analogy, the "end" of a gene body is the position where Pol II dissociates from the genomic DNA template. The gene body is entirely transcribed into single-stranded pre-mRNA, which is composed of exons (numbered green and brown cylinders) and intervening introns (**i**). The introns are removed by splicing and the 5'- and 3'-end of the mRNA molecule are protected against digestion by exonucleases via a nucleotide cap and the addition of hundreds of adenines (polyadenylation (poly(A)), respectively (**ii**). Mature mRNA is then exported by an active, i.e., ATP (adenosine triphosphate) consuming, process from the nucleus through nuclear pores to the cytoplasma (**iii**). Small ribosome subunits scan the mRNA molecule from its 5'-end for the first available AUG (the "start codon"), assemble then with large subunits and perform protein translation process until they reach the sequence UAA, UAG or UGA (the "stop codons") (**iv**). The mRNA sequences upstream of the start codon and downstream of the stop codon are not translated and referred to as 5'- and 3'-UTRs. The resulting polypeptide chains fold into proteins, most of which are further posttranslationally modified, in order to reach their full functional profile (**v**). The central dogma of molecular biology indicates that the flow of information from DNA to RNA to protein has one clear direction. Please note that for simplicity in this and in all following figures the nuclear envelope is drawn as single lipid bilayer and not as a double lipid bilayer. A = adenine, C = cytosine, G = guanine, T = thymine (occurs only in DNA), U = uracil (occurs only in RNA)

The initial step of gene expression is the transcription of the genomic DNA of the gene body (i.e., the DNA sequence between the TSS and the transcription termination site) into messenger RNA (mRNA), which after splicing and transport from the nucleus to the cytoplasma is translated into protein (Fig. 2.1). **Proteins are the "workers" within a cell and basically mediate all functions therein**, such as signal transduction (Box 2.1), catalysis and control of metabolic reactions, molecule transport and many more. In addition, proteins contribute to the structure and stability of cells and intracellular matrices. Gene expression patterns are cell-specific, but can also drastically change after exposure to intra- and extracellular signals and in response to pathological conditions, such as microbe infection or cancer. Thus, **gene expression determines the phenotype, function and developmental state of cell types and tissues**.

Box 2.1: Signal Transduction Cascades The cascades of signal transduction pathways have a common structure and typically start with an extracellular **ligand** binding to its cognate **membrane receptor**, which then at its cytoplasmatic part interacts with **adaptor proteins** and activates them. This activation status is mostly transferred to a cascade of **protein kinases** that eventually either stimulate a **transcription factor** or a **chromatin modifying enzyme** in the nucleus or other key enzymes located in the cytoplasma. This often orchestrates gene expression changes that result in functional alterations of the concerned cell. In this way, extracellular (and in part also intracellular) signals are perceived by dozens of **signal transduction pathways**. The latter form networks of cellular circuits that represent key lines of intracellular communication, i.e., mechanisms by which a cell integrates and transduces intra- and extracellular signals from their source, e.g., the **extracellular environment**, into an action of the cell, such as to move, excite granules, grow, differentiate or keep homeostasis, i.e., of its **fate**. Some of these pathways do not involve gene expression changes but directly activate an (metabolic) enzyme. Since proteins, which are encoded by genes regulating cell fate, cell survival and genome maintenance, often interact with each other, their respective pathways overlap, i.e., their classification may not be as distinct as indicated.

Transcription is carried out by RNA polymerases, i.e., by enzymes that catalyze the DNA-dependent synthesis of RNA. In the traditional definition of genes, only the transcription of mRNA is meant, i.e., the RNA template used for protein translation. However, nowadays many other forms of transcribed RNA, such as ribosomal RNA (rRNA), transfer RNA (tRNA), micro RNA (miRNA) and long ncRNA, are known that serve for other functions than carrying the information for the amino acid sequence of a protein (Chap. 10).

There are three types of DNA-dependent RNA polymerases, I, II and III, that are responsible for the synthesis of different types of RNA. RNA polymerase I exclusively transcribes the genes of the three rRNAs, 5.8S, 18S and 28S. These rRNAs are structural components of ribosomes and represent more than 80% of the RNA content of a cell. RNA polymerase III is specialized on the synthesis of small RNA molecules (80–340 nucleotides), such as 5S rRNA, all tRNAs and a number of other small nuclear RNAs (snRNAs), such as U6 snRNA used in splicing. Thus, both **RNA polymerase I and III are producing RNA molecules that are needed for the basic function of a cell**. The genes encoding for these RNAs therefore belong to the group of ubiquitously expressed genes, which are often called "housekeeping genes". Such genes are regulated in a rather straightforward fashion using a limited number of transcription factors, in order to support a constant activity of these two types of RNA polymerases.

In contrast, RNA polymerase II (Pol II) transcribes all approximately 20,000 protein-coding genes and most ncRNAs (Chap. 10). In contrast to RNA polymerase I and III target genes, most of these Pol II-transcribed genes are tightly regulated and are responsive to intra- and extracellular signals. There are many mechanisms how the activity of Pol II can be regulated by some 1600 different human transcription factors (Chap. 4) and other nuclear proteins, such as cofactors and chromatin modifying enzymes (Sect. 8.3).

The TSS of a gene is the first nucleotide that is transcribed into mRNA, i.e., it defines the 5'-end (the "start") of a gene (Fig. 2.1). In analogy, the 3'-end of a gene is the position where RNA polymerases dissociate from the genomic DNA template. For a given gene the TSS is the reference point and any sequence in front of the TSS is referred to as "upstream", while "downstream" means after the TSS (in the main direction of transcription). The sequences between genes are referred to as intergenic regions that can range from several hundred to million base pairs. In total, these intergenic regions represent some 85% of the sequence of the human genome (Box 1.3), i.e., **only some 15% of the genome are transcribed into pre-mRNA** (Table 2.1). In eukaryotic organisms genes are organized into exons and introns. Already while the process of transcription is ongoing, a second process, referred to as splicing, digests the pre-mRNA at the exon–intron borders and ligates only the exons, in order to form mature mRNA molecules. Since introns are in average some 10-times longer than exons, mature mRNAs are far shorter than their respective pre-mRNAs. Speaking in numbers, the approximately 20,000 human protein-coding genes have an average pre-mRNA length of more than 16,000 nucleotides, while the average human protein is composed of 460 amino acids, for which only 1380 nucleotides of mature mRNA are needed.

In additional processing steps, called capping and polyadenylation, the 5'- and 3'-end of the mRNA molecules are protected against the action of exonucleases, i.e., the stability of mRNA increases (Fig. 2.1). In this form the mRNA molecules are transported by an active, i.e., ATP consuming, transport process through nuclear pores

Table 2.1 The human genome in numbers

Number of chromosomes	22 + X + Y
Genome size (nt)	3,054,815,472
Number of genes	58,037
Number of transcripts	203,835
Number of protein-coding genes	19,901
Number of protein-coding transcripts	80,087
Number of long ncRNA transcripts	15,779
Number of pseudogenes	14,723
Number of small RNAs	7258
Number of miRNAs	2588
Number of tRNAs	631

The size, number or genes and transcripts of the latest version of the human genome (hg38) are indicated. Pseudogenes are regions of the genome that contain defective copies of genes. These numbers may still change a bit based on new annotation analyses

into the cytoplasm. In the cytoplasma the small subunits of ribosomes are scanning the mRNAs from their 5'-end for the first available AUG start codon, assemble then with the large subunit of the ribosome, start the protein translation process and progress with it until they reach one of three possible stop codons (UAA, UAG or UGA). The mRNA sequence upstream of the start codon and downstream of the stop codon are not translated and referred to as 5'- and 3'-untranslated regions (UTRs). This means that **only a minor proportion of a gene's sequence** (some 5–10%, representing less than 2% of the human genome) **are finally used for coding proteins**.

During the different steps of transcription, various protein complexes are deposited along the mRNA forming a mature messenger ribonucleoprotein that is subsequently exported to the cytoplasm. These steps were traditionally thought to occur independently, but there is extensive coupling between them, including the recruitment of both splicing and export factors during transcription, as well as interdependence between polyadenylation and export. The first step (transcription) in the flow of information from genomic DNA to a functional protein is the most controlled and regulated one. This seems to be logical, as it is most economic and safe to tightly control the first step of a regulatory process than a later step. However, this does not imply that the later steps are not controlled at all. Mechanisms that stop gene expression, such as in situations, in which the initial stimulus for the activation of a gene has disappeared, are as important as activation mechanisms. In this context, ncRNAs play an important regulatory role (Chap. 10).

2.2 Nucleosomes: Central Units of Chromatin

Due to its phosphate backbone, at physiologic pH genomic DNA is negatively charged. The electrostatic repulsion between adjacent DNA regions makes it impossible to fold the long DNA molecules (46 to 249 Mb) of individual chromosomes into the limited space of the nucleus. This problem is solved by combining genomic DNA with histone proteins forming nucleosome complexes. The general feature of the four core histones H2A, H2B, H3 and H4 is their small size of some 11–15 kD (kilo Dalton) and their disproportional high content of the of the positively charged amino acids lysine (K) and arginine (R), in particular at their amino-termini (Table 2.2).

Nucleosomes are the subunits of chromatin and in every cell of the human body the diploid genome is covered by approximately 30 millions of these complexes. Two copies each of H2A, H2B, H3 and H4, the so-called histone octamer, and 147 base pairs genomic DNA, which is wrapped nearly twice around the octamer, form the nucleosome (Fig. 2.2). The bending of genomic DNA is primarily enabled through the attraction between the positively charged histone tails and the negatively charged DNA backbone. In addition, at some genomic regions, the bending is supported by the natural curvature of DNA that is achieved by AA/TT dinucleotides repeating every 10 base pairs and a high CG content. Together with the linker histone H1 the nucleosome forms the so-called chromatosome (however, this term is not used very often in the literature). Each nucleosome is connected with the following one via linker DNA (20–80 base pairs). This forms a repetitive unit approximately every 200 base pairs of genomic DNA. The regular positioning of nucleosomes has the effect that the position of a given nucleosome determines the location of its neighbors. Through the investment of ATP, chromatin remodeling protein complexes are able modulate the position and composition of nucleosomes (Sect. 9.1).

Table 2.2 Types and properties of human histones

Histone	Molecular weight [kDa]	Number of amino acids	Content of basic amino acids	
			Lys [%]	Arg [%]
H1	22.1	223	29.5	1.3
H2A	14.0	129	10.9	9.3
H2B	13.8	125	16.0	6.4
H3	15.3	135	9.6	13.3
H4	11.2	102	10.8	13.7

Histone H1 binds to linker DNA, while histones each a pair of H2A, H2B, H3 and H4 form the nucleosome core

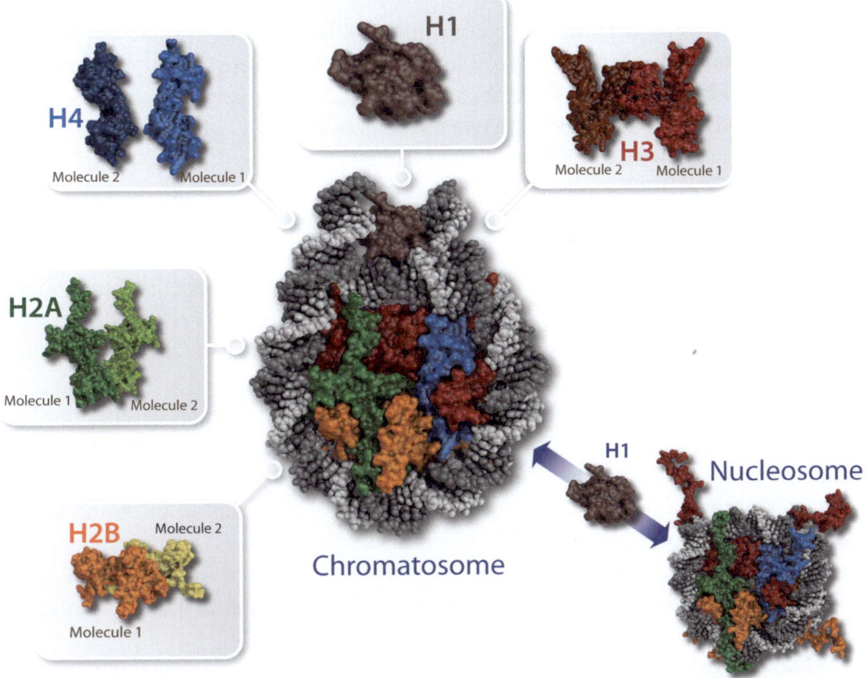

Fig. 2.2 The nucleosome. This space-filling surface representation of a nucleosome contains two copies each of the four core histone proteins H2A (green), H2B (orange), H3 (red) and H4 (blue) and 147 base pairs of genomic DNA (gray) being wrapped 1.8-times around the histone core. In complex with the linker histone H1 (brown) the nucleosome is referred to as chromatosome

The phosphate backbone of 200 base pairs genomic DNA carries 400 negative charges, which are in part neutralized by the approximately 220 positively charged lysine and arginine residues of the histone octamer. However, higher order folding of chromatin requires the neutralization of the remaining 180 negative charges by the positively charged linker histone H1 and further positively charged nuclear proteins associating with chromatin.

Nucleosomes are the regularly repeating units of chromatin, but they can vary from one genomic region to the other by different posttranslational modifications of the amino acid residues of their histones (Sect. 8.1) and the introduction of histone variants (Box 2.2). Genomic locus-specific histone modifications are reversible and an important component of the epigenetic memory affecting transcription factor binding and differential gene expression (Sect. 8.2). Thus, **nucleosomes do not only pack the DNA and are therefore not simply barriers that block access to genomic DNA but serve as dynamic platforms linking and integrating many biological processes**, such as transcription and replication.

Box 2.2: Histone Variants The core histones H2A, H2B, H3 and H4 represent the majority of histone proteins. In addition, there are eight variants of H2A (H2A.X, H2A.Z.1, H2A.Z.2.1, H2A.Z.2.2, H2A.B, macroH2A1.1, macroH2A1.2 and macroH2A2), two variants of H2B (H2BFWT and TSH2B) and six variants of H3 (H3.3, histone H3-like centromere protein A (CENPA), H3.1 T, H3.5, H3.X and H3.Y), while humans have no variants of H4. Core histones are assembled into nucleosomes behind the replication fork to package newly synthesized genomic DNA. By contrast, the incorporation of histone variants into chromatin is independent of DNA synthesis and occurs throughout the cell cycle. Interestingly, core histones have no introns, i.e., they have no splice variants while most of the genes for the histone variants do have introns and thus alternative splice variants. Histone variants are often subjected to the same modifications as core histones, but there are also variant-specific modifications on residues that differ from their canonical counterparts. Accordingly, histone variants also directly influence the structure of nucleosomes. For example, H2A.Bbd lacks acidic amino acids at its carboxy terminus, as a consequence of which only 118–130 base pairs (*versus* 147 base pairs) of genomic DNA are wrapped around the respective histone octamer. This leads to the formation of less compact and more accessible chromatin, which facilitates gene expression.

The same nucleosome may contain multiple histone variants. There are homotypic nucleosomes, which carry two copies of the same histone and heterotypic nucleosomes, which contain a core histone and a variant histone or two different histone variants. This allows for greater variability in nucleosome formation, stability and structure. For example, nucleosomes that contain H2A.Z and H3.3 are less stable than core nucleosomes and are often found at nucleosome-depleted regions of active promoters, enhancers and insulators (Sect. 9.1). These labile H2A.Z/H3.3-containing nucleosomes serve as "place holders" and prevent the formation of stable nucleosomes around regulatory genomic regions. They can be easily displaced by transcription factors and other nuclear proteins that are not able to bind genomic DNA in the presence of a nucleosome composed of core histones. Thus, **variable composition of nucleosomes can directly influence gene expression**.

2.3 Chromatin Structure and Epigenetics

Most of us probably had the first contact with epigenetics, when we were looking at chromosomes, either directly through a microscope or in a textbook. Chromosomes are formed of chromatin, which packs the human genome, i.e., the 16 to 85 mm long DNA molecules of each one chromosome, in the nucleus of a cell with a diameter of only 6–10 μm. However, chromosomes are only visible during a special phase of the cell cycle, referred to as the metaphase of mitosis. During mitosis the prime importance is that the genome is divided equally to the two daughter cells. Therefore, within the metaphase 2×46 DNA molecules need to be in highest compaction, referred to as chromosomes. At that time all genes are temporally switched off for approximately 1 h, i.e., **mitosis represents the most extreme case of epigenetic regulation of the human genome**.

More than 99% of the approximately 30 trillion (3×10^{13}) cells of our body are terminally differentiated, i.e., they do not divide anymore and are in the interphase. In this phase, within the nucleus only lighter and darker areas can be distinguished, which represent loosely packed euchromatin and tightly packed heterochromatin, respectively (Fig. 2.3). In the "beads-on-a-string" model of euchromatin (Fig. 2.3, bottom right), nucleosomes are regularly positioned every 200 bp leaving 50 base pair-sized gaps of freely accessible genomic DNA between them. Euchromatin becomes condensed only during mitosis and has a higher gene density than heterochromatin. **Genes can only be transcribed into RNA, i.e., they get expressed, when they are located within euchromatin**. While the euchromatin fiber has a diameter of 11 nm, more compacted heterochromatin forms a 30 nm fiber (Fig. 2.3, bottom left) or even higher order structures of 100 nm in diameter. For comparison, the diameter of a chromosome is even 700 nm.

The phenomena of genomic imprinting (Sect. 7.3) and X chromosome inactivation (XCI) (Sect. 10.3) were the first indications that identical genetic material (from individual genes to whole chromosomes) can be in the same nucleus in an "on" as well as in an "off" state. Thus, genes can either be actively expressed, i.e., their information is copied into RNA, or they are inactive, i.e., not expressed. In analogy to the term "epigenesis" (i.e., morphogenesis and development of an organism) the word "epigenetics" was created for describing changes in the phenotype that are not based changes in the genotype. This definition was extended to "**epigenetics is the study of changes in gene function that are mitotically and/or meiotically heritable and that do not entail change in DNA sequence**".

Epigenetic changes are functionally relevant when they result in changes in mRNA levels and initiate the production of proteins. However, there are also epigenetic changes that do not have an effect on transcription and may be related primarily to the creation of memory (Sect. 11.5). Gene expression patterns may stay persistent throughout the following cell divisions for the remainder of the cell's life. Interestingly, they can even last for multiple generations (Sect. 12.1). This implies that **epigenetics can inherit the information which genes are expressed in which cells**.

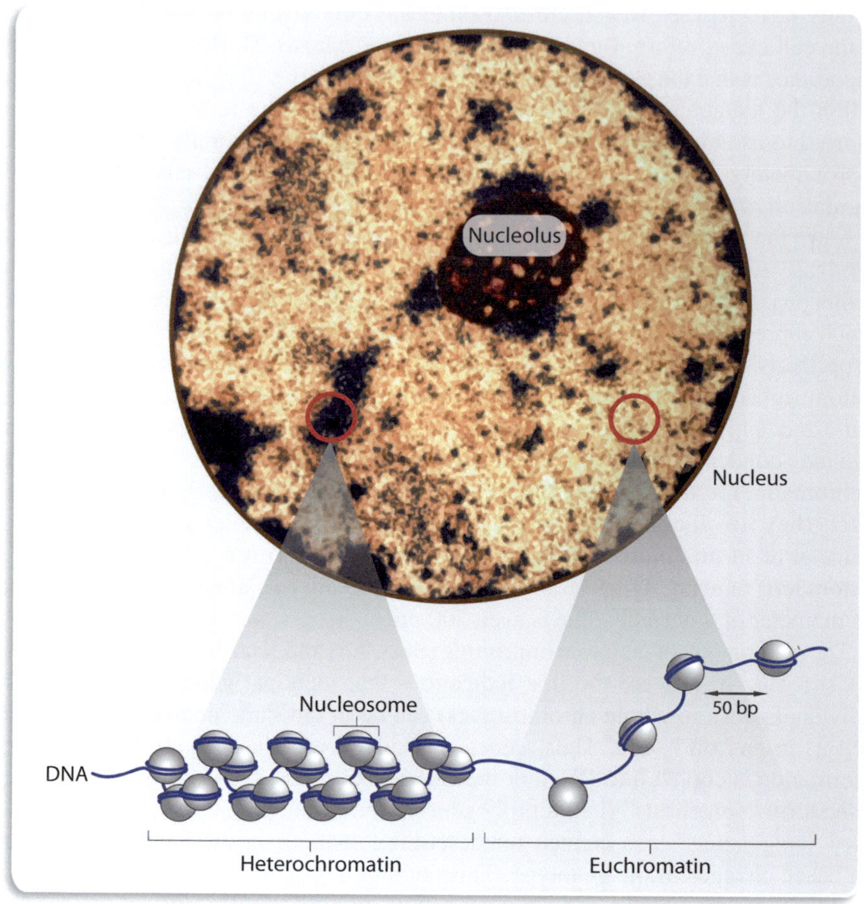

Fig. 2.3 Eu- and heterochromatin. An electron microscopic picture of a nucleus during interphase is shown. The darker areas, located mostly in the periphery of the nucleus, represent constitutive heterochromatin (inactive), whereas the lighter areas in the center are euchromatin (active). Please note that euchromatin takes far more space than heterochromatin. The nucleolus is a nuclear substructure, where rRNA genes are transcribed. A schematic drawing (**bottom**) monitors dense nucleosome packing in heterochromatin (**left**, also referred to as closed chromatin) and loose nucleosome arrangement in euchromatin (**right**, open chromatin)

2.4   Epigenetics Enables Gene Expression

Chromatin acts as a filter for the access of DNA-binding proteins to functional elements of the human genome, such as TSS regions and enhancers. Genes can only be transcribed into mRNA, when their TSS regions are accessible to the basal transcriptional machinery, which is a large protein complex containing Pol II (Sect. 3.1). However, even with given DNA access, mRNA transcription is often weak in the absence of stimulatory transcription factors. Therefore, the second condition for efficient gene expression is that enhancer regions in relative vicinity to the TSS are not buried in heterochromatin and can be recognized by transcription factors. Thus, **in order to activate and transcribe a gene, the chromatin at both TSS and enhancer regions, which control the gene's activity, needs to be accessible**. This implies that, in most cases, gene activation requires the transition from heterochromatin to euchromatin.

Enhancers are genomic regions that contain binding sites for sequence-specific transcription factors, which recruit coactivator and chromatin modifying proteins (Sect. 8.3) to the respective genomic loci. Thus, enhancers are stretches of DNA that function via the cooperative binding of multiple types of proteins. Since this often happens in less than one nucleosome length, nucleosome eviction is not essential for enhancer function (Sect. 9.1). **Enhancer activity is determined by epigenome stages**, which can be described by histone markers of accessible chromatin, such as H3K4me1 and H3K27ac (Sect. 8.2). When enhancers are close ($\pm$ 100 base pair) to the TSS, they are also often referred to as promoters. Thus, **there is no functional difference between enhancers and promoters besides their distance relative to the TSS of the gene that they are regulating**.

Enhancers that regulate the activity of a given gene need to be located within the same **topologically associated domain** (TAD) (Sect. 2.5). Since TADs have an average size of 1 Mb, this limits the maximal linear distance between an enhancer and the TSS(s) that it regulates (Fig. 2.4, top). Complexes of the proteins cohesin and CTCF (CCCTC-binding factor) mediate these DNA looping events. These 3D structural arrangements bring transcription factors that bind to enhancers into close vicinity of TSS regions. In this way, transcription factors, which bind to distant enhancers, can contact and activate the basal transcriptional machinery via intermediary complexes, such as the Mediator (MED) complex (Sect. 3.3).

The DNA looping mechanism also implies that enhancer regions are as likely upstream as downstream of TSS regions and may have tissue-specific usage and effects for transcription. For example, in tissue A enhancer A is used for activation, whereas in tissue B it mediates repression (Fig. 2.4, center and bottom). Results of the *Encyclopedia of DNA elements (ENCODE)* project (Sect. 6.1) demonstrated that basically all regulatory proteins have a Gaussian-type distribution pattern in relation to TSS regions, i.e., the probability to find an active TFBS symmetrically declines both up and downstream of the TSS. Thus, **the classical definition of a promoter as a sequence being located only upstream of the TSS is outdated**.

The structure and organization of chromatin can be interpreted as a number of superimposed epigenetic layers that lead either to open euchromatin and active gene expression ("on", Fig. 2.5, right) or to closed heterochromatin and no gene expression ("off", Fig. 2.5, left):

- The core of chromatin is genomic DNA that can be modified at cytosines, in particular at CpGs (CG dinucleotides, Sect. 7.1). Therefore, the first epigenetic layer is the DNA methylation status where hypermethylation stimulates the formation of heterochromatin.

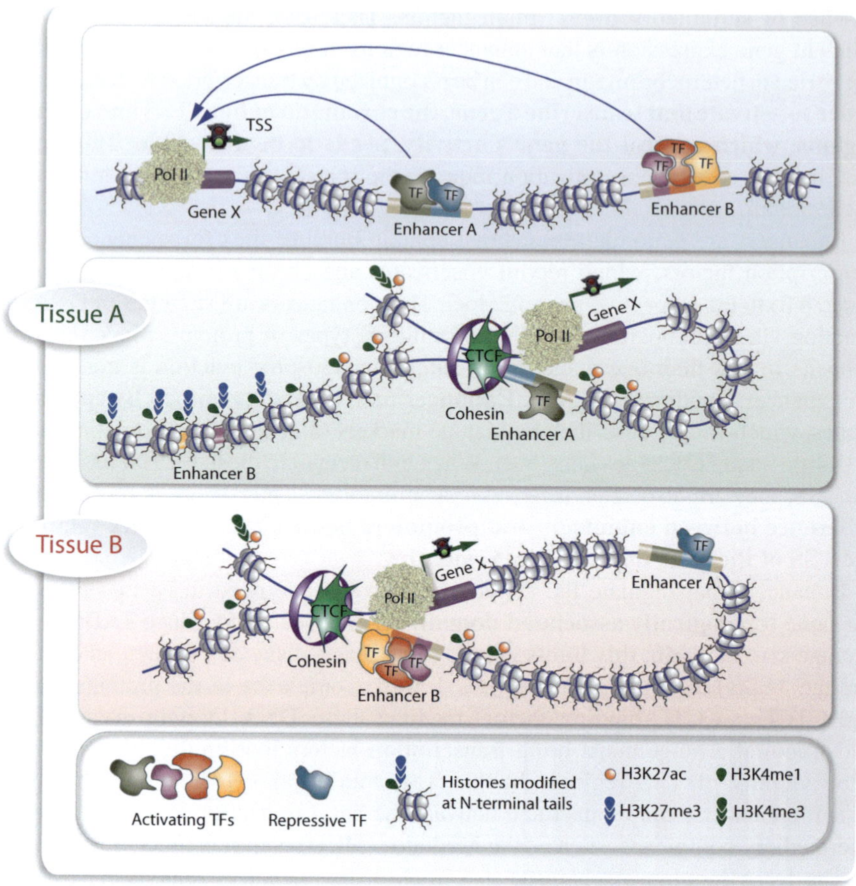

Fig. 2.4 Enhancer function. Enhancers are stretches of genomic DNA that contain binding sites for one or multiple transcription factors (TFs) stimulating the activity of the basal transcriptional machinery (Pol II and associated proteins) bound to the TSS of a gene. Enhancers are located both upstream and downstream of their target genes in linear distances of up to 1 Mb (**top**). Transcription factor-bound, active enhancers are brought into proximity of TSS regions by DNA looping, which is mediated by a complex of cohesin, CTCF and other proteins. Active TSS regions show depletion of nucleosomes while nucleosomes flanking active enhancers have specific histone modifications, such as H3K27ac and H3K4me1 (**center, tissue A**). In contrast, inactive enhancers are silenced by a number of mechanisms, such as repressing Polycomb proteins binding to H3K27me3 marks or by binding of repressive transcription factors (**bottom, tissue B**)

- The packing of nucleosomes represents level 2 where more dense arrangements indicate heterochromatin.
- Histone modifications (Sect. 8.1) at specific positions are level 3 and mark for either active chromatin (mainly acetylated) or inactive chromatin (mainly methylated).

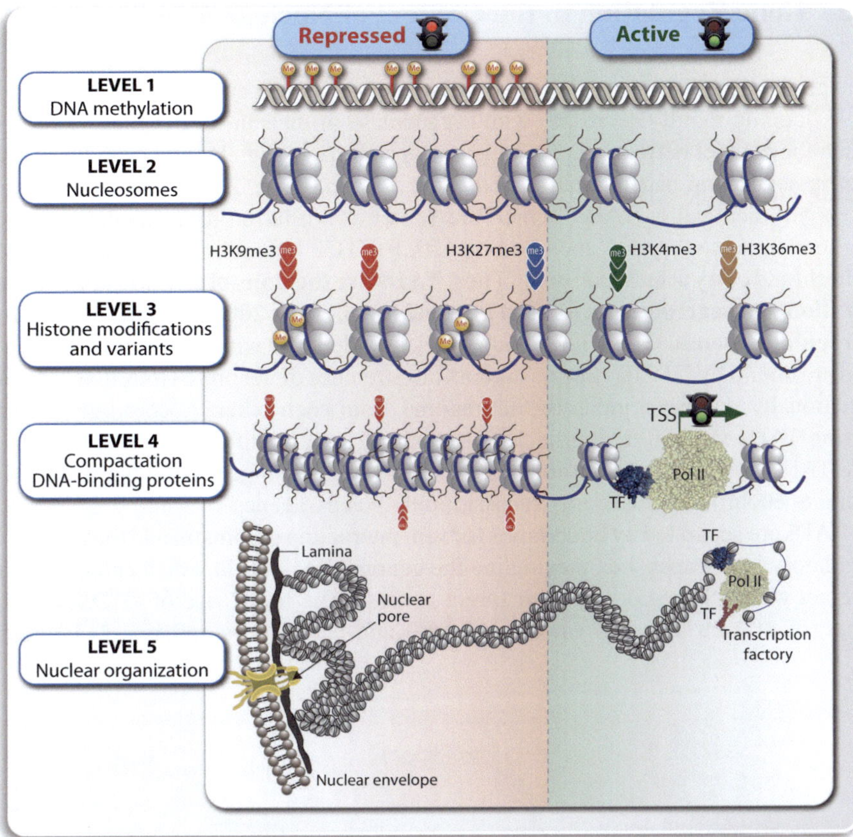

Fig. 2.5 Epigenetic layers of chromatin organization. There are at least five different layers of chromatin organization that are associated with inactive ("off", **left**) or active ("on", **right**) transcription

- The resulting accessibility of genomic DNA for the binding of transcription factors is considered as level 4.
- Finally, the complex formation and relative position of chromatin, such as active transcription factories (Sect. 9.4) in the center of the nucleus and inactive chromatin in lamin-associated domains (LADs) attached to the nucleoskeleton at the nuclear periphery, represent level 5 (Sect. 2.5).

Importantly, also during the interphase, 90% of the genomic DNA of terminally differentiated cells is not accessible to transcription factors. Thus, **heterochromatin is the default state of chromatin**.

2.5 Gene Regulation in the Context of Nuclear Architecture

The probability that two regions of a chromosome contact each other by chance via DNA looping rapidly decreases with the increase of their linear distance. However, when the contact between the two regions is stabilized, e.g., by associated proteins, then architectural and regulatory loops are forming (Fig. 2.6). Most architectural loops are identical to TADs (which are sometimes referred to as insulated neighborhoods), since they are anchored by CTCF-CTCF homodimers in complex with cohesin and carry at least one gene. Thus, **TADs are the units of chromosomal organization and segregate the human genome into at least 2000 domains containing coregulated genes**. Often TAD boundaries are identical with insulators (Sect. 7.3) and are bound by CTCF. Thus, insulators are stretches of genomic DNA that separate functionally distinct regions of the genome from each other. Accordingly, neighboring TADs can differ significantly in their histone modification pattern, such as one TAD being in heterochromatic state containing silent genes and the other TAD being in euchromatin carrying transcriptionally active genes.

TADs are separated by boundaries for self-interacting chromatin and thus organize regulatory landscapes, i.e., they define the genomic regions, in which enhancers can interact with TSS regions of their target gene(s). The linear size of TADs is in the range of 100 kb to 5 Mb (median: 1 Mb) and each TADs contains 1–10 genes

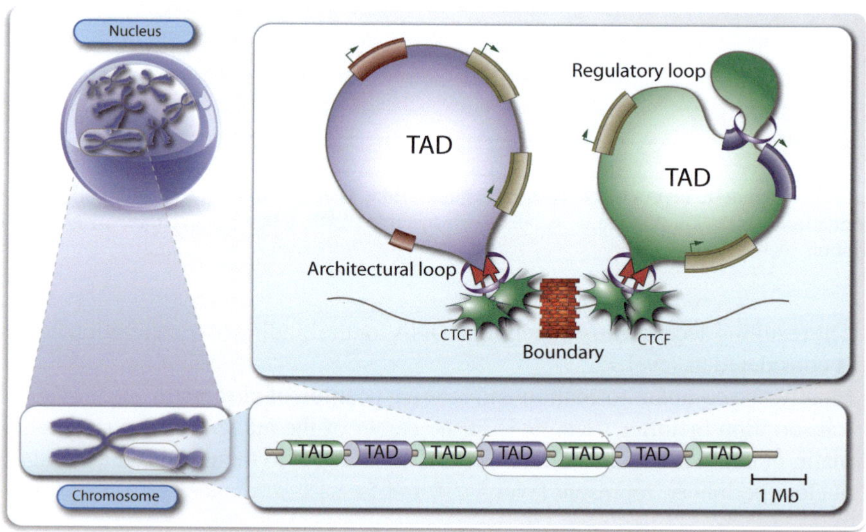

Fig. 2.6 Organization of chromosomes into TADs. The human genome is subdivided into a few thousand TADs defining genomic regions in which most genes have their specific regulatory elements, such as TSS regions/promoters and enhancers. TADs are architectural loops of chromatin that are insulated from each other by anchor regions binding complexes of CTCF and cohesin. Within TADs smaller regulatory loops between enhancers and TSS regions are formed

(median: 3 genes). Accordingly, most TADs comprise a number of genes that may be regulated by the same set of enhancers, such as often observed for clusters of genes of the same family. Regulatory loops are formed between enhancers and TSS regions that are located within the same TAD, i.e., they are smaller than TADs (Fig. 2.6). The formation of regulatory loops relies on the binding of transcription factors to the enhancer regions and its functional result is the stimulation of gene expression (Fig. 2.4).

The inner surface of the nuclear envelope is coated with nuclear lamina, which is a complex of lamins and a number of additional proteins (Fig. 2.7). Lamins maintain the shape and mechanical properties of the nucleus and serve as attachment points for LADs. **LAD-lamin interactions form a nucleoskeleton**, i.e., they serve as a structural backbone for the organization of interphase chromosomes. Like TADs, LADs vary in size from 0.1–10 Mb and cover up to 40% of the human genome. In contrast to TADs, LADs have a low gene density and are primarily formed by heterochromatin. However in total LADs still contain thousands of genes, most of which are not expressed. Accordingly, **the nuclear periphery is enriched for hete-rochromatin, whereas euchromatin is found more likely in the center of the nucleus** (Fig. 2.3). This suggests that the location of a gene within the nucleus is a functionally important epigenetic parameter.

The clustering of heterochromatin at the nuclear periphery creates silencing foci, so-called Polycomb bodies. These complexes are formed by members of the Poly-comb family, such as Polycomb repressive complex (PRC) 1 and 2 (Sect. 8.3). PRCs act as transcriptional repressors that are essential for maintaining tissue-specific gene expression programs, i.e., they ensure the long-term repression of specific target genes.

The position of chromatin and genes is not fixed, as there are dynamic changes in the contacts between the nucleoskeleton and genomic DNA involving single genes or small gene clusters. These changes are most pronounced during development. Of all human cell types, ES cells have the most accessible genome, i.e., the chromatin of these cells is largely open. During the differentiation process, cells change their chromatin structure and larger compaction of their genome occurs (Sect. 11.3). Thus, embryonic development proceeds from a single cell with dispersed chromatin to differentiated cells with nuclei that show compact chromatin domains being located in the periphery (Sect. 11.1). Accordingly, the physical relocation of a gene from the nuclear periphery to the center would unlock it to be expressed in a future developmental stage.

Another level of chromatin architecture in the interphase nucleus is the location of whole chromosomes in separate chromosome territories, which are separated by an interchromosomal compartment (Sect. 9.4). Chromosomes fold in their territories in such a way that active and inactive TADs are found in distinct nuclear compart-ments. Active regions are preferentially located in the nuclear interior, whereas inac-tive TADs accumulate at the periphery. In addition, TADs that are heavily bound by

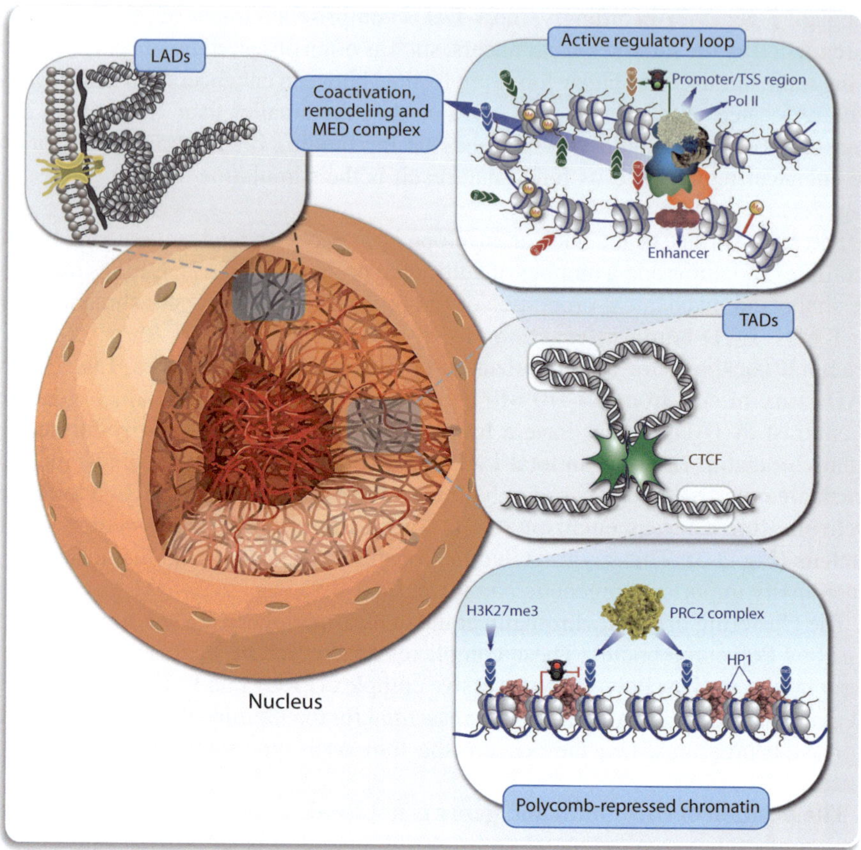

Fig. 2.7 Chromatin architecture. Mediated by structural proteins, chromatin forms a 3D architecture the nucleus (**center left**). Heterochromatin is composed of stably repressed, inaccessible genomic elements and is located closer to the nuclear lamina (**top left**). Two CTCF proteins bound at adjacent chromatin boundaries form a complex with cohesin and other MED proteins (**center right**). In this way, regulatory genomic regions, such as enhancers and promoters/TSS regions, which are separated by a genomic distance, can get into physical contact within DNA loops (**top right**). TADs distinguish such genomic regions with active enhancers from chromatin tracts that are silenced by PRCs (**bottom right**)

tissue-specific transcription factors are in different neighborhoods than those interacting with repressive PRCs (Fig. 2.7, right). To some extent, chromosome territories intermingle, which could explain interchromosomal interactions. Nevertheless, interactions between loci on the same chromosome are much more frequent than contacts between different chromosomes. Since the volumes of chromosome territories depend on the linear density of active genes on each chromosome, **chromatin with higher transcriptional activity occupies larger volumes in the nucleus than silent chromatin**.

(Clinical) conclusion: An overview of the impact of chromatin on gene expression is essential, in order to obtain an understanding of the molecular basis of epigenetics. This insight is important for realizing the role of epigenetics in nearly all processes of our life, many of which are discussed in the following chapters.

Additional Reading

Buccitelli, C. and Selbach, M. (2020). MRNAs, proteins and the emerging principles of gene expression control. Nat Rev Genet *21*, 630–644.

Buchwalter, A., Kaneshiro, J.M. and Hetzer, M.W. (2019). Coaching from the sidelines: The nuclear periphery in genome regulation. Nat Rev Genet *20*, 39–50.

Isbel, L., Grand, R.S. and Schubeler, D. (2022). Generating specificity in genome regulation through transcription factor sensitivity to chromatin. Nat Rev Genet *23*, 728–740.

Klemm, S.L., Shipony, Z. and Greenleaf, W.J. (2019). Chromatin accessibility and the regulatory epigenome. Nat Rev Genet *20*, 207–220.

Michael, A.K. and Thoma, N.H. (2021). Reading the chromatinized genome. Cell *184*, 3599–3611.

Pang, B., van Weerd, J.H., Hamoen, F.L. and Snyder, M.P. (2023). Identification of non-coding silencer elements and their regulation of gene expression. Nat Rev Mol Cell Biol *24*, 383–395.

Schoenfelder, S. and Fraser, P. (2019). Long-range enhancer-promoter contacts in gene expression control. Nat Rev Genet *20*, 437–455.

van Steensel, B. and Furlong, E.E.M. (2019). The role of transcription in shaping the spatial organization of the genome. Nat Rev Mol Cell Biol *20*, 327–337.

Zhou, K., Gaullier, G. and Luger, K. (2019). Nucleosome structure and dynamics are coming of age. Nat Struct Mol Biol *26*, 3–13.

Chapter 3
The Basal Transcriptional Machinery

Abstract In this chapter, we will discuss the precise spaciotemporal formation of the basal transcriptional machinery on TSS regions, which are also known as core promoters. This is a prerequisite for understanding how transcription by Pol II is controlled. Pol II is the core of the basal transcriptional machinery that contains a large number of general transcription factors (GTFs), such as the TATA box-binding protein (TBP), many of which are summarized as the transcription factor (TF) IID complex. The TATA box is the prototype of a site-specific TFBS determining the position of Pol II on the TSS. However, genome-wide analysis showed that the majority of human genes use alternative binding sites for GTFs. Diversity and complexity of the human transcriptome are based on that most genes have multiple TSS regions and that the TSS of many genes is not a single defined nucleotide. The basal transcriptional machinery interacts via another multiprotein complex of coactivators, termed the MED complex, with a large variation of transcription factors. In parallel, the MED complex coordinates the action of cofactors of transcription factors, some of which are chromatin modifying enzymes.

Keywords RNA polymerase II · TBP · GTFs · TATA box · Core promoter · TSS · Basal transcriptional machinery · TFIID · Sequence logo · TFBS · MED complex

3.1 The Core Promoter and Its Elements

Most protein-coding genes show a tissue- and signal-specific expression pattern that is mediated by a large set of some 3200 site-specific transcription factors (encoded by approximately 1600 genes, Chap. 4). These transcription factors bind to enhancers, most of which are located in some distance to the TSS region(s) of the gene that they are regulating (Fig. 2.4). Distal binding transcription factors recruit in a precisely orchestrated way a large set of coactivator proteins, in order to have an effect on the transcriptional activity of their target gene. They take advantage of the fact that genomic DNA can loop effectively into any desired direction (Sect. 2.5). This is an universal mechanism how **transcription factors can contact the basal transcriptional machinery and affect the activity of Pol II** (Fig. 3.1).

C. Carlberg, *Gene Regulation and Epigenetics*,
https://doi.org/10.1007/978-3-031-68730-3_3

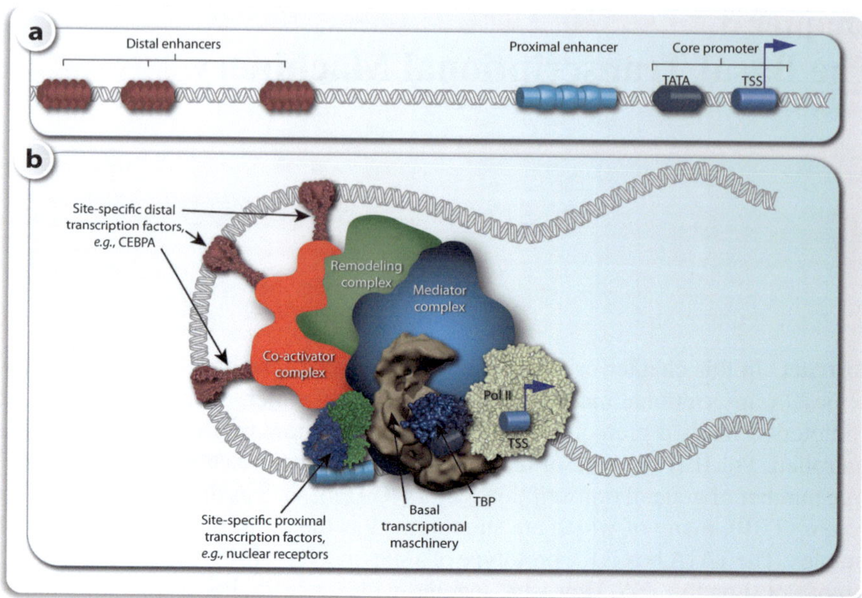

Fig. 3.1 Components of transcriptional regulation. In a linear schematic picture of the regulatory region of a gene **a**, the core promoter (TSS region), proximal TFBSs (proximal enhancers) and distal enhancers are distinguished. For simplicity only elements upstream of the TSS are indicated but, with exception of the TATA box, these TFBSs are also found downstream of the TSS. A more realistic DNA looping model **b**, in which also transcription factors, coactivators, other chromatin modifying proteins and Pol II are shown, suggests that all protein-bound TFBSs are connected via several multiprotein complexes, such as the coactivator complex, the remodeling complex, the MED complex and the basal transcriptional machinery. Different complexes are distinguished here, because they can be separately purified or assembled in vitro, but it is likely that they all together form a large super-complex, also called the transcription factory (Sect. 9.4)

The core promoter is the genomic region ± 50 base pairs of a TSS. This stretch of genomic DNA contains all essential elements, in order to allow the assembly of the basal transcriptional machinery (also called the preinitiation complex), which as a total size of 4 MDa. This process **places the catalytic site of Pol II on a suitable position of the genome and defines in this way a TSS**. The activity of the basal transcriptional machinery is modulated by proximal and distal activator and repressor proteins via the MED complex (Sect. 3.3). In humans, more than 50 different proteins bind to the core promoter and form the components of the basal transcriptional machinery (Table 3.1). These include the 12 subunits of Pol II and other multiprotein complexes, such as TFIID.

The most studied element of the core promoter is the TATA box, which is the binding site for the GTF TBP. When TBP has found an accessible core promoter, i.e., when this genomic region is sufficiently depleted from nucleosomes (Sect. 9.1), it binds to the TATA box and associates then with some 20 different TBP-associated factors (TAFs), 13 of which are forming the 1.3 MDa complex TFIID. Each RNA

Table 3.1 Components of the basal transcriptional machinery

General transcription factor TFII#	Subunit	Function/activity
D	TBP + TAFs	DNA binding to TATA box (core promoter), coactivation phosphorylation, ubiquitination and HAT activity
A	GTF2A1, GTF2A2	TBP-DNA stabilization, co-activation
B	GTF2B	TBP-DNA stabilization, Pol II and TFIIF recruitment TSS targeting
F	GTF2F1, GTF2F2	Pol II interaction and recruitment to promoter cooperation with TFIIB in TSS targeting recruitment of TFIIE and H enhances Pol II transcription start and elongation
E	GTF2E1, GTF2E2	Facilitation of the Pol II initiation-competency helping in promoter clearance recruitment of TFIIH
H	GTF2H1, GTF2H2, GTF2H3, GTF2H4, GTF2H5, MNAT1, CCNH, CDK7, ERCC2, ERCC3	Helping in promoter clearance and transcriptional initiation ATPase, helicase and E3 ubiquitin ligase activity transcription-coupled nucleotide excision repair phosphorylating Pol II C-terminal repeat domain (CTD)
Pol II	POLR2A-M	Initiation, elongation and termination of transcription recruitment of mRNA capping proteins recruitment of transcription-coupled splicing and 3' end processing factors CTD phosphorylation, glycosylation and ubiquitination

MNAT1 = MNAT1 component of CDK activating kinase, POLR2 = RNA polymerase II

polymerase type has its own set of TAFs, i.e., Pol II is interacting with TAFIIs. Interestingly, TAFIIs are also found in chromatin remodeling complexes (Sect. 9.2). The significant homology between TAFIIs and histones suggests that TFIID may mimic nucleosome function concerning stabilization of DNA. In fact, genomic DNA can be wrapped around TFIID similar to as it is wrapped around a nucleosome, so that the latter can be displaced while genomic DNA is alleviated during transcription complex assembly. TFIID modifies then the surrounding chromatin via the histone acetyltransferase (HAT) activity of TAF1. With TBP in its core, TFIID is the main GTF that directly binds to DNA. Therefore, **DNA-bound TFIID is the landmark for the core promoter** and the sign for other GTFs, such as TFIIA, B, E, F and H and Pol II to assemble in an ordered fashion at this genomic locus (Fig. 3.2). In contrast to some bacteriophage RNA polymerases, Pol II itself is not able to recognize any specific DNA-binding sequence. Thus, **the TSS is determined solely by steric constrains of the position of Pol II in relation to that of TFIID.**

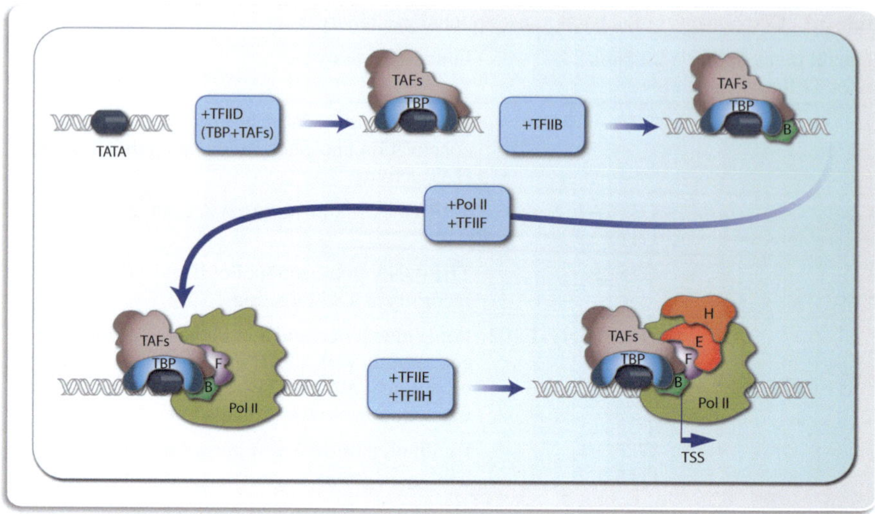

Fig. 3.2 Assembly of the basal transcriptional machinery. The TATA box of a core promoter (TSS region) is specifically bound by TBP that forms together with TAF proteins the multiprotein complex TFIID. In an ordered fashion further GTFs, such as TFIIB, F, E and H, as well as Pol II are recruited to the DNA-bound TFIID complex. Within this basal transcriptional machinery, the catalytic site of Pol II is in a defined distance of the TATA box, i.e., the binding of TBP determines the start of transcription

The multisubunit enzyme Pol II depends on a large number of additional proteins, in order to initiate, elongate and terminate transcription. Transcription initiation begins with the formation of the complex of the basal transcriptional machinery. Isomerization of this closed promoter complex to an open complex involves separation of the DNA strands, since the RNA synthesizing activity of Pol II needs partially single-stranded DNA as a template. DNA opening is mediated by the DNA translocase ERCC3 (ERCC excision repair 3, TFIIH core complex helicase subunit), which is a subunit of TFIIH and binds DNA downstream of Pol II. ERCC3 hydrolyses ATP to unwind DNA and propel it into the active center of the polymerase. TFIIE then binds and stabilizes the melted DNA. TFIIH has a dual role as it participates both in transcription and in the DNA repair process nucleotide excision repair (NER). In addition to its helicase subunit, TFIIH also contains the kinase subunit CDK7 (cyclin-dependent kinase 7), which phosphorylates the carboxy-terminal domain of Pol II. This phosphorylation step is necessary to dissociate Pol II from the GTFs. The transcribing Pol II complex is initially unstable and abortive initiation can create a number of short RNAs, such as enhancer RNAs (eRNAs) (Sect. 10.4). Nevertheless, from a critical length of the mRNA molecule on, initiation factors are released from the Pol II complex and a stable elongation complex is formed. This complex is facing obstacles, such as nucleosomes. With the help of elongation factors and chromatin remodelers (Chap. 9) the Pol II complex peels off the DNA from the nucleosome, in order to allow further pre-mRNA elongation. In parallel, the RNA gets processed through 5'-end capping, cotranscriptional splicing and 3'-end processing. The latter is tightly coupled with Pol II transcription termination. Most mRNAs are polyadenylated at their 3' ends by the CPF (cleavage and polyadenylation factor) complex, which recognizes an A-rich polyadenylation signal in the 3'-UTR of the mRNA and binds the Ser2-phosphorylated Pol II. Subsequently, the RNA is cleaved and the poly(A) tail is added, which enables export of the polyadenylated mRNA to the cytoplasma.

The TATA box is a prototype binding motif of site-specific transcription factors. The first nucleotide of a TATA box is located approximately 30 base pairs upstream of the TSS of a gene (Fig. 3.3a). The name TATA is a short form of its consensus sequence TATAWADR (Fig. 3.3b, for nucleotide abbreviations see Table 3.2) and it is specifically recognized by a homodimer of the transcription factor TBP (Fig. 3.3c).

Consensus sequences have been used in the past to represent the properties of known TFBSs. The binding sites are aligned below each other and a consensus nucleotide letter (Table 3.2) is assigned, in order to indicate the nucleotide composition in each column. Although consensus sequences represent a TFBS better than a

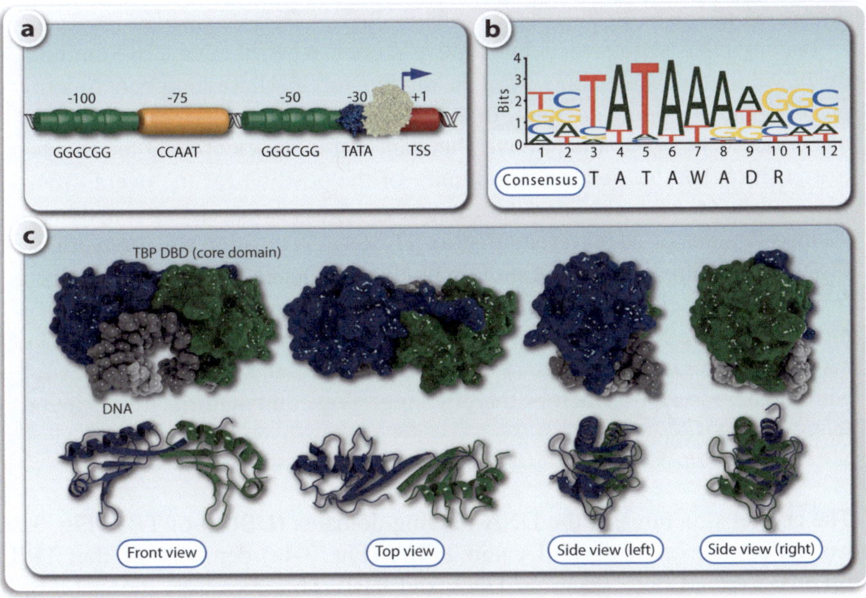

Fig. 3.3 TATA box in complex with TBP. The TATA box is found some 30 base pairs upstream of the TSS of a subset of human genes and is specifically recognized by TBP **a**. Other possible proximal TFBSs are CG-rich motifs being recognized by the transcription factor SP1 or CCAAT boxes bound by the transcription factors of the CEBP (CCAAT enhancer binding protein) family. All these elements belong to the core promoter (TSS region). The TATA box is the prototype of a TFBS. It can be represented either by a consensus sequence or more accurately by a sequence logo **b**. Two nearly identical DBDs (blue and green) of TBP are shown in a Connolly surface model (**c, top**) in complex with DNA (gray) or as a ribbon model (**c, bottom**) in the absence of DNA

single sequence, they do not accurately reflect the quantitative characteristics of this protein-DNA interaction. Thus, **sequence logos** (Fig. 3.3b) **are more appropriate**, since they are based on position frequency and position weight matrices (Box 3.1). Moreover, they allow a fast intuitive visual assessment of the characteristics of a TFBS (Table 4.2).

Box 3.1: Sequence Logos and de Novo Motif Analysis In order to reflect more accurately the characteristics at each position of a TFBS, a position frequency matrix is created that describes the number of nucleotides observed at each position. This frequency matrix is often converted to a position weight matrix, where normalized frequency values are indicated in a log scale (this makes computational analysis more efficient). Targets of a given transcription factor can be predicted by screening genomic DNA locally or genome-wide for regions, in which the local sequence fits with the position weight matrix. However, this approach does not address any redundancy in recognition by related transcription factors, the accessibility of the sequence within chromatin structure or contributions of other transcription factors binding up or downstream.

For any DNA sequence, a quantitative score can be calculated by summing up the values for each nucleotide of the binding motif. These scores are roughly proportional to binding energies. In sequence logos (Fig. 3.3b) the scale of each nucleotide is based on the relative abundance of the nucleotide at the respective position and the relative importance of the position for the overall transcription factor binding. Therefore, sequence logos are better suited and more intuitively understood representations of TFBSs than position weight matrices. Comparing a large number of protein binding sequences, such as determined from ChIP-seq (chromatin immunoprecipitation sequencing) data via de novo motif finding, allows the most reliable description of a TFBS (Sect. 6.1). Moreover, the same method can also reveal the presence of binding sites for additional transcription factors, thereby suggesting combinatorial transcription factor complexes.

The crystal structures of the DNA-binding domains (DBDs) of TBP (Fig. 3.3c) and of its complexes with TFIIA and TFIIA (Fig. 3.4a) demonstrate that TFIIA and TFIIB contact both genomic DNA and TBP. This increases the stability of the TBP-DNA complex. Moreover, these structures show that the DNA is dramatically bent and unwound. TAFIIs in conjunction with TFIIA induce conformational changes in the complex leading to wrapping of the core promoter around TFIID. **Core promoter regions are often nucleosome-depleted**, i.e., in comparison to TFBSs at other genomic regions, such as enhancers, TSS regions represent the most accessible form of genomic DNA (Sect. 9.1).

There are different types of general TFBSs within TSS regions. In analogy to prokaryotes, it was initially assumed that every core promoter contains a TATA box

Table 3.2 The nucleotide code

Base	Meaning	Origin of designation
A	A	**A**denine
B	G or T or C	Not-A
C	C	**C**ytosine
D	G or A or T	Not-C
G	G	**G**uanine
H	A or C or T	Not-G
K	G or T	**K**eto
M	A or C	a**M**ino
N	G or A or T or C	a**N**y
R	G or A	pu**R**ine
Y	T or C	p**Y**rimidine
S	G or C	**S**trong interaction
T	T	**T**hymine
V	G or C or A	Not-T (not-U)
W	A or T	**W**eak interaction

The following abbreviations for nucleotides are internationally used

sequence. However, in fact **only 10–20% of mammalian core promoters carry a functional TATA box**. Therefore, alternative binding sites for GTFs have to take over the role of the TATA box. The initiator (Inr) element is functionally analogous to the TATA box as it is directing the formation of the basal transcriptional machinery, determining the location of the TSS and mediating the action of upstream activator proteins. The consensus sequence of Inr is YYANWYY and it directly overlaps with the TSS. The Inr element is bound by a complex of TAF1 and TAF2 and then recruits the other subunits of TFIID (Fig. 3.4b). After the stable binding of TFIID to the core promoter, the remaining steps of the formation of a functional basal transcriptional machinery and transcription initiation follow a similar mechanism than for TATA box-containing promoters.

The downstream promoter element (DPE) bears the consensus sequence RGWYV and is located approximately 30 bp downstream of the TSS. The DPE is found in TATA box-lacking core promoters and often acts in conjunction with the Inr element to direct specific initiation of transcription (Fig. 3.4b). In contrast, the TFIIB recognition element (BRE) binds TFIIB, has the consensus sequence SSRCGCC and is often found upstream of the TATA box (Fig. 3.4c). Members of a class of core promoters being often found with housekeeping genes lack both TATA and Inr elements but instead contain several transcription initiation sites, have a high CG content and multiple binding sites for the ubiquitously expressed mammalian transcription factor SP1 (specificity protein 1) (Fig. 3.4d). SP1 directs the formation of the basal transcriptional machinery to a region 40–100 base pairs downstream of its binding sites involving TAF1, TAF2 and TAF4.

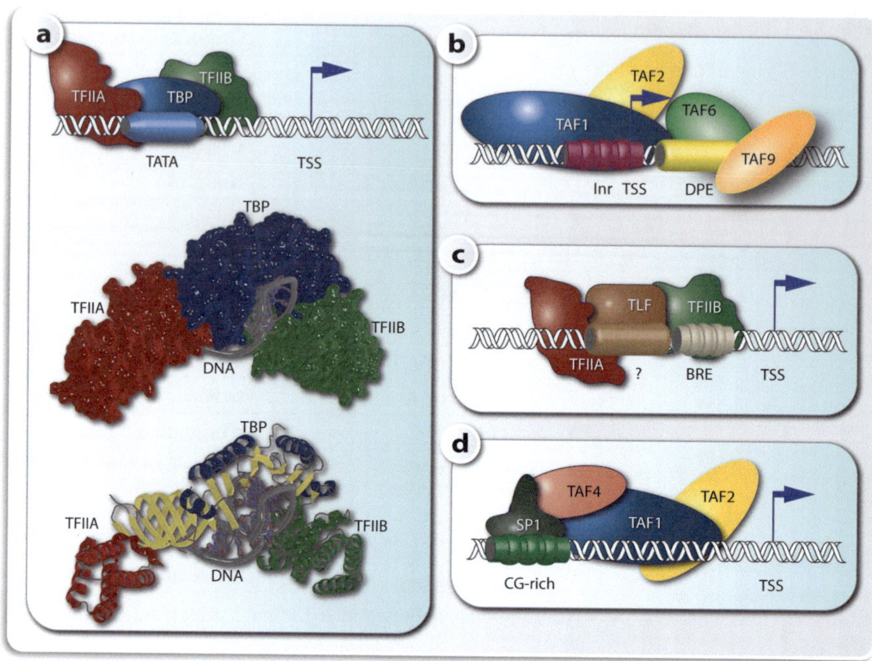

Fig. 3.4 Different protein complexes on TSS regions. Core promoters that contain a TATA box are bound by TBP in complex with TFIIA and TFIIB **a**. The complex is shown as a schematic drawing (**top**), as a Connolly surface model (**center**) or as a ribbon model (**bottom**). The unwound DNA is visible best in the ribbon model. On TATA-lacking core promoters the Inr element is used alone or in combination with DPE to attract TAF1 and TAF2 (to Inr) and TAF4 and TAF9 (to DPE) **b**. Alternatively, TBP-like factor (TLF) can form a complex with TFIIA and TFIIB on a BRE element **c** or SP1 binding to a CG-rich sequence directs complex assembly of TAF1, TAF2 and TAF4 **d**

Sequence elements of core promoters are commonly conserved across orthologous genes, but the complete set of mammalian promoters is too diverse to allow reliable prediction of TSS regions without reference to the experimental data (Chap. 6). For example, one of the main characteristics of human TSS regions is that approximately 60% of them are situated in proximity to CpG islands (Sect. 7.1).

3.2   Genome-Wide Core Promoter Identification

The availability of whole genome sequences of humans and other species led to the development of new high-throughput methods, some of which are targeted toward locating the 5'-ends of mRNAs or active TSS regions (Sect. 6.2). Next-generation sequencing (NGS) methods, such as RNA sequencing (RNA-seq, Box 3.2), indicate on a genome-wide level relative mRNA expression. The database DBTSS (http://dbtss.hgc.jp) describes the exact position of experimentally validated TSS regions for a number of species. It integrates RNA-seq data and ChIP-seq data of histone modifications as well as the binding of Pol II and several transcription factors. This also includes public data, such as from the *ENCODE* project (Sect. 6.1). Interestingly, many of the newly identified TSS regions are not associated with a protein-coding gene but lead to the production of ncRNAs (Chap. 10). The *FANTOM* (functional annotation of the mammalian genome) *5* project (Sect. 6.1) systematically used the method CAGE (cap analysis of gene expression, Box 3.2) with samples from nearly 1000 primary human tissues and cell lines and identified some 185,000 TSS regions throughout the human genome. Many of these clusters are core promoters.

Box 3.2: Transcriptome Profiling Methods RNA-seq is nowadays the standard method for transcriptome profiling and uses NGS technologies. In this method, a population of RNA molecules, such as total RNA or a poly(A)$^+$ mRNA subset, is converted into a library of cDNA fragments. The library is then sequenced in a high-throughput approach (Sect. 6.2). This provides short sequence tags (comparable to those produced in ChIP-seq, Fig. 6.3) from either one end (referred to as "single-end" sequencing) or both ends (named "pair-end" sequencing). RNA-seq allows a more precise measurement of transcript levels than previously used methods, such as microarrays, that are based on nucleic acid hybridization. This allows to quantify differential gene expression in comparison of two or more conditions, such as stimulation with a signaling molecule. The method CAGE is a special version of RNA-seq that focuses on the 5'-end of the RNA population of a biological sample. In this technique small fragments from the 5'-ends of capped mRNA transcripts are extracted, reverse-transcribed to DNA, polymerase chain reaction (PCR) amplified and sequenced. This method was extensive used by the *FANTOM5* project (Sect. 6.1).

For example, ChIP-seq analyses identified Pol II bound to the TSS regions of active genes. Pol II is recruited to these TSS regions depending on the studied gene and differentiation status of the cell. This implies that the recruitment and mRNA transcript elongation by Pol II is regulated differently at different genes. Furthermore, Pol II is also associated with enhancer elements, which supports the model presented in Fig. 3.1 that distal binding transcription factors are connected via protein–protein interactions with the basal transcriptional machinery. Further genome-wide studies indicated also for a number of histone modification and DNA methylation marks a correlation to active TSS regions.

Genome-wide approaches also demonstrated that most human core promoters lack a distinct TSS to be located at one specific nucleotide position, but they consist of an array of closely located TSSs that have a median spread of some 70 base pairs. **This distinguishes "broad" core promoters from "sharp" ones.** However, the various individual TSSs within a broad promoter are located within the same nucleosome-depleted region (Sect. 9.1). Variant hybrids between these two core promoter types also exist. Interestingly, sharp core promoters more likely contain TATA boxes, while broad promoters often are close to CpG islands (Sect. 7.1). Moreover, sharp promoters are used preferentially for tissue-specific expression, whereas broad promoters are often associated with housekeeping genes. The use of multiple TSSs over an extended genomic region in genes with broad core promoters requires that the respective genes exclude ATG translation start codons close to the TSS. Accordingly, more than 80% of human genes have a long 5'-UTR. Furthermore, this implies that the TFIID complex binds relatively non-specifically to these broad core promoters.

A third type of core promoters are those of key developmental transcription factors involved in patterning and morphogenesis. These resemble housekeeping gene core promoters, but are distinctly bivalently marked with both H3K4me3 and H3K27me3 (Sect. 8.2), which primes them for activation in the correct cell lineage and for silencing in all other cells. These **poised promoters are associated with CpG islands** (Sect. 7.2).

Importantly, in humans **most protein-coding genes have more than one TSS region**. These alternative core promoters are generally used in different contexts or tissues, in order to produce distinct protein products. In many cases, the different TSS regions generate alternative 5'-exons that sometimes contain alternative translation start codons. Moreover, the same gene locus can carry both sharp and broad core promoters. **The use of alternative core promoters substantially contributes to the complexity of the human transcriptome and proteome**.

3.3 TFIID and MED as Paradigms of Multiprotein Complexes

The schematic drawings of the different protein complexes on TSS regions (Fig. 3.4) focus only on the key proteins and are not in scale. For a better illustration of a multi-protein complex, we display TFIID in two different ways. In the schematic drawing shown in Fig. 3.5a all subunits of TFIID (TBP and TAFs 1–13, Table 3.3) are shown in correct stoichiometry and are scaled according to their relative molecular mass. This illustrates the size of the complex in relation to the core promoter and demonstrates that the different subunits can simultaneously contact different binding sites, such as a TATA box, an Inr element or a DPE site that spread over more than 50 base pairs in distance. Furthermore, the figure suggests that irrespective of the exact composition of binding sites within a given TSS region, the same large protein complex can be formed. Nevertheless, all protein complexes involved in transcriptional regulation have a dynamic structure, i.e., the different subunits assemble and dissociate, so that the detailed composition of the complex varies over time. The degree of this variance may depend on the binding sites found in the respective core promoter and may influence its interaction with other protein complexes.

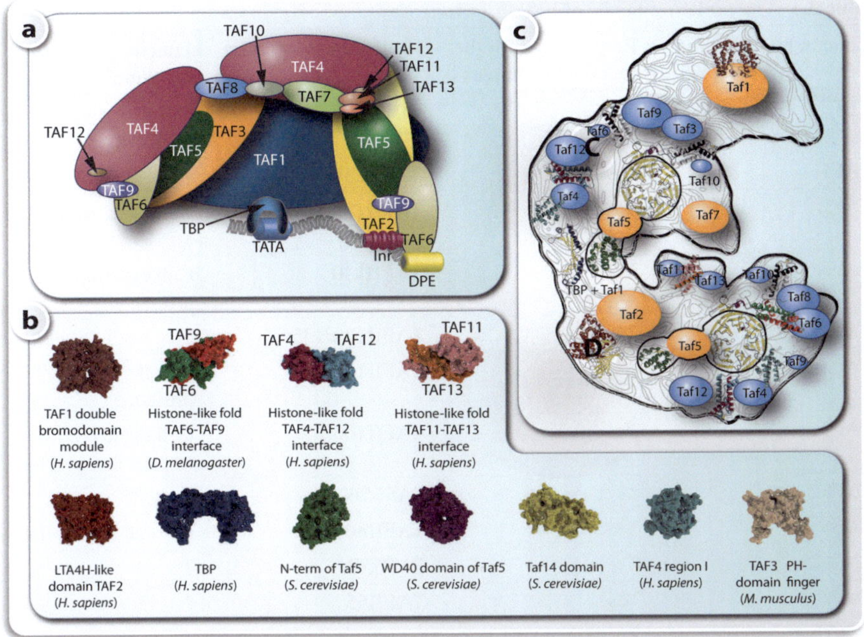

Fig. 3.5   TFIIB as a paradigm of a multiprotein complex. Schematic representation of the multi-protein complex TFIID, where the size of the different subunits is relative to their molecular mass **a**. Crystallized domains and folds of TAFs from different species **b**. With the exception of the histone fold domains, there is the same fold in TAF10-TAF8 and TAF10-TAF3 interacting surfaces. The LTA4H (leukotriene A4 hydrolase)-like domain that is homologous to TAF2 is based on human M1 aminopeptidase (PDB identifier 3B7S) and the characteristic WD40 propeller domain found in TAF5 is based on the carboxy-terminal domain of Tup1, which is a transcriptional corepressor in yeast (PDB identifier 1ERJ). Subunits of the yeast TFIID complex **c**. The known 3D structures of yeast Taf domains or their homology models are roughly positioned according to the available data on protein–protein interactions into an electron density map obtained from electron microscopic images. Tafs containing histone folds are displayed in blue. TBP is complexed with the TAND domain of Taf1

Figure 3.5b and c provide an even more realistic view on the TFIID complex. Crystal structure data of individual TFIID subunits were combined with an electron microscopic density map of the whole complex. The complex shown is from yeast, but the high evolutionary conservation of GTFs suggests that also the human TFIID complex has a comparable structure. The surface of this large multiprotein complex has a number of contact points for DNA (TBP, TAF1 and TAF4) that were already indicated in the schematic pictures of Fig. 3.4. In addition, the complex provides numerous interfaces for the interaction with other proteins, such as other GTFs, Pol II and members of the MED complex.

Hundreds of coactivator proteins are involved in the transfer of information from activated transcription factors, which bind to distal enhancers, to the basal transcriptional machinery. However, only a limited number of these coactivators directly

Table 3.3 The protein subunits of human TFIID

General nomenclature	Requirement for the functional complex	Human nomenclature	Function, activity or structural similarity
TBP	No	TBP	DNA binding to TATA box
TAF1	Yes	TAFII250	HAT
TAF2	Yes	TAFII150	Cell cycle (G1/S arrest)
TAF3*	Yes	TAFII140	Cell cycle (G2/M arrest)
TAF4*	?	TAFII130/135	?
TAF4B*	?	TAFII105	B-cell specific presence
TAF5	Yes	TAFII100	Cell cycle (G2/M arrest)
TAF5L	?	PAF65B	?
TAF6*	Yes	TAFII80	Histone H4 similarity
TAF6L	?	PAF65B	?
TAF7	Yes	TAFII55	?
TAF7L	?	TAF2Q	?
TAF8*	?	TAFII43	?
TAF9*	Yes	TAFII31/32	Histone H3 similarity
TAF9B	?	TAFII31L	?
TAF10*	Yes	TAFII30	Cell cycle (G1/S arrest)
TAF11*	Yes	TAFII28	Histone H3 similarity
TAF12*	Yes	TAFII20/15	Histone H2B similarity
TAF13*	Yes	TAFII18	Histone H4 similarity
TAF15	?	TAFII68	?

* TAFs with histone-like fold

interact with components of the basal transcriptional machinery, some of which are the subunits of the MED complex. Specific protein–protein interactions occur both between individual subunits of the MED complex and site-specific transcription factors as well as between the MED complex and Pol II. This suggests that **gene regulatory signals are processed through the MED complex**. Since the MED complex senses a multitude of different signals, it integrates them and consecutively delivers a properly calibrated output to the basal transcriptional machinery.

Most of the 26 subunits of the core MED complex (total size in humans 1.4 MDa) are evolutionarily conserved from yeast to humans. Based on their position within the complex the proteins belong to the head, middle, tail and kinase module (Fig. 3.6).

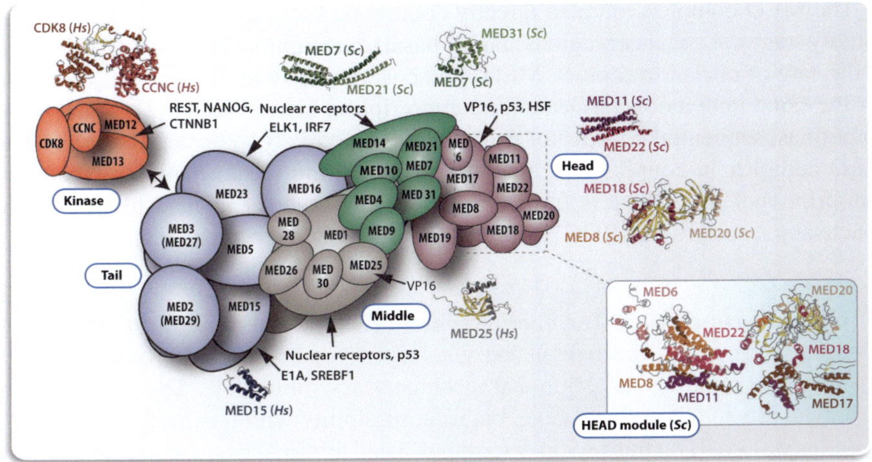

Fig. 3.6   The MED complex. The schematic structure of the human MED complex is displayed. The relative position of the subunits in the modules kinase (orange), tail (blue), middle (brown and green) and head (red) is based on the displayed cocrystal structures

The relatively stable core structure of the MED complex is formed by the modules head, middle and tail, while the components of the kinase module, CDK8, cyclin C (CCNC), MED12 and MED13, associate reversibly with the complex. Under these conditions MED26 dissociates, i.e., the active MED complex has 29 subunits and mediates then via recruitment of the super-elongation complex the activation of Pol II. The head and middle modules of the MED complex are involved in interactions with the basal transcriptional machinery, whereas all module subunits interact with various transcription factors. Since the kinase module interacts with Pol II, in its absence the MED complex rather exerts a repressive function on gene transcription.

The subunits of the MED complex show preference for different transcription factor classes. For example, MED1 is the major interaction partner of nuclear receptors, such as thyroid hormone receptor (THR) (Sect. 5.1), but members of this transcription factor superfamily can also bind to MED14. Moreover, MED1 interacts also with other transcription factors, such as GATA-binding protein 1 (GATA1). MED23 is the main sensor for mitogen-activated protein kinase (MAPK) signaling by interacting with the transcription factor ELK1 (ETS transcription factor ELK1) and in parallel one of the end points of signal transduction initiated by insulin. MED15 interacts with the cholesterol-sensing transcription factor SREBF1 (sterol regulatory element-binding transcription factor 1) and therefore belongs to the key regulators of lipid homeostasis. The tumor suppressor and transcription factor tumor protein p53 (encoded by the gene *TP53*) (Sect. 15.4) and the viral activator protein VP16 both interact with MED17. In addition, developmental and neuronal pathways interact with subunits of the kinase module. Taken together, **the MED complex is a signal-sorting center that is involved in the regulation of the transcription of nearly all human genes** and in parallel mediates the transactivation effects of most transcription factors (Sect. 4.2).

The MED complex can also directly coordinate between changes in chromatin activity stages of enhancer regions and the basal transcriptional machinery. However, in the case of nuclear receptors, MED1 and coactivators with HAT activity compete for the same interaction surface on the transcription factor (Sect. 5.2). Under these conditions, sequential coactivator exchange is more likely to occur. The role of the MED complex in coupling chromatin remodeling and the formation of the basal transcriptional machinery is further fine-tuned by other gene- and tissue-specific coactivators, such as PPARGC1A (proliferator-activated receptor γ, coactivator 1A).

(Clinical) conclusion: The core promoter (TSS region) is the most essential region of a gene. Throughout the whole genome on every TSS region the same type of basal transcriptional machinery assembles. The TSS region not only marks the 5'-end of a gene, but its accessibility within chromatin and the activity of Pol II on this region determines if and how much a gene is expressed.

Additional Reading

Cramer, P. (2019). Organization and regulation of gene transcription. Nature *573*, 45–54.

Girbig, M., Misiaszek, A.D. and Muller, C.W. (2022). Structural insights into nuclear transcription by eukaryotic DNA-dependent RNA polymerases. Nat Rev Mol Cell Biol *23*, 603–622.

Malik, S. and Roeder, R.G. (2023). Regulation of the RNA polymerase II pre-initiation complex by its associated coactivators. Nat Rev Genet *24*, 767–782.

Richter, W.F., Nayak, S., Iwasa, J. and Taatjes, D.J. (2022). The Mediator complex as a master regulator of transcription by RNA polymerase II. Nat Rev Mol Cell Biol *23*, 732–749.

Stark, R., Grzelak, M. and Hadfield, J. (2019). RNA sequencing: the teenage years. Nat Rev Genet *20*, 631–656.

Wang, H., Schilbach, S., Ninov, M., Urlaub, H. and Cramer, P. (2023). Structures of transcription preinitiation complex engaged with the +1 nucleosome. Nat Struct Mol Biol *30*, 226–232.

Chapter 4
Transcription Factors and Signal Transduction

Abstract In this chapter, we will present transcription factors as the key controllers of gene expression. The activities of these proteins critically influence how a cell functions and responds to environmental perturbations. The most characteristic domain of a transcription factor is its DBD, but the proteins also contain domains for homo- and heterodimerization and for contacts with cofactors and other nuclear proteins. The structural and functional understanding of site-specific transcription factors provides insight how they link to signal transduction and the sensing of intra and extracellular lipophilic molecules via nuclear receptors. Thus, a central characteristic of life, the response to molecules of the extracellular environment, is mediated by signal transduction cascades that mostly start with an extracellular signaling molecule and end with an activated transcription factor, i.e., with a change in gene expression.

Keywords DBD · Zinc finger · Helix-turn-helix · Homeodomain · Leucine zipper · Homodimer · Heterodimer · TFBS · Signal transduction

4.1 Site-Specific Transcription Factors and Their Domains

The binding of GTFs, such as TFIID, to TSS regions usually results in low transcriptional activity, i.e., **on its own the basal transcriptional machinery does not initiate any substantial mRNA production**. In contrast, gene transcription significantly increases when site-specific transcription factors contact their specific DNA binding sequences within enhancers, which are located proximal or distal to the gene's TSS(s). When both TSS region and enhancer are within open, accessible chromatin, the activity of site-specific transcription factors is critical in determining, whether and to what extent a given gene is expressed. Moreover, transcription can also be downregulated by transcription factors with repressive functions. The latter interfere with the activity and binding capacity of activating transcription factors (Sect. 4.2). In this way, they prevent the recruitment of the basal transcriptional machinery (Chap. 3) or recruit chromatin modifying enzymes that create repressive chromatin structures (Chap. 8).

C. Carlberg, *Gene Regulation and Epigenetics*,
https://doi.org/10.1007/978-3-031-68730-3_4

In the past, site-specific transcription factors were distinguished into those binding close to TSS regions and others being preferentially associated with distal enhancer, silencer or insulator regions. However, genome-wide analyses of TFBSs via ChIP-seq and similar methods (Sect. 6.1) indicate that this distinction is not appropriate. Binding sites of basically all site-specific transcription factors are found in any distance from TSS regions and DNA looping mechanisms allow them to come into close contact with the basal transcriptional machinery, i.e., the linear distance between the TSS and the enhancer is not critical (Fig. 2.1). Nevertheless, **the likelihood of a sequence-specific transcription factor to be involved in the control of the transcription of a given gene symmetrically decreases with its distance from the TSS**. Exceptions are, e.g., members of the E2F family, which are mostly found in core promoter regions, i.e., their binding pattern resembles that of components of the basal transcriptional machinery.

A typical transcription factor is composed of multiple domains that allow the functions:

- contacting sequence-specifically DNA (DBD)
- dimerizing with other transcription factors (dimerization domain)
- being activated via ligands or signal transduction pathways (transactivation domain), in order to contact via cofactors and the MED complex the basal transcriptional machinery.

For most transcription factors the DBD and the transactivation domain can be clearly distinguished, while dimerization activity is often attributed to both types of domains.

DBDs interact specifically with genomic DNA by recognizing base-specific surface features on the DNA molecule. Hydrogen-bond donor and acceptor groups exposed to the major groove of the DNA are the chemical groups that differ among the four bases (A, T/U, C and G) and permit discrimination between them. **Most of the protein-DNA contacts that mediate sequence-specific binding of transcription factors are hydrogen bonds**. An exception is the non-polar surface close to the C5 position of pyrimidines, where thymine can be distinguished from cytosine by its bulky methyl group. Protein-DNA contacts are also possible in the minor groove of the DNA, but the hydrogen-bonding patterns mostly do not allow base-specific contacts. Therefore, the dimension of the major groove limits the number of bases that are contacted by the DBD of a site-specific transcription factor to six, i.e., **the DNA recognition sequence of a single transcription factor normally is in maximum a hexameric motif**.

Furthermore, DBDs are rather small protein domains with an average size of 60–90 amino acids. However, only a few of these residues are used to interact with bases in the major groove of the DNA. These amino acids are often stably protruding from the protein surface and show positive (lysine (K) and arginine (R)) or negative (asparagine (N), glutamine (Q) and glutamic acid (E)) charge. Nevertheless, each base pair can be recognized in multiple ways by a transcription factor, i.e., **there is no simple amino acid-to-base code**.

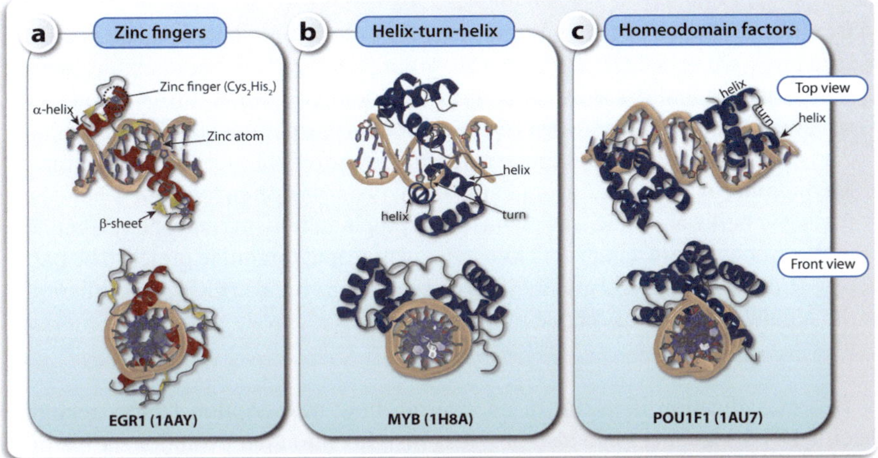

Fig. 4.1 Three main classes of DBDs. Representatives of the three main DBD classes, zinc finger **a**, helix-turn-helix **b** and homeodomain **c**, are displayed in two different orientations. Please note that homeodomains are a subgroup of helix-turn-helix motifs

The most common classification of transcription factors is based on the structure of their DBDs (Box 4.1). The major types of DBDs are:

- **Zinc finger** (Fig. 4.1a): A typical zinc finger consists of about 30 amino acid residues, four of which (either four cysteines (C4) or two cysteines and two histidines (C2H2)) coordinate a single zinc ion that stabilizes the 3D structure of the small motif. Since the interaction of a single zinc finger with DNA is weak, transcription factors have multiple zinc fingers that cooperatively contact DNA with a significantly enhanced affinity (see the example of CTCF in Sect. 7.3). The precise manner, in which zinc finger proteins bind to DNA, varies a lot and not all zinc fingers contain amino acids that recognize DNA in a sequence-specific way. Moreover, zinc fingers can also serve as RNA-binding motifs.
- **Helix-turn-helix** (Fig. 4.1b): This motif is formed by about 20 amino acids in two short α-helical segments, each 7–9 amino acid residues long that are separated by a β-turn. In order to form a stable structure, the two α-helices have to be supported by other helices of the DBD. One of the two α-helices, the recognition helix, protrudes from the DBD surface, so that it fits into the major groove of the DNA and makes there sequence-specific contacts.
- **Homeodomain** (Fig. 4.1c): This form of a DBD is a subtype of helix-turn-helix motifs. Its name derives from the regulation homeotic genes, such as those of the homeobox (*HOX*) family, that are critical for body pattern formation during development.

Box 4.1: Secondary Protein Structures The most common secondary protein structures are α-helices and β-sheets. Tight β-turns and loose, flexible loops link α-helices and β-sheets. In contrast, random coil is not a true secondary structure, but it is formed when regular secondary structures are absent. Amino acids have an individual preference to form the one or other secondary structure. Methionine, alanine, leucine, glutamate and lysine are often found in α-helices, while the bulky aromatic side chains of tryptophan, tyrosine and phenylalanine or the branched side chains of isoleucine, valine and threonine prefer to be part of β-strands. Proline and glycine often serve as helix breakers, since they disrupt the regularity of the backbone of α-helices.

This classification is useful in uncovering how transcription factors recognize specific DNA sequences and also provides insights into their evolutionary histories. **The three DBD classes account for a large part of all human transcription factors**. The superfamily of C2H2-type zinc finger proteins has 675 members and is probably so large, because this structural motif is rather insensitive against mutations happening during evolution. Moreover, zinc finger DBDs can be linked in a sequential manner, in order to extend its capacity to recognize a larger diversity of DNA-binding sites. Moreover, there are some 90 helix-turn-helix transcription factors and some 250 homeodomain transcription factors, for which the DBD provides clues to their function.

Furthermore, transcription factors can be distinguished via the domains that they use for protein–protein interactions, such as:

- **Leucine zipper** (Fig. 4.2a): This motif is formed by a pair of amphipathic α-helices carrying a series of hydrophobic amino acid residues on one side that provide with their hydrophobic surfaces the contact between two helices of the dimer. Very often leucine residues occur at every 7th position (please note that the helical repeat of an α-helix is 3.5 amino acids), forming a straight line along the hydrophobic surface. Leucine zipper proteins often have a separate DBD with a high concentration of positively charged amino acids (lysine and arginine) that interact with the negatively charged DNA backbone.
- **Basic helix-loop-helix (BHLH)** (Fig. 4.2b): A conserved region of about 50 amino acid residues is important for both DNA binding and protein dimerization. Two short amphipathic α-helices are linked to a loop of variable length, the helix-loop-helix. DNA binding is mediated by a neighboring short amino acid sequence that is rich in positively charged residues.
- **β-scaffold factors with minor groove contacts** (Fig. 4.2c): Some transcription factors, such as TBP, distort the DNA at their binding site by inserting amino acid side chains between the base pair, partially unwinding the helix and kinking it. The distortion is accomplished through a great amount of surface contact between the protein and the DNA. The transcription factors bind to the negatively charged DNA backbone through positively charged lysine and arginine residues. In case of TBP, the sharp bend in the DNA is produced through projection of four bulky phenylalanine residues into the minor groove.

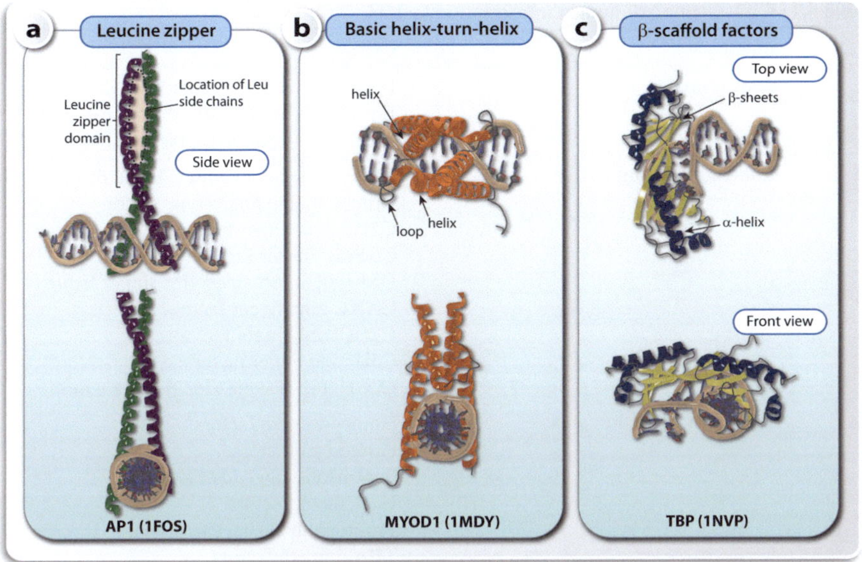

Fig. 4.2 Three main classes of protein–protein interaction modes of transcription factors. The DNA interaction of transcription factors is often directed by their mode of dimerization. Representatives of the three main groups, leucine zipper **a**, helix-loop-helix **b** and β-scaffold factors with minor groove contacts **c**, are displayed in two different orientations

Site-specific transcription factors have been related to a number of human diseases. At present, **more than 150 transcription factors are known to be directly responsible for some 300 diseases**, but far more transcription factor-disease associations will be identified. Many oncogenes, such as *MYC* (MYC proto-oncogene, BHLH transcription factor), *FOS* (FOS proto-oncogene, AP1 transcription factor subunit) or *JUN* (JUN proto-oncogene, AP1 transcription factor subunit), or tumor suppressor genes, such as *TP53*, code for transcription factors (Chap. 15). Importantly, **one third of human developmental disorders are attributed to dysfunctional transcription factor genes and proteins**. Furthermore, alterations in the activity and regulatory specificity of transcription factors majorly contribute to the phenotypic diversity of humans.

There are approximately 1600 human genes encoding for transcription factors, i.e., some 8% of all protein-coding genes. A classification of transcription factors based on their shared DBDs is provided in Table 4.1. A subset of all transcription factors, such as nuclear factor κB (NFκB) (Sect. 13.4), p53 (Sect. 15.4), JUN/FOS and the nuclear receptors estrogen receptor (ESR) 1 and 2 as well as androgen receptor (AR) (Sect. 5.1), have been most intensively studied. For example, there are more publications on p53, ESR1 and FOS than on the sum of all other transcription factors.

Table 4.1 Classification of human transcription factors

#	Superclasses	Classes
I	Basic domains	Basic leucine zipper factors (bZIP) Basic helix-loop-helix factors (bHLH) Basic helix-span-helix factors (bHSH)
II	Zinc-coordinating DBDs	Nuclear receptors with C4 zinc fingers Other C4 zinc finger-type factors C2H2 zinc finger factors C6 zinc cluster factors DM-type intertwined zinc finger factors CXXC zinc finger factors C2HC zinc finger factors C3H zinc finger factors C2CH THAP-type zinc finger factors
III	Helix-turn-helix domains	Homeodomain factors Paired box factors Fork head/winged helix factors Heat shock factors Tryptophan cluster factors TEA domain factors ARID domain factors
IV	Other all-α-helix DBDs	High-mobility group (HMG) domain factors Heterodimeric CCAAT-binding factors
V	α-helix exposed by β-structures	MADS box factors E2-related factors SAND domain factors
VI	Immunoglobulin fold	Rel homology region (RHR) factors STAT domain factors p53 domain factors Runt domain factors T-box factors NDT80 domain factors Grainyhead domain factors
VII	β-Hairpin exposed to an a/b-scaffold	SMAD/NF-1 DNA-binding domain factors GCM domain factors
VIII	β-Sheet binding to DNA	TATA-binding proteins A.T hook factors
IX	β-Barrel DBDs	Cold-shock domain factors
X	Not yet defined DBDs	AXUD/CSRNP domain factors NonO domain factors Leucine-rich repeat flightless-interacting proteins NFX1-type putative zinc finger factors GTF2I domain factors CG-1 domain factors Uncharacterized

This classification is based on information of the database TRANSFAC (www.edgar-wingender.de/huTF_classification.html)

Bioinformatic methods for the identification of critical genomic elements, such as TFBSs, are significantly improving in quality when they are "trained" by experimental data. In the past, in vitro approaches, such as gelshift or reporter gene assays, were used to define TFBSs being necessary for both basal transcriptional activity in TSS regions and for cell type-specific, hormonal or environmental transcriptional responses via enhancer regions. However, **nowadays most data on transcription factors and their binding sites are obtained by genome-wide approaches**, such as ChIP-seq (Sect. 6.2). For example, the de novo motif analysis of sequences below ChIP-seq peak summits provides far more reliable data on the specificity of the DNA recognition of a transcription factor than previous position weight matrix analysis based on DNA sequence comparison (Box 4.2). Therefore, the results of Big Biology projects, such as *ENCODE*, *FANTOM5* or *Roadmap Epigenomics*, are very important for the systematic analysis of transcription factor binding, locations of histone modifications and other genome-wide features of gene regulation (Sect. 6.1).

> **Box 4.2: Bioinformatic Identification of TFBSs** The large size of the human genome (3.05 Gb) and the huge number of site-specific human transcription factor genes (some 1600) can only be handled by the use of bioinformatic methods. Most binding site screening studies take the assumption that transcription factors recognize in vivo the same binding motifs as those identified by in vitro studies. Table 4.2 lists the sequence logos of a number of other important transcription factors. However, any in silico screening tends to over-predict binding sites by a factor of up to 1000 (called the futility theorem). In fact, the vast majority of the predicted binding sites will not be used in vivo although the transcription factor would bind to them in vitro, i.e., **in most genomic regions containing a transcription factor consensus binding motif the respective site may not be accessible due to tight chromatin packing**. Moreover, DNA methylation of a crucial cytosine within the binding motif can change the affinity for the transcription factor (Sect. 7.3).

Table 4.2 Sequence logos of key TFBSs

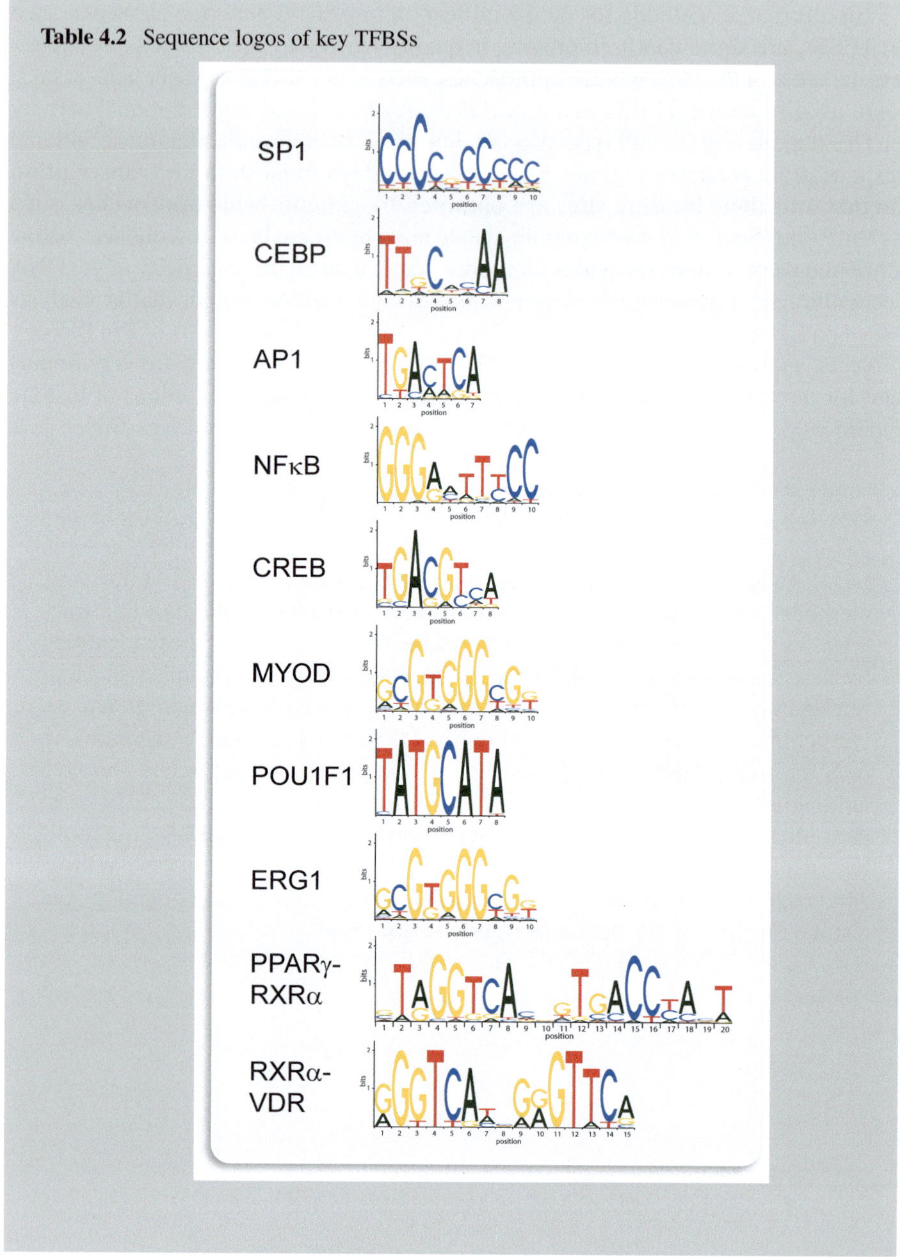

4.2 Classification of Transcription Factors

The approximately 3200 human transcription factors (encoded by some 1600 genes) are often classified by the mode of their activation (Fig. 4.3):

- **Constitutively active transcription factors**. These transcription factors can be subdivided into two main groups.
- **Ubiquitous transcription factors**. These are a smaller group of site-specific transcription factors that are always located in the nucleus, such as SP1, CEBPs and nuclear factor 1 (NF1). These proteins are primarily involved in the regulation of housekeeping genes, such as structural proteins like actin or metabolic enzymes like glyceraldehyde phosphate dehydrogenase.
- **Cell type-specific transcription factors**. The process of development is critically dependent on sequential waves of activating cell type-specific transcription factor genes. These are genes for developmental transcription factors, such as the members of the *HOX* gene cluster and the helix-loop-helix protein MYOD1 (myoblast determination protein 1) that is central in muscle differentiation. The expression of these transcription factors is mostly limited in time, such as during embryogenesis, but they do not need any additional signals to be active. However, their activity is often regulated by posttranslational modifications, such as phosphorylations. The expression of an individual developmental transcription factor is not necessarily tissue-specific, but the combinatorial distribution of multiple of such proteins contributes to cell type determination and differentiation.

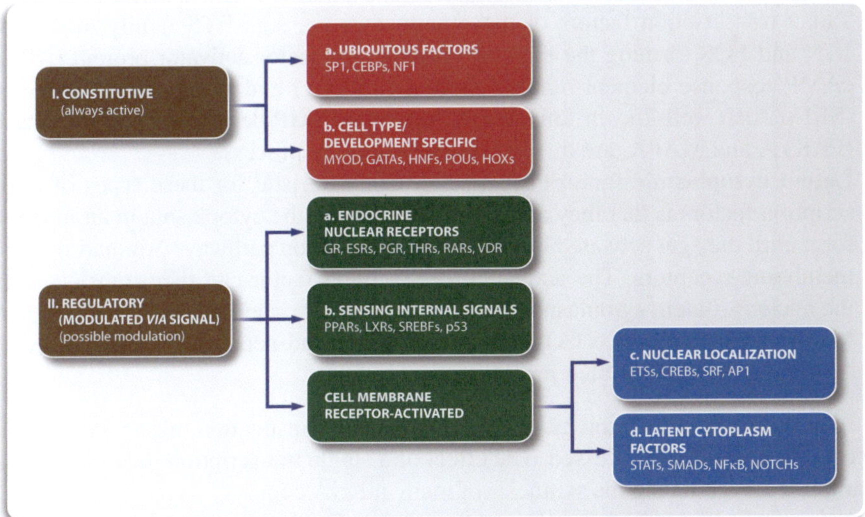

Fig. 4.3 Functional classification of positive-acting transcription factors. Most transcription factors can be classified by the way of their activation

- **Signal-dependent transcription factors**. These transcription factors (or their precursors) are inactive (or minimally active) until the cell is exposed to an appropriate intra- or extracellular signal. They can be subdivided into four main groups:
- **Endocrine nuclear receptors**. Some members of this transcription factor super-family (Sect. 5.1) can get activated by small lipophilic endocrine ligands, such as cortisol, the steroid hormones estrogen, progesterone and testosterone, the vitamin A and vitamin D derivatives all-*trans* retinoic acid (RA) and $1\alpha,25$-dihydroxyvitamin D_3 ($1,25(OH)_2D_3$), respectively, and the thyroid hormone triiodothyronine (T_3). Some of these endocrine nuclear receptors settle down on genomic DNA even before they bind their ligand.
- **Transcription factors activated by internal signals**. These transcription factors are activated by intracellular signaling molecules. In case of SREBF1, internal sterol concentrations regulate the proteolysis of a membrane protein precursor of SREBF1. Also adopted orphan nuclear receptors, such as peroxisome proliferator-activated receptor (PPAR) and liver X receptor (LXR) and the sensor for DNA damage, p53, belong to this group.
- **Constitutive transcription factors activated by serine phosphorylation**. When small hydrophilic signaling molecules, such as epinephrine, or peptide hormones bind to their respective G-protein coupled receptor (GPR) proteins, intracellular second messengers, such as cAMP (cyclic adenosine monophosphate), diacyl-glycerol (DAG) and Ca^{2+}, trigger serine kinase cascades and phosphorylation of transcription factors. Similarly, the activation of receptor tyrosine kinases (RTKs) by smaller proteins, such as growth factors and cytokines, or the peptide insulin finally also leads to serine kinase cascades and transcription factor activation. Target transcription factors of these pathways are, e.g., ETS family members, JUN and FOS forming the dimeric transcription factor activator protein (AP1), cAMP response element-binding protein (CREB1) and serum response factor (SRF). Also well-known kinases, such as the cAMP-dependent protein kinase (PRK) A and MAPK, are involved in the signaling process.
- **Latent cytoplasmic factors** (Sect. 4.3). Characteristic for these types of tran-scription factors is that they are initially located in the cytoplasma in an inactive form until they get activated by signaling transduction pathways originating from membrane receptors. The activated transcription factor can then translocate to the nucleus. Latent cytoplasmic transcription factors, such as SMADs (Sma- and Mad-related proteins), STAT (signal transducer and activator of transcription), NFκB and NOTCH (Notch receptor).

Like for most other proteins involved in signal transduction, also transcription factors are not highly expressed. The effect of a single transcription factor molecule is amplified by inducing the synthesis of many mRNA copies of its target genes, i.e., there is no need for a high number of them per cell. Moreover, the low expression level for transcription factor genes allows an easier triggering of a regulatory event by altering transcription factor concentrations or activity. Nevertheless, the number of transcription factors range from approximately 100 molecules per cell for a highly specific proteins regulating only a few target genes up to more than 100,000 per cell for ubiquitous factors being involved in the control of most genes, such as SP1.

Transcription factors act not only as activators of gene expression but can also function as **repressors**. In fact, it is not easy to classify transcription factors as definitively activating or repressive. There are approximately 400 different domains in transcription factors that are not related to DNA binding, a number of which have an repressive effect on gene expression. These non-DBD domains vary significantly in their abundance. For example, the repressive KRAB (Krüppel associated box) domain is in roughly 350 human transcription factors. There are several mechanisms of transcriptional repression, such as the competition for the binding an activator or cofactor as well as the binding to TFBS within an enhancer. This will be illustrated at the example of the inhibition of NFκB signaling in Sect. 13.4 (Fig. 13.10). In addition to this short-range repression, there is also long-range repression, which involves the inactivation of enhancers by PRCs binding to H3K27me3 marks (Sect. 2.4) or the RCOR (REST corepressor) complex (Sect. 14.3). These gene repression-mediating enhancers are referred to as **silencers**.

4.3   Activation of Transcription Factors

The activation of membrane receptors by hydrophilic signaling molecules leads either to the stimulation of a kinase signaling pathway resulting in the phosphorylation of target transcription factors, such as CREB1, AP1, ETS and others (Fig. 4.4, left), or in the activation of latent transcription factors (Fig. 4.4, right). **Transcription factors are called latent, when they are activated through their translocation from the cytoplasma to the nucleus**. Latent transcription factors are involved in a number of critical signal transduction pathways, the most important of which are:

- **SMAD pathway**. The family of TGFβ (transforming growth factor β) contains 33 structurally related growth and differentiation factors that include TGFβs, activins, nodal and BMPs (bone morphogenetic proteins). TGFβ family members activate cells via a complex of two types of serine-threonine kinase membrane receptors. Ligand binding to this complex induces phosphorylation and activation of type I receptors by type II receptors that leads to the activation of SMAD transcription factors and their translocation to the nucleus. In response to the activation of the membrane receptors by ligand binding, the effector SMADs are phosphorylated (e.g., SMAD1, SMAD5, SMAD8 and SMAD9 in response to BMP SMAD2 and SMAD3 in response to activins or nodal). SMAD4 and SMAD10 are cofactors of the effector SMADs, while SMAD6 and SMAD7 block SMAD4 binding, i.e., they are negative regulators. SMADs form heterodimeric complexes with partner transcription factors, in which the partner primarily mediates the DNA contact and the SMADs the transactivation. The specificity of SMADs in response to different ligands is related to the selection for their heterodimeric partner proteins.

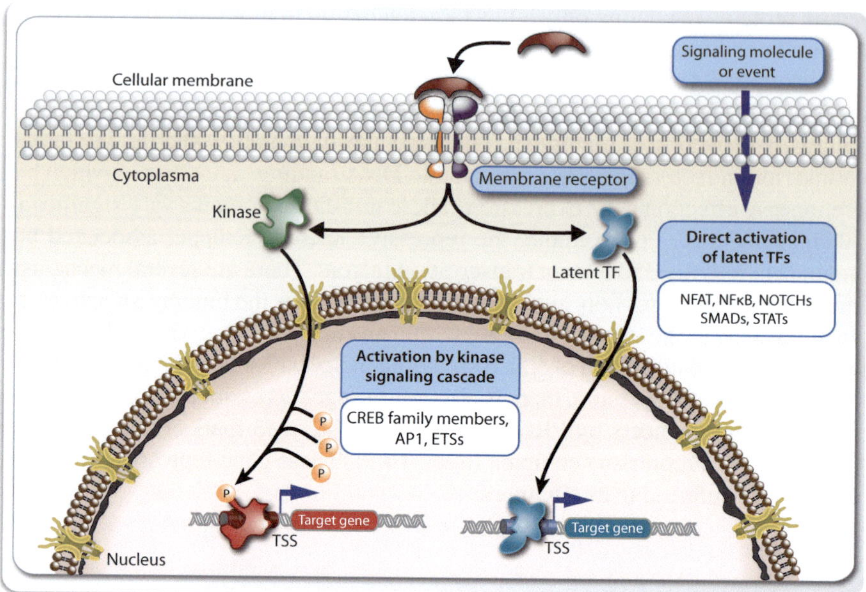

Fig. 4.4 Activation of transcription factors by membrane receptor signaling pathways. There are two major pathways, how activated membrane receptors can activate transcription factors. Either they stimulate kinase signaling cascades that result in phosphorylation of various resident nuclear transcription factors in the nucleus (**left**), or they induce the translocation of latent transcription factor from the cytoplasma to the nucleus (**right**)

- **STAT pathway**. Some 20 different cytokines activate via their membrane receptors Janus kinases (JAKs) that phosphorylate the ligand-bound receptor and the associated STAT transcription factor at tyrosines. Phosphorylated STATs translocate into the nucleus and bind as dimers to their genomic binding sites. There are seven different STATs forming a number of homo- and heterodimeric complexes that differ in the cytokine mediating their activation. STATs can also be activated by RTKs, like epidermal growth factor receptor (EGFR), by non-RTKs, such as SRC (SRC proto-oncogene, non-receptor tyrosine kinase) and ABL1 (ABL proto-oncogene 1, non-receptor tyrosine kinase) as well as through GPRs.
- **NFκB pathway**. The five members of the NFκB transcription factor family are activated by a variety of extracellular products, most of which are related to inflammation, such as tumor necrosis factor (TNF), interleukin (IL) 1β, growth factors, infections by bacteria and viruses, oxidative stress and a number of synthetic compounds. More details about this pathway will be provided in Sect. 13.4.
- **Hedgehog pathway**. Hedgehog is a lipid-anchored cell surface ligand that binds to the patched receptor (PTCH), relieving PTCH-mediated inhibition of the GPR Smoothened. Smoothened signaling leads to activation of the transcription factor GLI family zinc finger (GLI). The *PTCH* gene is also a target of GLI, forming a negative feedback loop.

- **Wingless-type (WNT) pathway**. The more than 30 members of the WNT (wingless-type MMTV integration site family member) family act as ligands to receptors of the Frizzled family. The first intracellular step in WNT signaling is a phosphorylation of the protein dishevelled segment polarity protein (DVL) through GPR activation induced by WNT-Frizzled interaction that inhibits the kinase glycogen synthase kinase 3 (GSK3). GSK3 controls a proteolytic cascade that prevents nuclear accumulation of the cofactor protein CTNNB1 (catenin β1). When WNT binds Frizzled, activated DVL blocks GSK3 phosphorylation and subsequent the proteolysis of CTNNB1. Cytoplasmic CTNNB1 levels rise, the protein enters the nucleus, where it participates in gene activation via binding to the transcription factor TCF7L2 (transcription factor 7-like 2, also named TCF4).
- **NOTCH pathway**. The NOTCH signaling pathway is essential for proper embryonic development. The four NOTCH proteins are single-pass receptors that are activated by Delta, Delta-like and Jagged ligands. Interaction with these ligands leads to proteolytic cleavage of NOTCH so that NICD (NOTCH intracellular domain) is liberated from the cellular membrane. NICD translocates to the nucleus, where it forms a heterodimer with the helix-loop-helix transcription factor RBPJ (recombination signal-binding protein for immunoglobulin kappa J region L). Through interaction with NICD, RBPJ changes its interacting cofactors from a corepressor-histone deacetylase (HDAC) complex to a coactivator-HAT complex that then leads to the activation of NOTCH target genes.
- **Nuclear factor activated T cells (NFAT) pathway**. The family of NFAT transcription factors shows homology with those of the NFκB family, but their members are differently regulated. Cytoplasmic NFAT is heavily phosphorylated in resting immune cells, but activation of the T cell receptor (TCR) results in fluctuations of the internal concentrations of the second messenger Ca^{2+} in a cyclical fashion. Ca^{2+} activates the phosphatase calcineurin, resulting in dephosphorylation of NFAT and the accumulation of the transcription factor in the nucleus. NFAT is a rather weak DNA-binding protein that in most cases needs the support of other transcription factors, such as AP1.

(Clinical) conclusion: Transcription factors bind sequence-specifically to genomic DNA and are often the endpoints of signal transduction cascades. In this way, intra- and extracellular signals to cells are translated to activities of enhancer regions and in changes of gene expression. Therefore, in health and disease transcription factors play a key role in most physiological processes, such as inflammation, differentiation and cancer.

Additional Reading

de Boer, C.G. and Taipale, J. (2024). Hold out the genome: a roadmap to solving the cis-regulatory code. Nature *625*, 41–50.

Lambert, S.A., Jolma, A., Campitelli, L.F., Das, P.K., Yin, Y., Albu, M., Chen, X., Taipale, J., Hughes, T.R. and Weirauch, M.T. (2018). The human transcription factors. Cell *172*, 650–665.

Lim, B., Domsch, K., Mall, M. and Lohmann, I. (2024). Canalizing cell fate by transcriptional repression. Mol Syst Biol *20*, 144–161.

Marazzi, I., Greenbaum, B.D., Low, D.H.P. and Guccione, E. (2018). Chromatin dependencies in cancer and inflammation. Nat Rev Mol Cell Biol *19*, 245–261.

Stadhouders, R., Filion, G.J. and Graf, T. (2019). Transcription factors and 3D genome conformation in cell-fate decisions. Nature *569*, 345–354.

Chapter 5
A Key Transcription Factor Family: Nuclear Receptors

Abstract In this chapter, a highly interesting family of ligand-inducible transcription factors, referred to as nuclear receptors, will be presented. Nuclear receptors serve since some 30 years as paradigms for many functional and structural aspects of transcription factors, such as the interaction with coactivator and corepressor proteins. Many of the 48 human members of the largest family of transcription factors in metazoans have the unique property to be specifically activated by small lipophilic ligands in the size of cholesterol (approximately 400 Da). Some of these ligands are known as endocrine hormones, such as estradiol and testosterone, while others are metabolites of dietary compounds, such as fatty acids and cholesterol. Both types of lipids are of large physiological impact in health and disease, such as metabolic control and the modulation of immunity. This made nuclear receptors especially attractive for both basic and applied research.

Keywords Nuclear receptor · Superfamily · DNA binding · REs · Dimerization · Ligand-binding domain · Ligand-binding pocket · Transactivation · Repression · Coactivator · Corepressor · Nutrient sensor

5.1 The Nuclear Receptor Superfamily

Many signals to which a cell is exposed to, such as growth factors, cytokines and other hydrophilic signaling molecules, cannot pass the cellular membrane. Therefore, these signaling molecules need to interact with a membrane receptor, in order to stimulate a signal transduction cascade that eventually leads to the activation of a transcription factor. In contrast, lipophilic signaling molecules, such as steroid hormones, have more straightforward signal transduction pathways, since these compounds can pass cellular membranes and bind directly to a transcription factor that is often already located in the nucleus (Fig. 5.1). Therefore, these transcription factors are called **nuclear receptors**.

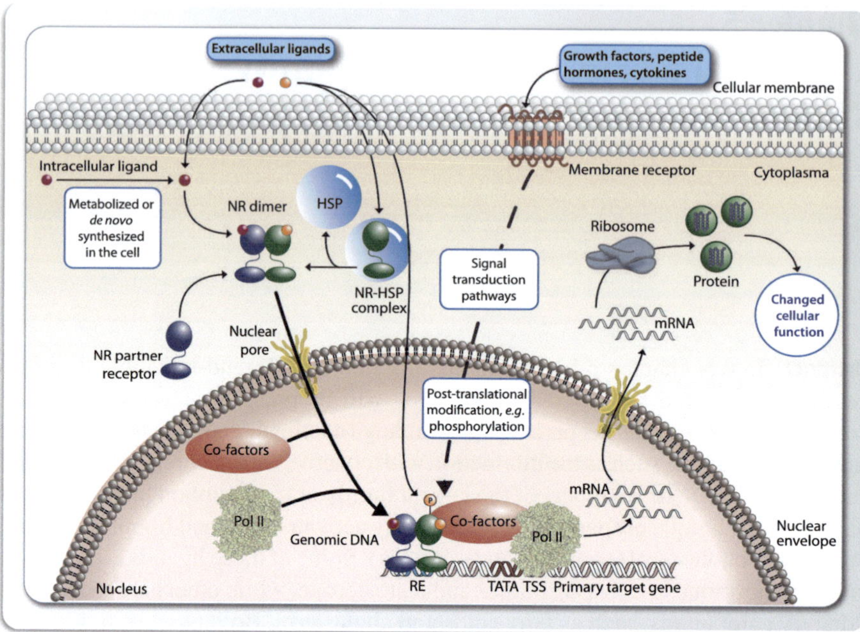

Fig. 5.1 Principles of nuclear receptor signaling. Nuclear receptors (NRs) reside either in the cytoplasma (e.g., AR and GR) in a complex with chaperone proteins, such as HSP, or are already located in the nucleus (like most nuclear receptors), when they are activated through the binding of their specific lipophilic ligand. The ligand is either of extracellular origin and has passed the cellular membrane or is a metabolite that was synthesized inside of the cell. Cytoplasmic nuclear receptors dissociate then from their chaperones and translocate to the nucleus, where they bind to their genomic binding sites (REs) in the vicinity of their target genes. Ligand-activated nuclear receptors interact with cofactors that build a bridge to the basal transcriptional machinery with Pol II in its core. This leads then to changes in the mRNA and protein expression of these target genes

Some nuclear receptors, such as GR or AR, wait in the cytoplasma for the arrival of their specific ligands, while most other nuclear receptors reside already in the nucleus and get activated there (Fig. 5.1). Nuclear receptors locating in the cytoplasm, yet in the absence of their specific ligand, are complexed with chaperones, such as heat shock proteins (HSPs). The dissociation of these chaperones after ligand binding allows the translocation of the transcription factors into the nucleus. Nuclear receptors bind as homo- or heterodimers to their specific genomic binding sides, often referred to as response elements (REs), in enhancer regions being located within the same TAD as the TSS regions of their primary target genes. Ligand-activated nuclear receptors preferentially interact with coactivator proteins that then together with proteins of the MED complex build a protein bridge to the basal transcriptional machinery (Sect. 3.3). This transactivation process leads to expression changes in the respective primary nuclear receptor target genes and eventually into changes of cellular functions. The straightforward signal transduction process of nuclear receptors can interfere with other signaling pathways that start at membrane receptors. This often results in the posttranslational modification of nuclear receptors or their coactivators, i.e., **nuclear receptors are directly or indirectly also sensitive to membrane signaling pathways**.

The members of the nuclear receptor superfamily are defined by their very conserved DBD. In humans there are 48 protein-coding genes for nuclear receptors, 12 of which belong to the subfamily of endocrine receptors (Fig. 5.2). These nuclear receptors have been identified, when one was looking for receptors of the already well-characterized hormones, such as the steroids testosterone, estradiol, progesterone, cortisol and aldosterone, the vitamin derivatives all-*trans* RA, $1,25(OH)_2D_3$ and the thyroid hormone T_3. **All these lipophilic hormones circulate in the serum at low nanomolar concentrations and, accordingly, their specific nuclear receptors bind them with K_d-values in the same nanomolar range**. Interestingly, testosterone, progesterone, cortisol, aldosterone and $1,25(OH)_2D_3$ have each only one specific receptor, estrogens and T_3 have two receptor isoforms (α and β) and for all-*trans* RA there are even three receptor subtypes (α, β and γ).

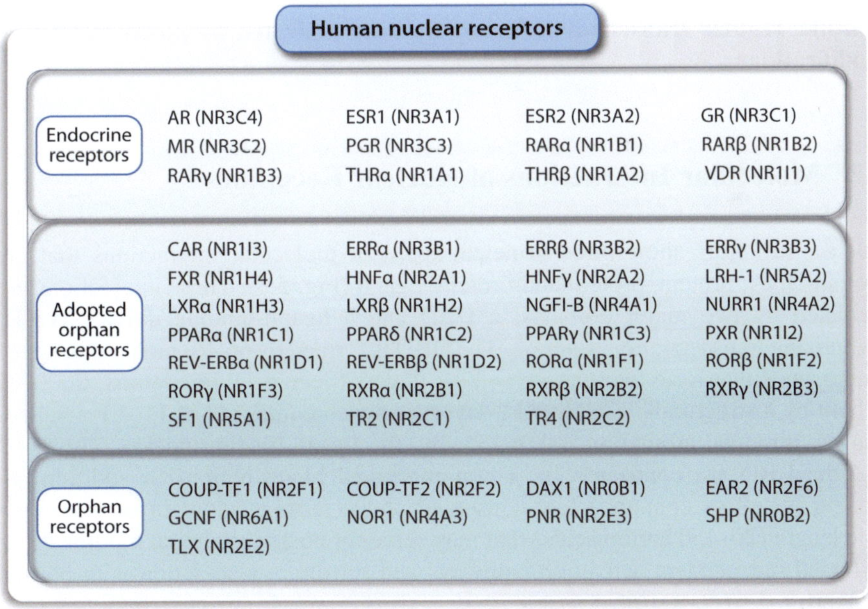

Fig. 5.2 The nuclear receptor superfamily. The 48 human members of the nuclear receptor superfamily are divided into three classes based on their nature and affinity for their ligand. The members of the family are defined by a highly conserved DBD. Please note that the receptors DAX and SHP lack a DBD

When the 36 remaining members of the nuclear receptor superfamily were cloned, they were termed "orphans", because instantly their ligand was not known. For some of these orphan nuclear receptors natural or xenobiotic compounds have been identified as ligands and the receptors were termed "adopted orphans". Interestingly, the natural ligands of adopted orphan receptors are all dietary lipids and their derivatives, such as bile acids for the farnesoid X receptor (FXR), oxysterols for LXRs, fatty acids for PPARs and 9-*cis* RA for retinoid X receptors (RXRs). Also for these receptors the K_d-values for their ligands were found to be in the same concentration range as the circulating concentrations, which, however, are in part in the low millimolar range. Thus, **some nuclear receptors represent true sensors for micro- and macronutrients.** In contrast, other nuclear receptors, such as HNF (hepatocyte nuclear factor) 4α and 4γ, LRH-1 (liver receptor homolog-1), REV-ERB (Reverse-Erb) α and β, ROR (RAR-related orphan receptor) α, β and γ as well as SF1 (steroidogenic factor 1), bind nutrient derivatives like fatty acids, phospholipids, heme and sterols, but **this interaction is constitutive and does not represent a sensing process.**

Simple eukaryotes, such as fungi, do not have genes encoding for nuclear receptors. It is assumed that the first nuclear receptors were orphans developing in metazoans as environmental sensors for nutritional compounds and toxins. In contrast, endocrine receptors are a rather recent evolutionary development. In this way, the three classes of the nuclear receptor superfamily (Fig. 5.2) represent the different stages of the family's evolutionary development. This implies that true orphan receptors have a too small ligand-binding pocket to harbor a ligand. Thus, **they function like regular transcription factors and are activated by posttranslational modifications.**

5.2   Molecular Interactions of Nuclear Receptors

Nuclear receptors show three principal types of molecular interactions that are protein-DNA, protein–protein and protein–ligand (Fig. 5.3). These interactions are mediated by two major domains, a DBD and a ligand-binding domain (LBD) comprising transactivation function. The DBD is formed by 66–70 highly conserved amino acids that form two Cys_4-type zinc fingers (Sect. 4.1). In contrast, the LBD is built by a structurally conserved 3-layer sandwich composed of 11–13 α-helices that are arranged around an internal cavity, the ligand-binding pocket. The DBD and the LBD are connected by a non-conserved hinge region. In addition, all nuclear receptors contain a low conserved amino-terminal domain of very variable length (20–450 amino acids) that may serve for posttranslational modifications in ligand-independent activation pathways and for direct association with partner proteins.

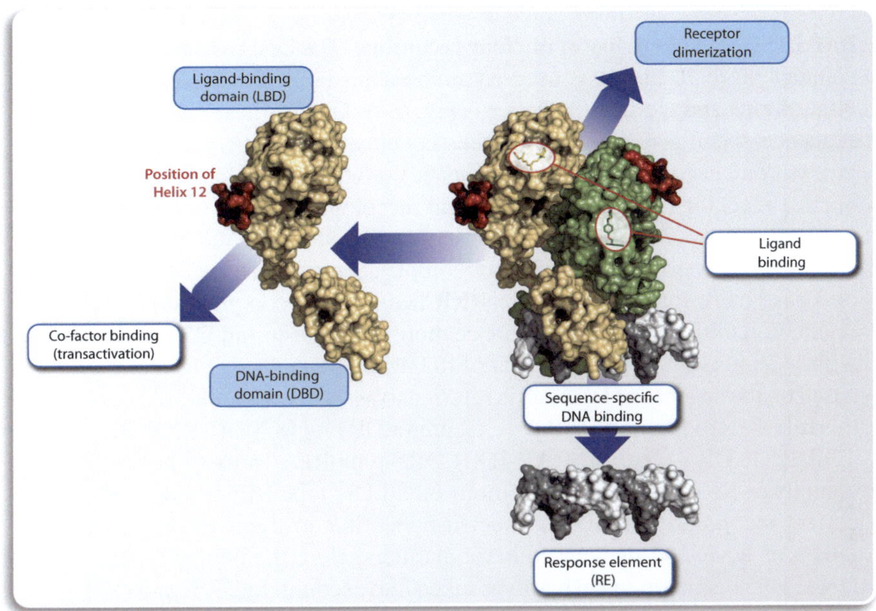

Fig. 5.3 Different molecular interactions of a nuclear receptor. Nuclear receptors are performing three types of molecular interactions. With their DBD they specifically contact DNA, with the inner surface of their LBD, the ligand-binding pocket, they bind their specific ligand and with the outer surface of their LBD they interact with partner nuclear receptors and cofactor proteins. DBD-DBD interactions direct the specificity of nuclear receptor dimers for their distinct REs. Ligand binding induces a conformational change within the LBD primarily affecting the most carboxy-terminal α-helix, helix 12. In this changed conformation the LBD has a significantly higher affinity for coactivator proteins. This ligand-induced protein–protein interaction is the core transactivation mechanism of nuclear receptors

Only a few members of the nuclear receptor superfamily, such LRH-1 and RORs, have already as a monomer sufficient affinity for DNA. Therefore, **most nuclear receptors have to interact with a partner receptor, in order to bind efficiently to DNA**. The dimerization partner is often the same type of receptor (e.g., in the case of all steroid receptors), so that the resulting complex is a homodimer. However, for 14 members of the nuclear receptor superfamily (RARs (retinoid acid receptors), THRs, VDR (vitamin D receptor), FXR, LXRs, PPARs, constitutive androstane receptor (CAR) and pregnane X receptor (PXR) (Fig. 5.2) the preferential coreceptor is RXR, i.e., they form heterodimers. Consecutively, nuclear receptor REs are two copies of the hexameric motif AGGTCA oriented as direct repeats (DRs), everted repeats or inverted repeats (Box 5.1).

Box 5.1: DNA specificity of nuclear receptors The DBD of nuclear receptors contains 66 to 70 highly conserved amino acids being composed of two zinc finger loops and a pair of α-helices (Fig. 5.4). One of these helices mediates sequence-specific recognition of the sequence AGGTCA via typical major groove contacts. Due to the high sequence conservation of the DBD within the nuclear receptor superfamily, individual receptor specificity and RE diversity is generated by the distance and relative orientation of the two AGGTCA sites. Homodimeric steroid receptor complexes prefer inverted repeats with 0 or 3 base pairs spacing, while for RXR heterodimer complexes the preferred orientation of the hexameric sequence motif is a head-to-tail DR arrangements with 1–5 intervening base pairs (DR1 to DR5). The pattern of RE selectivity is based on the spacing of DRs and is referred to as the "1-to-5 rule". According to this rule, heterodimers of PPAR-RXR prefer DR1-type, VDR-RXR DR3-type, THR-RXR DR4-type and RAR-RXR DR5-type REs. In these heterodimeric complexes RXR binds to the 5'-motif on all DR-type REs besides DR1. The correct recognition of REs is directed by steric constrains of the interacting DBDs of RXR and its heterodimeric partners. Here the helical repeat of the DNA, 10.5 base pairs/turn, has to be taken into account. In DR3- and DR1-type REs the DBDs are considerably tiled against each other (Fig. 5.4a and b), while in DR4- and DR5-type REs the DBDs of RXR and THR or RAR, respectively, are positioned to the same side of the DNA (Fig. 5.4c). In contrast, the nuclear receptors for steroids, such as GR, AR, mineralocorticoid receptor (MR) and progesterone receptor (PGR), form homodimers on two copies of AGAACA motifs in an inverted repeat orientation.

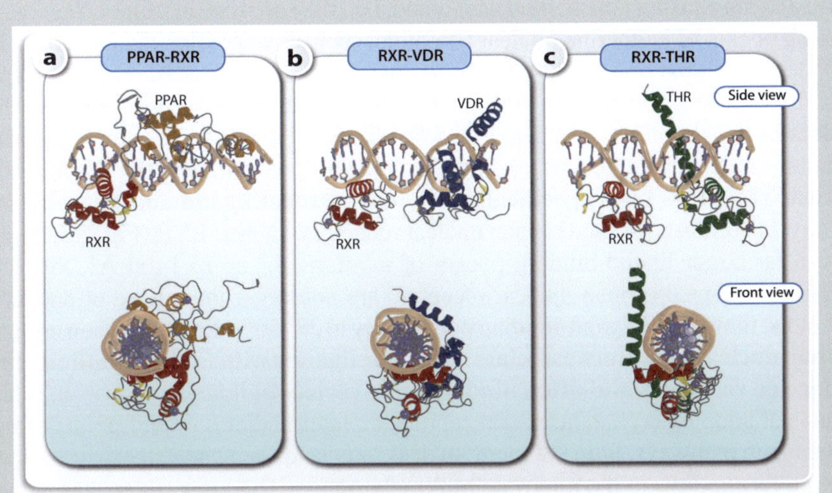

Fig. 5.4 Binding of RXR heterodimers to DR-type REs. Heterodimeric DBD complexes of RXR with PPAR on a DR1-type RE **a**, with VDR on a DR3-type RE **b** and with THR on a DR4-type RE **c** are displayed in two different orientations

The here described example of dimerizing nuclear receptors and the specific structure of their REs can be transferred to other transcription factors, i.e., it is a general principle that **most transcription factors act as dimers**. Monomeric transcription factor DBDs often recognize within the major groove of the DNA 3–6 base pairs in a sequence-specific way. Depending on the interaction of the DBDs of the dimerizing transcription factors, the individual binding motifs may be spaced by up to 5 base pairs. This explains why the length of TFBSs is in the range of 6–17 base pairs. In cases, where a transcription factor has multiple DBDs, such as CTCF (Sect. 7.3), or forms even tetrameric complexes, such as p53 (Sect. 15.4), these binding sites can even be longer.

The comparison of the crystal structures of the LBDs of the orphan receptor NOR1 (neuron-derived orphan receptor 1), the endocrine receptor VDR and the adopted orphan receptor PXR makes the **structural conservation** of this domain clearly visible (Fig. 5.5a). The lower part of each LBDs is more flexible than the top part and leaves space for an internal cavity, the ligand-binding pocket, of variant volume. Orphan nuclear receptors, such as NOR1 (Fig. 5.5a, left), lack this open space and thus are not able to bind any ligand, i.e., they are **true orphans**. The ligand-binding pocket of endocrine nuclear receptors, such as VDR (Fig. 5.5a, central), has a moderate volume of 300–700 Å^3. For comparison, the volume of nuclear receptor ligands is in the order of 250–400 Å^3 (Fig. 5.5b) that roughly corresponds to their molecular weight of 260–600 Da (Fig. 5.5c). They fill the ligand-binding pockets of endocrine nuclear receptors by 60–80%. This explains why **most of the 12 endocrine nuclear receptors bind specifically only one natural ligand** and this with high affinity. In contrast, adopted orphan nuclear receptors, such as PXR (Fig. 5.5a, right), have a far larger ligand-binding pocket of a volume of up to 1400 Å^3. Since the ligands of adopted orphan nuclear receptors are not larger than those of endocrine receptors, they fill the ligand-binding pocket only to 25–50%. For this reason **adopted orphan nuclear receptors associate with their ligands with far lower affinity than endocrine receptors and often bind a larger variety of ligands**. However, typical ligands of adopted orphan nuclear receptors are intermediates or end-points of lipid metabolism pathways. Some of them, such as fatty acids and cholesterol, have steady state concentrations in the micro- to millimolar range. Therefore, there was no need of their respective nuclear receptors to evolve a more specific ligand-binding pocket.

The binding of a specific ligand to amino acids within the ligand-binding pocket of a nuclear receptor results in a number of positional changes of α-helices of the LBD that affect also its outer surface. In case of endocrine nuclear receptors such conformational changes are visible via a reorientation of helix 12 (red in Fig. 5.6a). Like a mouse-trap the helix flips its position after ligand binding. However, in the absence of a ligand, corepressor proteins efficiently associate with the LBD, but in its changed position helix 12 prevents this interaction and favors a contact with coactivator proteins. In this way, **ligand binding changes the profile of interacting partner proteins and consequently also the function of the LBD**.

Figure 5.6b illustrates a 3-step transactivation process that is valid for nuclear receptors residing in the nucleus. In the absence of a ligand or in the presence of an antagonist, the DNA-bound dimeric nuclear receptor complex interacts with core-pressor proteins, such as NCOR (nuclear receptor corepressor) 1 or 2. The nuclear receptor is connected via these corepressors with a multiprotein complex that contains chromatin modifying enzymes, such as HDACs. The action of these nuclear enzymes leads to local chromatin condensation, i.e., the nuclear receptor target genes do not get transcribed. **The binding of an agonistic ligand to the nuclear receptor LBD leads**

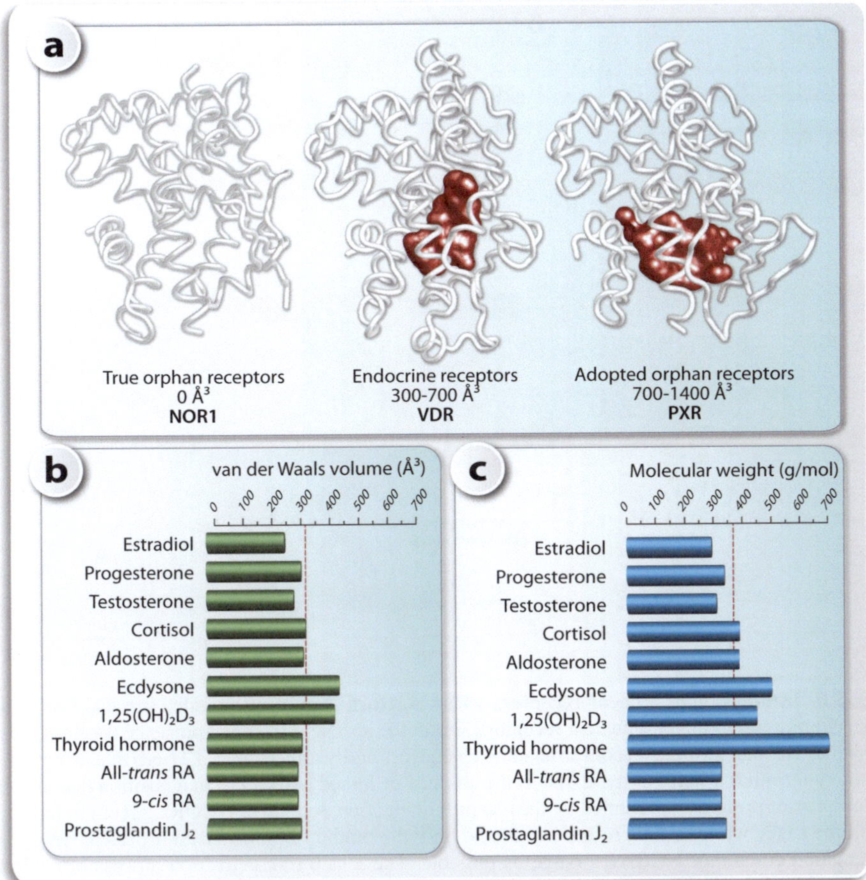

Fig. 5.5 The volume of ligand-binding pockets of nuclear receptors and their ligands. On the first view the LBDs of the true orphan receptor NOR1 (**left**), the endocrine receptor VDR (**center**) and the adopted orphan receptor PXR (**right**) appear very similar **a**. However, NOR1 has no ligand-binding pocket, while that of adopted orphan receptors in average has the double size compared to endocrine receptors. This explains the variability of adopted orphan in ligand affinities. For comparison, the volume **b** and the molecular weight **c** of important nuclear receptor ligands are indicated. The vertical red lines mark the average

to the dissociation of corepressor proteins and in turn to the association of coactivators, such as members of the nuclear receptor coactivator (NCOA) family. These coactivator proteins are connected with multiprotein complexes that are composed of chromatin modifying enzymes, such as HATs, leading to local chromatin decondensation. This process is also called de-repression. Furthermore, the local opening of chromatin is an essential but not a sufficient condition for the initiation of gene transcription. In the last step, coactivator proteins with HAT activity are replaced by components of the MED complex building a bridge to the basal transcriptional

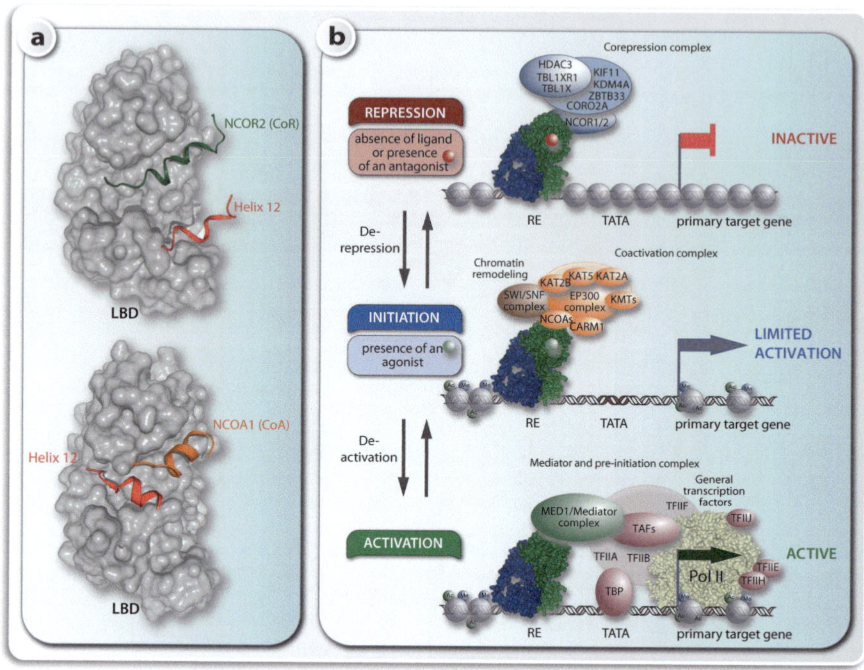

Fig. 5.6 Interaction of nuclear receptors with cofactors. A solvent excluded surface (Connolly surface) representation of a nuclear receptor LBD in the absence (**top**) and presence (**bottom**) of a ligand **a**. The ligand-induced conformational change primarily affects helix 12 (red) that is the most carboxy-terminal α-helix of the LBD. In the absence of ligand helix 12 is in a position that allows a corepressor protein (represented by the receptor interaction domain of NCOR2, green) to interact with the LBD, while in the presence of ligand only the binding of a coactivator protein (receptor interaction domain of NCOA1, orange) is possible. The 3-step transactivation process of nuclear receptors is shown in context of a target gene **b**. In the absence of a ligand the DNA-bound dimeric nuclear receptor complex is connected via corepressor proteins with a multiprotein complex with chromatin modifying activity that leads to local condensation of chromatin and repression of the target gene (**top**). Following the binding of an agonistic ligand the LBD of nuclear receptors is dissociating from corepressors and associating with coactivators that connect with a multiprotein complex having chromatin decondensation activity (**center**). The ligand-activated nuclear receptor is changing to another type of coactivator being a subunit of the MED complex (**bottom**). In this way, the basal transcriptional machinery and Pol II are activated and finally mRNA transcription starts. CORO2A = coronin 2A, KIF11 = kinesin family member 11, TBL1X = transducin β-like 1 X-linked, TBL1XR1 = TBL1X receptor 1, ZBTB33 = zinc finger and BTB domain containing 33

machinery that has assembled on the TSS region of the target gene (Sect. 3.3). This then leads to the activation of Pol II and gene transcription, i.e., the production of respective mRNA molecules. Taking all tissues and cell types together, a given nuclear receptor can have more than 1000 target genes (Box 5.2).

> **Box 5.2: Genome-wide nuclear receptor analysis** Different NGS methods, such as ChIP-seq, which were intensively used by the *ENCODE* project (Sect. 6.1), have also been applied for the genome-wide analysis of the action of nuclear receptors. The total sum of binding sites for an individual nuclear receptor, referred to as its **cistrome**, collectively for multiple tissues can exceed 20,000 loci. Moreover, transcriptome-wide methods, such as RNA-seq, identified in the sum of all tissues more than 1000 primary target genes for most nuclear receptors or their specific ligands. Not all of these binding sites and target genes are equally important, but their huge numbers indicate that **nuclear receptors and their ligands are involved in the control of more physiological processes than formerly assumed**. On many, if not on all of their genomic binding sites nuclear receptors colocate with other transcription factors, such as SPI1 (SPI1 (spleen focus forming virus proviral integration oncogene, also called PU.1), FOXA1 (forkhead box A1) or NFκB (Sect. 13.4), that either work as pioneer factors (Sect. 9.2) to open the local chromatin structure or to interact with other signal transduction pathways, of which these proteins are the end points (Sect. 4.3). In addition, nuclear receptors do not directly contact DNA on all of their genomic binding sites but can sometimes act as cofactors to other DNA-binding transcription factors (Sect. 6.3).

Although nuclear receptor signaling is per se independent from other signaling pathways that start at the cellular membrane, there are many occasions for an interference of both signal processes. Like any other cellular protein, nuclear receptors and their cofactors can be posttranslationally modified by phosphorylation, acetylation, methylation and ubiquitynation (Sect. 8.1). The origins of these modifications are classical signal transduction pathways starting at the cellular membrane. For example, members of the NCOA family are extensively posttranslationally modified.

5.3 Nuclear Receptors as Nutrient Sensors

Nuclear receptors and their ligands play an important role in the maintenance of homeostasis of the body that represents "health". The evolutionary oldest and likely still the most important role of nuclear receptors is the regulation of metabolism. There is an interrelationship of lipid metabolism (supplemented by micro- and macronutrients taken up by diet), metabolites and their converting enzymes, such as cytochrome P450s (CYPs), transporters and key representatives of the nuclear receptor superfamily. There are many examples (RAR, CAR, PXR, PPAR, VDR, LXR and FXR, differently color-coded in Fig. 5.7), where a metabolite activates a nuclear receptor, which in turn regulates the expression of the enzyme or transporter handling the metabolite. Nuclear receptor-controlled CYP enzymes

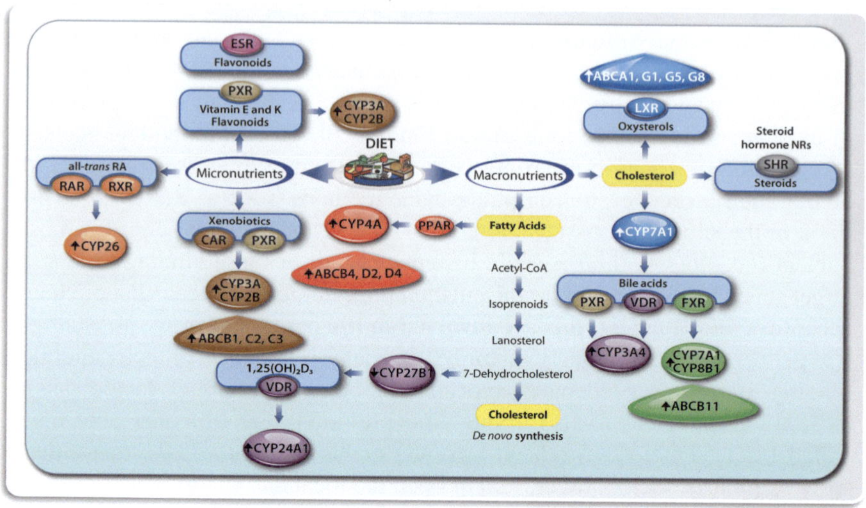

Fig. 5.7 Triangle regulatory circuits of nuclear receptors, their ligands and metabolite handling enzymes and transporters. The interrelationship between micro- and macronutrient metabolism involves enzymes, transporters and nuclear receptors. Only a selected number of metabolites and proteins are shown. There are many examples of triangle relationships (differently color-coded), in which the metabolite regulates its nuclear receptor, the receptor the expression of the metabolite converting enzyme and the enzyme the metabolite levels

have also a central role in ligand inactivation and clearance. These **triangle regulatory circuits are found at several critical positions in lipid metabolism pathways and allow a fine-tuned control on metabolite concentrations and thus nuclear receptor activity**. This suggests that dietary metabolites are ancestral precursors of endocrine signaling molecules, such as steroid hormones. In turn this emphasizes a nutrigenomics principle that **diet is not only a supply for energy but has also important signaling function**.

An immediate implication that followed from understanding the function of nuclear receptors is their potential as therapeutic targets. In fact, **nuclear receptor targeting drugs are widely used and commercially successful**. For example, bexarotene and alitretinoin (RXRs), fibrates (PPARα) and thiazolidinediones (PPARγ) are already approved drugs for treating cancer, hyperlipidemia and T2D, respectively. The nuclear receptor ESR1 belongs to the top 3 of the most studied transcription factors, mainly because of its role in the estrogen-dependent growth of breast cancer cells. Other nuclear receptor ligands, such as the physiologically active vitamin A and vitamin D metabolites all-*trans* RA and 1,25(OH)$_2$D$_3$ as well as their synthetic analogs, are known for their role in inducing cellular differentiation, e.g., of monocytes and peripheral nerval cells. This emphasizes the role of nuclear receptors in the control of cellular growth and differentiation. Moreover, synthetic GR agonists are very effective in the treatment of local and systemic inflammations and also other nuclear receptors, such as PPARs, LXRs and VDR, have an antiinflammatory potential. This supports the concept that nuclear receptors also play an important role in the control of the immune system (Sect. 13.3). Moreover, FXR and LXR agonists are in development for treating non-alcoholic fatty liver disease (NAFLD) and preventing atherosclerosis (Sect. 16.1). However, **natural nuclear receptor ligands that are taken up by healthy diet may avoid drug treatment**.

> **(Clinical) conclusion**: Nuclear receptors represent a very important superfamily of transcription factors that mediate the gene regulatory effects of steroid hormones (estrogen, testosterone, progesterone and cortisol), vitamins (vitamin A and vitamin D) as well as macronutrients and their derivatives (polyunsaturated fatty acid (PUFAs), oxysterols and bile acids) with their target genes. In this way, nuclear receptors have key roles in a large number of physiological processes, such as reproduction, differentiation, immunity and metabolism.

Additional Reading

Carlberg, C., Ulven, S.M. and Molnár, F. (2020). Nutrigenomics: how science works. Springer textbook.

Challet, E. (2019). The circadian regulation of food intake. Nat Rev Endocrinol 15, 393–405.

Dhiman, V.K., Bolt, M.J. and White, K.P. (2018). Nuclear receptors in cancer—uncovering new and evolving roles through genomic analysis. Nat Rev Genet *19*, 160–174.

Greco, C.M. and Sassone-Corsi, P. (2019). Circadian blueprint of metabolic pathways in the brain. Nat Rev Neurosci *20*, 71–82.

Reinke, H. and Asher, G. (2019). Crosstalk between metabolism and circadian clocks. Nat Rev Mol Cell Biol *20*, 227–241.

Chapter 6
Genome-Wide Principles of Gene Regulation

Abstract In this chapter, we will discuss about new principles of genome-wide gene regulation as obtained through data from Big Biology projects like *ENCODE*, *Roadmap Epigenomics* and *FANTOM5*. In this context, we will discuss epigenetic NGS methods used for the genome-wide assessment of accessible chromatin, histone modifications, transcription factor binding and 3D chromatin structure. The integration of these new data types provides further insight on the mechanistic and the human genome's functional landscape. In this way, we will obtain a genome-wide understanding of gene regulation and epigenetics.

Keywords Big biology projects · *ENCODE* · *Roadmap Epigenomics* · *FANTOM5* · NGS methods · ChIP-seq · Data integration · Epigenomics

6.1 Gene Regulation in the Context of Big Biology

With a delay of some 20 years, molecular biologists followed the example of physicists and realized that some of their research aims can only be reached by international collaborations of dozens to hundreds of research teams and institutions in so-called Big Biology projects (Fig. 6.1). The *Human Genome* project, which was launched in 1990 and completed in 2001 (Box 1.3), was the first example of a Big Biology project and has significantly changed the way of thinking in the bioscience community. In parallel, improved NGS methods allowed personal genome sequencing of both normal and cancer genomes. This made large-scale genome sequencing studies possible, such as in the projects *1000 Genomes* (www.1000genomes.org) (Sect. 1.3) and *TCGA* (www.cancer.gov/about-nci/organization/ccg/research/structural-genomics/tcga) (Sect. 15.3).

C. Carlberg, *Gene Regulation and Epigenetics*,
https://doi.org/10.1007/978-3-031-68730-3_6

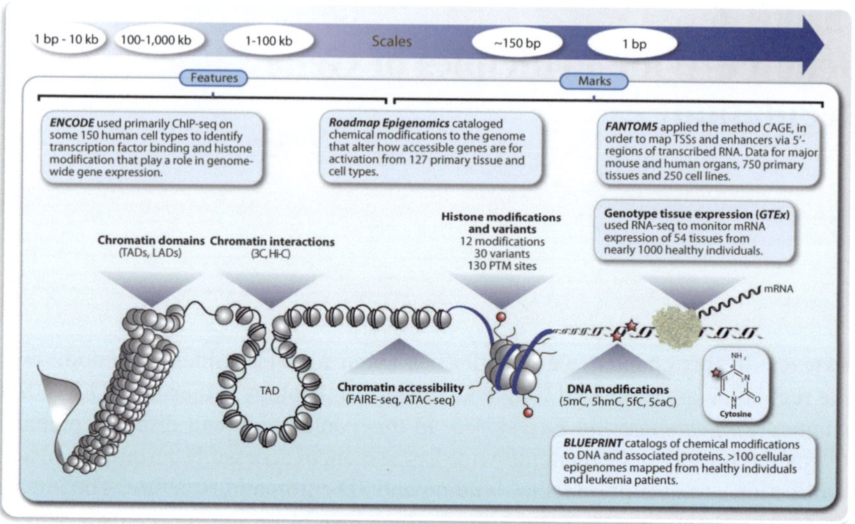

Fig. 6.1 Epigenetic Big Biology projects. Outline of the type of datasets collected by different Big Biology projects with an epigenetic focus. More details are provided in the text. PTM = posttranslational modification

The first Big Biology projects in the field of epigenomics and transcriptomics were *ENCODE* (www.encodeproject.org) and *FANTOM5* (http://fantom.gsc.riken.jp). The *ENCODE* project systematically mapped regions of transcription, transcription factor association, chromatin structure and histone modification. However, the project focused initially on some 150 human cell lines rather than on primary cells. The *FANTOM5* project used the method CAGE (Box 3.2), in order to map the 5'-end of de novo synthesized RNA at promoter and enhancer regions in some 750 primary tissues and 250 cell lines. Similarly, the *GTEx* (Genotype-Tissue Expression) project (https://gtexportal.org/home) measured mRNA expression from 54 non-diseased tissues from nearly 1000 individuals. The *ENCODE* follow-up project *Roadmap Epigenomics* (www.roadmapepigenomics.org) provided human epigenome references from 111 primary human tissues and cell types. In addition, the *BLUEPRINT* project (www.blueprint-epigenome.eu) focused on the epigenomes of human hematopoietic cells. Finally, *IHEC* (http://ihec-epigenomes.org) became the umbrella organization, under which international epigenome efforts are jointly coordinated. The key achievements of *IHEC* have been the introduction and implementation of quality standards for harmonizing epigenomic data collection, management and analysis. For example, based on *IHEC* standards each reference epigenome needs to be composed of at least nine profiles and assays, but typically reference epigenomes are composed of 20–50 genome-wide profiles and represent a multidimensional data matrix.

Epigenome profiling leads to maps of DNA methylation, histone marks, DNA accessibility and DNA looping that can be visualized with an appropriate web browser, such as the UCSC Genome Browser (Box 6.1). Although visualization can be highly illustrative and may induce hypotheses, epigenome maps are descriptive, i.e., primarily they are used for annotation. For example, enhancers, promoters and other genomic features have characteristic epigenomic signatures, such as H3K4me1 marks for enhancers and H3K4me3 marks for promoters (Sect. 8.2), on the basis of which they can be identified within epigenome maps.

Box 6.1: Visualizing Epigenomic Data A typical way of visualizing epige-nomic data, such as those from the *ENCODE* project, is to display a selected subset of them in a browser like the UCSC Genome Browser (http://genome. ucsc.edu/ENCODE). Datasets can be inspected without downloading them by creating a dynamic UCSC Genome Browser track hub that can be visualized on a local mirror of the UCSC Browser. Other visualization tools supporting the track hub format, such as Ensembl (www.ensembl.org), can also be used. For every given genomic position, a graphical display provides an intuitively understandable description of chromatin features, such as histone acetylation and methylation, that can be read in combination with experimentally proven information about transcription factor binding, as obtained from ChIP-seq experiments.

6.2 Epigenetic Methods

The rapid maturation of NGS technologies after finishing the *Human Genome* project led to the exponential development of methods for nearly all aspects of cellular processes, i.e., sequencing allows their detailed and comprehensive analysis. In these methods millions of DNA fragments are sequenced that origin either directly from genomic DNA (whole genome sequencing and bisulfide sequencing), the whole transcriptome (RNA-seq (Box 3.2) and related techniques) or subsets of these that origin from filtering for the interaction with transcription factors or accessible chro-matin (ChIP-seq and related methods). In general, **NGS methods have the advan-tage that they provide, in an unbiased and comprehensive fashion, information on the entire epigenome and genome**. Thus, global epigenomic profiling allows hypothesis-free exploration of new observations and correlations, i.e., conclusions drawn from a few isolated genomic regions might be extended to other parts of the genome.

The key epigenomic methods determine DNA methylation, transcription factor binding and histone modification, accessible chromatin and 3D chromatin architecture. The biochemical cores of these methods are:

- Different chemical susceptibility of nucleotides, such as bisulfite treatment of genomic DNA, in order to distinguish between C and 5mC (5-methylcytosine)
- Affinity of specific antibodies for chromatin-associated proteins, such as transcription factors, modified histones and chromatin modifying enzymes
- Endonuclease-susceptibility of genomic DNA within open chromatin compared to inert closed chromatin
- Physical separation of protein-associated genomic DNA (of fragmented closed chromatin) in an organic phase from free DNA (of accessible chromatin) in the aqueous phase
- Proximity ligation of genomic DNA fragments that via looping got into close physical contact.

The DNA methylome, i.e., a genome-wide map of 5mC patterns and its oxidized modifications (Sect. 8.2), is an essential component of the epigenome. Many datasets of different human tissues and cell types are already publicly available via the *IHEC* consortium. Global DNA methylation methods are either affinity-based methods, such as methylated DNA immunoprecipitation sequencing (MeDIP-seq), or base-resolution mapping methods, such as bisulfite sequencing (Fig. 6.2). Both types of methods consistently identify genomic regions, in which cytosines are frequently modified. Affinity-based assays use an antibody that enriches methylated fragments from sonicated genomic DNA. The resolution of these assays is highly dependent on the DNA fragment size and CpG density, i.e., they are more qualitative than quantitative. In contrast, bisulfite sequencing directly determines the methylation state of each cytosine of the whole genome. In this chemical conversion method, sodium bisulfite treatment of genomic DNA chemically converts unmethylated cytosines to uracils. Through PCR amplification all unmethylated cytosines become thymidines, i.e., the remaining cytosines correspond to 5mC. Whole genome sequencing provides then single base resolution of the methylation pattern. Interestingly, bisulfite sequencing is one of the first epigenomic methods being successfully applied on the single-cell level (Box 6.2). Since 5mC and 5hmC (5-hydroxymethylcytosine), but not 5fC (5-formylcytosine) or 5caC (5-carboxylcytosine), are resistant to bisulfite conversion, they cannot be distinguished from each other, i.e., antibody enrichment or more advance chemical conversion methods need to be applied.

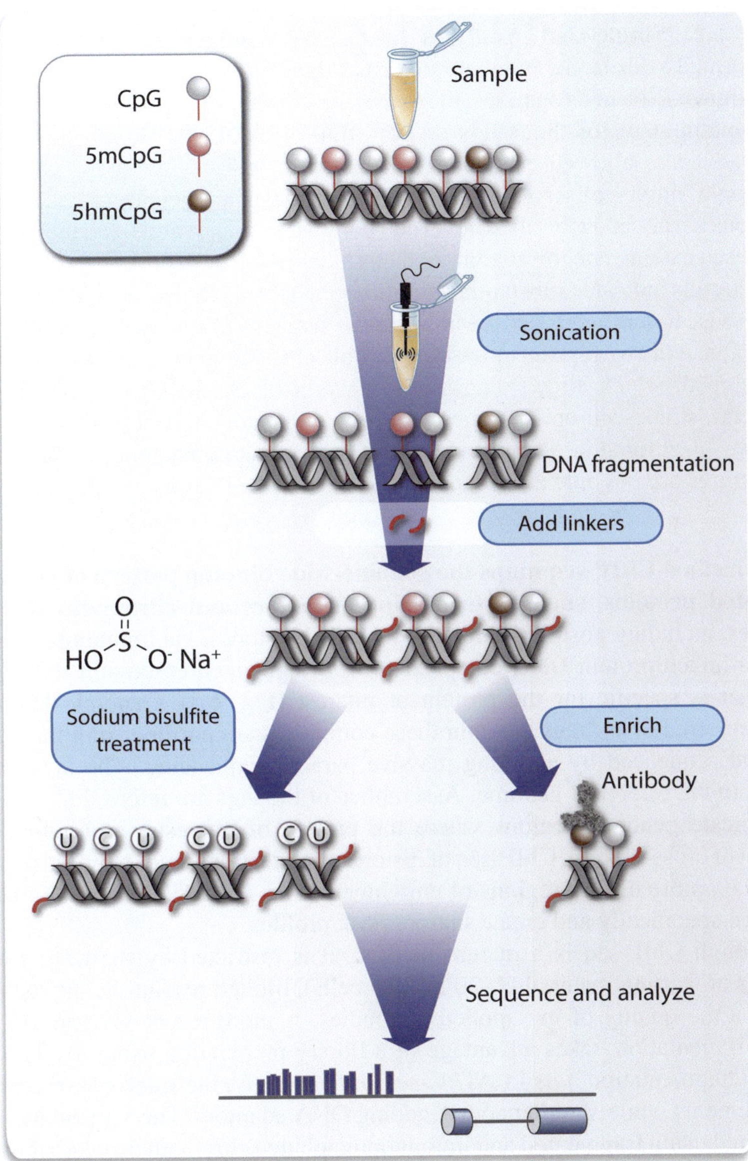

Fig. 6.2 Analysis of DNA methylation. Genome-wide 5mC and its oxidative derivative 5hmC are measured by conversion- and enrichment-based methods followed by sequencing. Bisulfite conversion allows quantification of 5mC but does not distinguish 5mC from 5hmC, while antibody enrichment enables qualitative measurement of 5mC and 5hmC. Bisulfite-converted or -enriched genomic DNA is purified, subjected to library construction and clonally sequenced. Finally, the sequencing reads are aligned to the reference genome

Box 6.2: Single-Cell Analyses In the past, epigenome methods were performed with larger numbers of cells ("bulk"). However, cell populations are known to be heterogenous, the respective results only represent the average chromatin state for thousands or even millions of cells. Recent technological advances allowed executing genome-wide analyses on single cells. For example, single-cell RNA-seq showed substantial heterogeneity of cell types in various tissues and identified novel cell populations. The single-cell technology has been extended to the genome and DNA methylome. Bisulfite sequencing of single cells indicates substantial variations in DNA methylation patterns across otherwise homologous cells residing in the same tissues. Moreover, single-cell ATAC-seq (assay for transposase-accessible chromatin using sequencing) was developed, which allows single-cell analyses of chromatin accessibility. In general, single-cell epigenomics will provide insights into the combinatorial nature of chromatin, such as which combinations of epigenetic marks and structures are possible and what mechanisms control them.

The method **ChIP-seq maps the genome-wide binding pattern of chromatin-associated proteins, such as transcription factors and chromatin modifying enzymes**, including posttranslationally modified histones, via immunoprecipitation of cross-linked protein-DNA complexes from fragmented chromatin with an antibody that is specific for the protein of interest (Fig. 6.3). Genomic DNA fragments, referred to as "tags", within these complexes are purified from the enriched pool and sequenced by applying massive parallel sequencing. The tags are then aligned to the reference genome. Assemblies of the tags are referred to as "peaks" and indicate genomic regions where the protein of interest was binding at the moment of cross-linking. ChIP-seq of histone modifications tends to produce broader peaks, i.e., more diffuse regions of enrichment, than transcription factors that bind sequence-specifically and create sharper peak profiles.

Although ChIP-seq is a mature method, it is restricted by the need for large amounts of starting material (1–20 million cells), limited resolution and the dependence on the quality of the applied antibodies. A more recent variation of ChIP-seq, ChIPmentation, takes advantage of a library preparation using the Tn5 transposase ("tagmentation") as in ATAC-seq. Interestingly, the transposase cuts DNA into fragments while simultaneously adding DNA adaptors. The sequencing library is prepared using fragmented and immunoprecipitated chromatin instead of the standard purified, i.e., protein-free, immunoprecipitated genomic DNA. This tagmentation step reduces the number of cells needed in the experiments by a factor of 10–100. Finally, single-cell approaches (Box 6.2) allow even more powerful analyses of chromatin states and their associated gene regulatory networks.

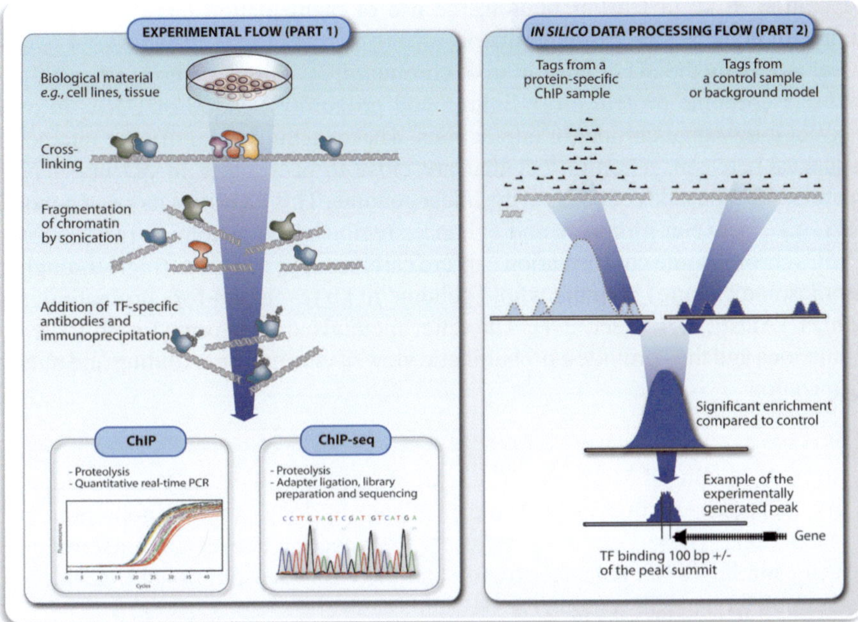

Fig. 6.3 ChIP-seq and its analysis. Short chromatin fragments are prepared from cells in which nuclear proteins are covalently attached to genomic DNA by short-term formaldehyde cross-linking. After chromatin fragmentation, immobilized antibodies against a protein of interest, such as a transcription factor (TF) or a histone mark, are used to immunoprecipitate the chromatin fragments associated with the respective protein (**left**). All genomic fragments are subjected to massive parallel sequencing, e.g., by the use of an Illumina Genome Analyzer. Typically, sequencing runs provide tens of millions of sequencing tags (small arrows) that are uniquely aligned to the reference genome (**right**). Clusters of these tags form peaks that represent transcription factor binding loci when they show significantly higher binding than the control sample. One has to take into account that regular ChIP-seq assays measure averaged protein binding events based on chromatin templates being obtained from millions of cells, i.e., a weak ChIP-seq peak may represent strong binding that is only observed in a small subset of cells

In the past, the main assays for genome-wide mapping of open chromatin were DNase-seq (DNase I hypersensitivity followed by sequencing) and FAIRE-seq (formaldehyde-assisted identification of regulatory elements followed by sequencing). The experimental principle of the DNase-seq method is a limited digestion of chromatin with the endonuclease DNase I that releases nucleosome-depleted fragments of genomic DNA. In contrast, FAIRE-seq takes advantage of the fact that genomic DNA within open chromatin regions is particularly sensitive to shearing by sonication. Chromatin is isolated from formaldehyde cross-linked DNA, sonicated and subjected to a phenol–chloroform extraction. Protein-free genomic DNA can be isolated from the aqueous phase, while protein-bound DNA remains in the organic phase. However, the present state-of-the-art method for mapping chromatin accessibility is ATAC-seq, which uses the Tn5 transposase for sequencing library preparation. Like in ChIPmentation, the use of tagmentation largely reduces the number of cells needed per experiment.

For assessing the 3D organization of chromatin, 3C-based methods (Box 6.3) are used that combine protein cross-linking and proximity ligation of DNA, in order to detect long-range chromatin interactions. These methods quantify the interaction frequency between genomic loci that are close to each other in 3D but may be separated by thousands of bases in the linear genome. This identifies loops of genomic DNA, e.g., between promoter and enhancer regions. The genome-wide version of 3C, 5C (chromosome conformation capture carbon copy) and Hi-C (high-throughput chromosome capture) map the whole genome in kb resolution for chromatin loops, such as TADs/LADs (Sect. 2.4). The latter methods are performed with large cell populations and thus provide a probabilistic view of chromosome folding and nuclear organization.

Box 6.3: 3C-Based Methods 3C is a proximity ligation method that can identify loops of genomic DNA being mediated by long-range protein–protein interactions. These loops may represent a connection between a transcription factor binding to an enhancer region and the basal transcriptional machinery assembled on a TSS region. The 3C method involves:

- cross-linking of segments of genomic DNA to proteins and of proteins with each other like in ChIP (Fig. 6.3)
- restriction digestion of the cross-linked DNA, in order to separate non-cross-linked DNA from the cross-linked chromatin
- intramolecular ligation of neighboring, previously cross-linked DNA fragments with the corresponding junctions
- reverse cross-linking resulting in linear DNA fragment with a central restriction site corresponding to the site of ligation
- quantitative PCR using primers and Taqman probes against the site of ligation to measure quantitatively the fragment of interest.

The frequency with which two restriction fragments become ligated indicates how often they interact in the nucleus. In genome-wide versions of the 3C method, such as circularized chromosome conformation capture (4C), 5C and Hi-C, the 3C protocol is combined with high-throughput genomic methods, which is greatly enhancing the power of discovery.

6.3 Integrating Epigenome-Wide Datasets

Individual research teams as well as large consortia, such as the *ENCODE* project and the *Roadmap Epigenomics* project, have already produced thousands of epigenome maps from hundreds of human tissues and cell types. The integration of these data, e.g., transcription factor binding and characteristic histone modifications, allows the prediction of enhancer and promoter regions as well as monitoring their activity and many additional functional aspects of the epigenome.

There are a variety of approaches to integrate epigenomics data within or across omics layers, such as correlation or co-mapping. When the datasets, which are to be compared, have a common driver, or if one regulates the other, correlations or associations should be observed. In most cases, more than two datasets, which often derive from different omics layers, are integrated (often referred to as modeled) in gene regulatory networks. Thus, in order to infer function from epigenomics data, they need to be integrated within the same layer (Fig. 6.4), such as histone marks with DNA accessibility, or across layers, such as transcription factor binding with mRNA expression as obtained by the *GTEx* project or DNA methylation with genetic variation described in the GWAS catalog.

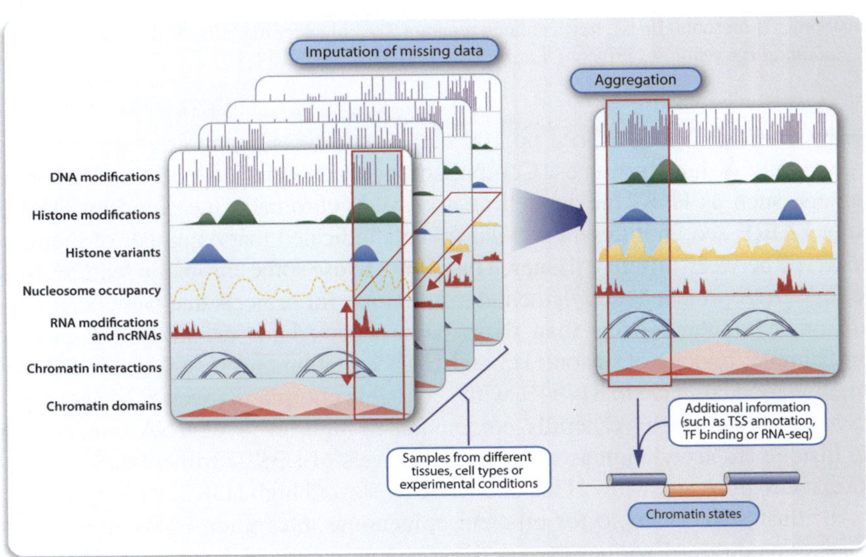

Fig. 6.4 Multidimensional integration of epigenome profiles. Epigenome data integration is achieved through imputation of missing data via profiles from the same and/or closely related samples and addition of non-epigenomic data, such as transcriptomic data (e.g., gene expression levels and TSS use). This allows the aggregation and segmentation of the datasets into a number of different chromatin states

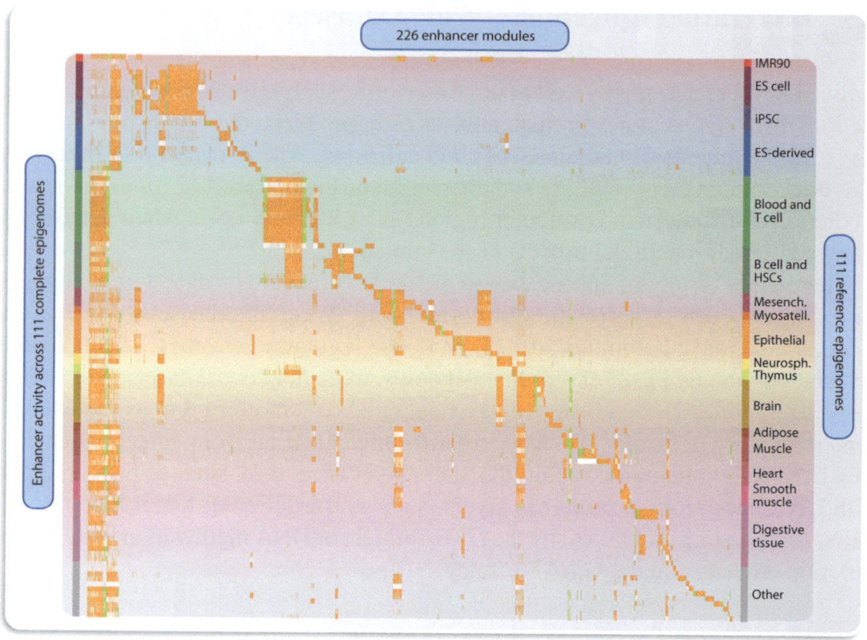

Fig. 6.5 Mapping enhancer modules. Regulatory modules, such as enhancers, can be identified based on activity-based clustering of 2.3 million accessible genomic regions across 111 reference epigenomes (horizontal lines). Vertical lines separate 226 enhancer modules. Data were taken from the *Roadmap Epigenomics* project (*Nature* 518, 317–330)

The *ENCODE* project used up 100 human cell lines as representatives for the large variety of human tissues. Comparison between data for the same chromatin markers, such as H3K27ac as well as accessible chromatin measured by DNase-seq or FAIRE-seq, in different cellular models indicated that a number of them are conserved between different tissues. This allows to use some chromatin features from the *ENCODE* project as supplemental information for projects that were performed with other cellular models than those being selected for the *ENCODE* project. For example, regions of histone H3 and H4 acetylation and H3K4 mono-, di- and trimethylation coincide to 81–94% with accessible chromatin (Sect. 8.2). Moreover, active genomic regions generally correspond to high levels of RNA transcription and histone H3 acetylation as well as to low levels of H3K27 trimethylation, while repressed regions show low H3ac and RNA levels and high H3K27me3 signal.

An illustrative example for efficient epigenome integration is the mapping of enhancer modules (describes as accessible genomic regions determined by DNase-seq) over the 111 reference epigenomes of *Roadmap Epigenomics* (Fig. 6.5). It shows that a limited set of enhancer modules are active in nearly all tissues and cell types, while the majority of the enhancers are specific to same lineage, such as stem cells or mature blood cells. This demonstrates that **the activity of enhancer modules**

can be used to monitor the similarity and relationship of cell types, including common functions and phenotypes. For the interpretation of human genetic variation and disease this is very useful, as the enrichment of respective traits is strongest for enhancer-associated marks.

Disease- and trait-associated genetic variants are often show tissue-specific enrichment for enriched in epigenomic marks. This demonstrates the central role of epigenome-wide information for understanding gene regulation as well as embryogenesis (Sect. 11.1) and cellular differentiation (Sect. 11.3). Moreover, **the broad coverage of epigenomic annotations improves the understanding of common diseases, such as cancer, Alzheimer disease, autoimmune diseases and T2D, beyond the level of protein-coding genes**.

Data integration reduces the list of affected epigenomic regions to a subset with inferred function. Many thousand epigenomes from hundreds of human tissues and cell types already significantly improved insight into epigenomics. However, much more remains to be investigated, such as better characterizations of epigenome variations in human populations and in patients, in order to fully appreciate the potential of epigenomics in human health and disease. Moreover, modern technologies like **CRISPR genome editing methods** (Box 6.4) are applied, e.g., for the knockdown or excision of gene regulatory regions, such as enhancers and promoters, in order to prove their impact on gene regulation and epigenetics.

> **Box 6.4: CRISPR Methodologies** The CRISPR method is based on a defense system of bacteria against the integration of the DNA of bacteriophages (viruses infecting bacteria) into their genome. An array of sequences that recognize the integrated virus DNA (CRISPR = clustered regularly interspaced short palindromic repeats) is guiding a set of Cas (CRISPR-associated) endonucleases, at which positions of the bacterial genome they should cut, in order to excise the invaded DNA. Engineering in particular the nuclease Cas9 and the use of synthetic guiding oligonucleotides created a toolbox that enabled the programmed generation of site-specific double-strand DNA breaks for genome editing. Additional engineering of the Cas9 enzyme by fusion with additional protein domains can change the functional profile from a sequence-specific nuclease to basically any type of protein, such as an activating or repressing transcription factor or a chromatin modifying enzyme and others, and allows the precise and targeted modification of the epigenome. Thus, CRISPR technologies can also be applied for **epigenome editing**, such as the validation of enhancers, promoters, insulators and silencers. Nowadays, CRISPR is used in basic and applied research ranging from agriculture via biotechnology to medicine (gene therapy).

6.4 Genome-Wide Understanding of Epigenetics

Pure in silico screening for consensus sequences of TFBSs, which are typically 6–17 base pairs in length, has only rather low information content, since it largely overrepresents the sites used in vivo. This provides the chromatin structure with a critical role in determining, whether a suitable TFBSs is accessible. Numerous ChIP-seq studies indicated that the number of genomic binding sites vary greatly between transcription factors. Moreover, **the number of direct target genes is far lower than those of binding events**, since only a subset of the sequences below the summits of the ChIP-seq peaks contain binding sites for the selected transcription factors. Thus, the understanding of the action and function of transcription factor has to be adapted to these new genome-wide insights.

For example, ChIP-seq indicated that in erythroblasts the transcription factor GATA1 has over 15,000 binding sites, while for TAL1 (TAL BHLH transcription factor 1, erythroid differentiation factor) only 3000–6000 binding sites were identified. Most of the TAL1 binding sites colocate with GATA1 sites, i.e., GATA1 acts as a pioneer factor for TAL1. In contrast, ChIP-seq on the transcription factor MYOD1 in skeletal muscle cells identified some 30,000–60,000 binding sites. MYOD1 is the most important transcription factor in muscle cells, as it controls via a feed-forward circuit the temporal expression pattern of genes important for skeletal muscle differentiation. Interestingly, both TAL1 and MYOD1 heterodimerize with an E-box protein and the respective heterodimers recognize the same binding site. Therefore, the tenfold difference in amount of experimentally proven binding events cannot be related to a difference in their DNA binding sites. However, the accessibility of these binding sites may be significantly different between erythroblasts and myocytes. MYOD1 can initiate chromatin opening at otherwise inaccessible sites, i.e., it can bind independently of other factors, whereas TAL1 requires GATA1 or other proteins, in order to get access to its binding sites. Thus, the difference in the number of genomic binding events of MYOD1 and TAL1 reflects their ability to act as a pioneer factor or as a following factor, respectively. Accordingly, **the actions of MYOD1 are rather independent from other transcription factors, while TAL1 needs the help of other proteins**.

Accessible chromatin and TSS regions both reflect genomic regions that are intensively used for gene regulation, in particular, when they overlap. In combination with data on RNA transcripts that are now typically obtained by genome-wide approaches, *ENCODE* data provide substantial experimental evidence for the different promoter types used for human genes. For example, TSS regions close to CpG islands display a broader distribution of histone modification than those not being colocated with CG-rich sequences (Sect. 3.2). Importantly, distal regulatory regions show characteristic patterns of histone modification being clearly different from TSS regions that show high H3K4me1 levels combined with lower levels of H3K4me3 and H3ac. Moreover, many proteins with high occupancy at TSS regions, such as the transcription factors

E2F4 and YY1 (YY1 transcription factor), are seldomly found at enhancer regions, whereas other transcription factors, such as MYC or CTCF, are enriched at both TSS and enhancer regions. Moreover, some transcription factors, such as JUND and ESR1, show considerable cell type-specific binding. Such **differential behavior of sequence-specific transcription factors points to biological differences between enhancer and TSS regions**.

TFBSs that occur outside of genomic regions directly involved in gene regulation may be non-functional or random. Many of these experimentally validated TFBSs are only of low-affinity and may contribute to gene expression only at low levels that, however, is sufficient enough to allow evolutionary conservation. Alternatively, accessible genomic DNA may serve as a low-affinity reservoir for transcription factors that are not directly regulating gene transcription in vicinity to their binding site. For example, in mouse ES cells there are approximately 3700 binding sites for the pluripotency transcription factor OCT4 (octamer-binding transcription factor 4, also called POU5F1), 4500 for SRY-box 2 (SOX2) and 10,000 for NANOG (nanog homeobox). However, only a few genomic regions were bound simultaneously by all three embryonal transcription factors (Sect. 11.2), i.e., **functionally effective sites may only be achieved by cooperative binding**. This can be achieved either by direct interaction between the transcription factors or by indirect interaction through cofactors.

Some transcription factors are recruited by a common motif to their genomic binding sites, while others use a number of different recruitment mechanisms. For example, de novo motif analysis after ChIP-seq indicated that some transcription factors, such as p63 and STAT1, show high enrichment for a specific motif, while E2F family members seem not to require a specific DNA sequence for their binding in vivo. The lack of a consensus motif can be explained by binding of the transcription factor to a distal site with the consensus motif and looping to the proximal site via the MED complex or other cofactors (Fig. 6.6a), "piggyback" binding to a second transcription factor that contacts DNA directly (Fig. 6.6b), the use of a different dimerization partner that results in significantly different DNA binding specificity (Fig. 6.6c) or the stabilization via interaction with chromatin factors (Fig. 6.6d). Thus, **the more protein–protein interactions are involved in the complex formation, the more difficult it is to use a pure bioinformatic approach for the identification of TFBSs.**

In most cases first the histone marks of a genomic region are changed before transcription factors are binding. Therefore, specific chromatin modifications, such as H3K4me1 for enhancer regions, may enhance transcription factor recruitment while others prevent it, i.e., certain transcription factors may have an affinity for a specific histone modification (Fig. 6.6d).

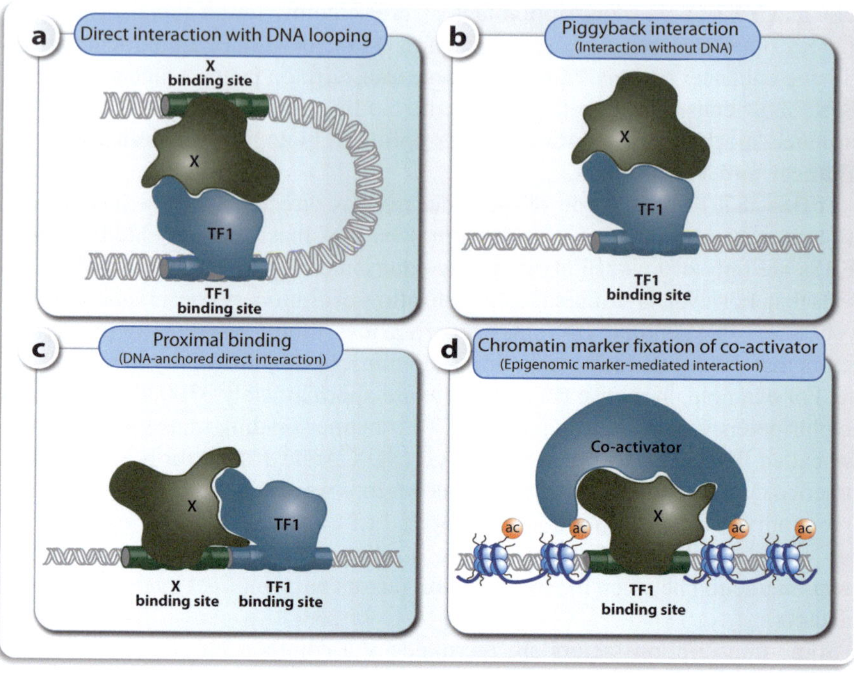

Fig. 6.6 Alternative binding modes of transcription factors. ChIP-seq results on the genomic binding sites of the uncharacterized transcription factor X can be explained via looping **a**, piggyback interaction with a partner transcription factor (TF1) **b**, proximal binding **c** or chromatin marker fixation of coactivator proteins **d**

In order to understand how a gene is expressed in its chromosomal environment, one should ideally be able to identify all TFBSs that are required for its regulation under all physiological conditions. The bioinformatic method of **comparative genomics** is based on the fundamental assumption that sequence similarity between orthologous sequences of different species results from selective pressure during evolution (Box 6.5). Comparative genomics with the goal to identify functional TFBSs is called **phylogenetic footprinting**. For example, a genome-wide comparison of TSS regions and their surrounding sequence between the mammalian species human, mouse, rat and dog suggests that the substitution rate at each site is lowest within the 50 base pairs upstream of the TSS, i.e., within the classical core promoter and increases linearly further upstream. Interestingly, TATA box-containing sharp promoters evolve more slowly than CpG island-containing broad promoters (Sect. 3.2). This suggests that **the more constrained architecture of the TATA box containing sharp promoters is needed to ensure efficient transcription initiation**, so that any change in the sequence is likely to have significant consequences on the function of the respective TSS region.

Box 6.5: Orthologous Genes and Sequence Alignment Genes are called orthologous with each other, when they originate from the same ancestral gene and are diverged by a speciation event. Phylogenetic footprinting assumes that orthologous genes are under common evolutionary constraints. At every particular position the genes are either analyzed for substitutions compared to neutral base exchange rates based on a multisequence alignments or alternatively the presence and frequency of intraspecies polymorphisms is determined. Both approaches are independent of any specific function that the analyzed sequence may confer. Duplication and/or deletion of genes during evolution complicates the determination of orthologs. Suitable sequences are aligned, in order to identify segments of similarity. Once the alignments are defined, the data interpretation is assisted by tools, such as the VISTA browser (pipeline.lbl.gov). The latter creates graphs of nucleotide identity over a sliding window along a pairwise alignment. Such a graphical display helps in the visualization of the alignment results, but for the analysis of long sequences additional computational analysis of the conservation patterns is needed.

In general, regulatory genomic regions, such as promoters, enhancers and insulators, are composed of TFBSs that should show a high level of interspecies conservation. However, not all TFBSs are equally well conserved, as some were just recently recruited in evolution and therefore may be species-specific. Moreover, not all conserved non-coding sequences are proven regulatory regions. **A search for regulatory elements is most efficient when it incorporates a comparison between different species and includes data about open chromatin and histone modifications, i.e., on the epigenome**.

(Clinical) conclusion: Big Biology projects, such as ENCODE, FANTOM5 and others, provide huge amounts of epigenome and transcriptome datasets from many human tissues and cell types. These data are the basis for a genome-wide understanding of gene regulation and epigenetics.

Additional Reading

Andersson, R. and Sandelin, A. (2020). Determinants of enhancer and promoter activities of regulatory elements. Nat Rev Genet *21*, 71–87.

Badia, I.M.P., Wessels, L., Muller-Dott, S., Trimbour, R., Ramirez Flores, R.O., Argelaguet, R. and Saez-Rodriguez, J. (2023). Gene regulatory network inference in the era of single-cell multi-omics. Nat Rev Genet *24*, 739–754.

Gasperini, M., Tome, J.M. and Shendure, J. (2020). Towards a comprehensive catalogue of validated and target-linked human enhancers. Nat Rev Genet *21*, 292–310.

Haniffa, M., Taylor, D., Linnarsson, S., Aronow, B.J., Bader, G.D., Barker, R.A., Camara, P.G., Camp, J.G., Chedotal, A., Copp, A., *et al.* (2021). A roadmap for the Human Developmental Cell Atlas. Nature *597*, 196–205.

Heumos, L., Schaar, A. C., Lance, C., Litinetskaya, A., Drost, F., Zappia, L., Lucken, M. D., Strobl, D. C., Henao, J., Curion, F., *et al.* (2023). Best practices for single-cell analysis across modalities. Nat Rev Genet *24*, 550–572.

Jerkovic, I. and Cavalli, G. (2021). Understanding 3D genome organization by multidisciplinary methods. Nat Rev Mol Cell Biol *22*, 511–528.

Lappalainen, T., Scott, A.J., Brandt, M. and Hall, I. M. (2019). Genomic analysis in the age of human genome sequencing. Cell *177*, 70–84.

Minnoye, L., Marinov, G.K., Krausgruber, T., Pan, L., Marand, A.P., Secchia, S., Greenleaf, W.J., Furlong, E.E.M., Zhao, K., Schmitz, R.J., *et al.* (2021). Chromatin accessibility profiling methods. Nature Reviews Methods Primers, *1*, 10.

Moshitch-Moshkovitz, S., Dominissini, D. and Rechavi, G. (2022). The epitranscriptome toolbox. Cell *185*, 764–776.

Pacesa, M., Pelea, O. and Jinek, M. (2024). Past, present and future of CRISPR genome editing technologies. Cell *187*, 1076–1100.

Stark, R., Grzelak, M. and Hadfield, J. (2019). RNA sequencing: the teenage years. Nat Rev Genet *20*, 631–656.

Villiger, L., Joung, J., Koblan, L., Weissman, J., Abudayyeh, O.O. and Gootenberg, J.S. (2024). CRISPR technologies for genome, epigenome and transcriptome editing. Nat Rev Mol Cell Biol *25*, 464–487.

Wang, Y., Zhao, Y., Bollas, A., Wang, Y. and Au, K.F. (2021). Nanopore sequencing technology, bioinformatics and applications. Nat Biotechnol *39*, 1348–1365.

Zhang, Y., Boninsegna, L., Yang, M., Misteli, T., Alber, F. and Ma, J. (2024). Computational methods for analysing multiscale 3D genome organization. Nat Rev Genet *25*, 123–141.

Chapter 7
DNA Methylation

Abstract In this chapter, the best-understood epigenetic mark, cytosine methylation of genomic DNA, will be introduced. DNA methylation is performed by DNA methyltransferases (DNMTs) and primarily results in 5mC within CpGs. It is a prominent epigenetic mechanisms, which has an impact on genome stability, gene expression and development. In most cases, DNA methylation leads to the formation of heterochromatin and subsequent gene silencing. Coordinated DNA methylation and its recognition via methylation-sensitive DNA-binding proteins have a large impact on health, such as genomic imprinting, i.e., the expression of a gene in a parent-of-origin-specific manner. In contrast, aberrant DNA methylation is a well-established marker of diseases, such as cancer. This can lead to the inactivation of tumor suppressor genes, a disturbance in genomic imprinting and genomic instabilities through reduced heterochromatin formation on repetitive sequences. In addition, we will learn about the main function of the transcription factor CTCF which mediates intra- and interchromosomal contacts. In this way, CTCF stabilizes 3D complexes of chromatin loops. CTCF-mediated loops at some 100 developmentally regulated loci provide a mechanistic explanation of genomic imprinting in health.

Keywords DNA methylation · CpG islands · DNMTs · TET proteins · 5mC modifications · Gene silencing · Insulator · CTCF · Genomic imprinting

7.1 Cytosines and Their Methylation

The identity of each of the at least 400 human tissues and cell types is based on their respective unique gene expression patterns, which in turn are determined by differences in their epigenomes. **For the proper function of tissues, it is essential that cells memorize their respective epigenetic status** and pass it to daughter cells when they are proliferating. The main mechanism for this long-term epigenetic memory (Sect. 11.5) is the methylation of genomic DNA, in order to yield 5mC preferentially within CpGs. Since CpGs are the only dinucleotides that can be symmetrically methylated, they are the exclusive methylation marks that remain after DNA replication on both daughter strands.

The average CG base pair percentage of the human genome is 42% (Fig. 7.1). In principle, each cytosine can shift into 5mC, but those of CpGs are functionally most important, in particular, if they occur in clusters. However, less than 10% of all CpGs are found in genomic regions of at least 200 base pairs in length with a CG density of more than 55%. **These regions are referred to as CpG islands.**

The human genome contains approximately 28,000 CpG islands. Interestingly, many of the 20,000 protein-coding genes have a CpG island close to their TSS region. In fact, genes are distinguished into those with and without CpG islands in the vicinity of their promoters (Sect. 3.2). Interestingly, **actively transcribed gene bodies carry both 5mC and 5hmC marks, whereas active promoters are unmethylated**. Most of CpGs remain methylated during development, but CpG islands located close to the TSS regions of housekeeping or developmentally regulated genes have a very low

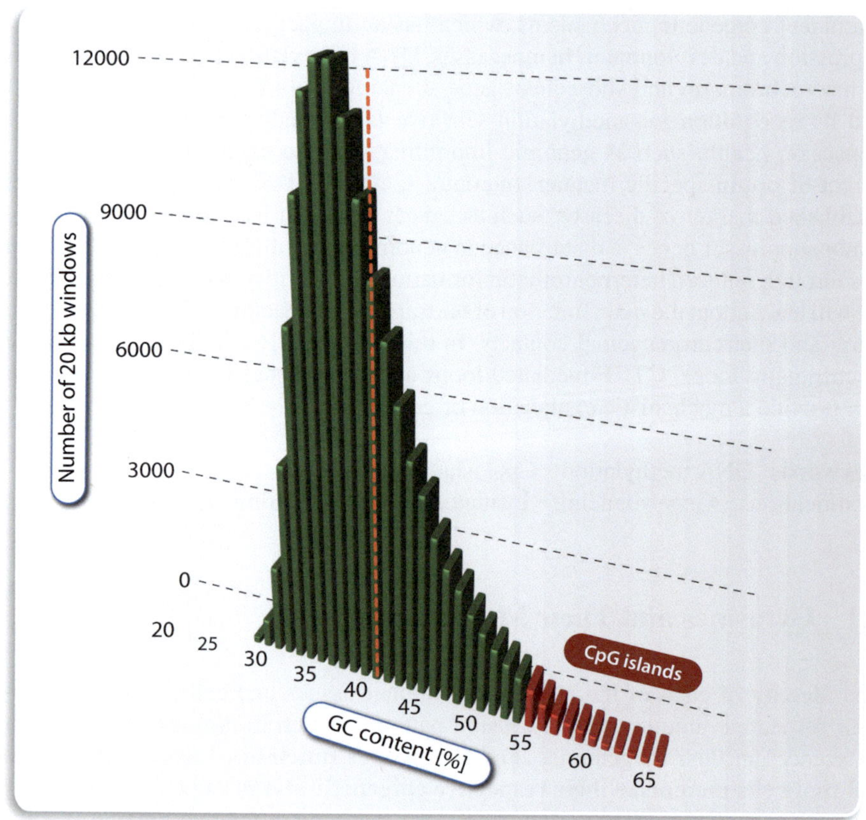

Fig. 7.1 CG content of the human genome. Due to subsequent passive deamination of genomic DNA, the average CG base pair content in the human genome is not 50% but only 42% (red line). CpG islands (red) have a CG percentage of 55%, i.e., only a minority of CpGs belong to CpG islands

methylation status. Interestingly, a C to T transition at the location of CpG islands is one of the most frequent mutations found in human diseases. This implies that DNA methylation reduces the efficiency of DNA repair resulting in the accumulation of mutations at these sites.

Cytosine methylation does not only occur at CpGs, but also at CpH dinucleotides (H = A, C or T). **Non-CpG methylation (mCH) occurs in all human tissues, but is most common in long-lived cell types, such as stem cells and neurons**. This type of methylation contributes to memory function and is negatively correlated with gene activity (Sect. 11.5). Proteins that specifically bind methylated DNA, such as MECP2 (methyl-CpG-binding protein 2), do not only interact with methylated CpG sites but also with CpH loci. MECP2 belongs to the family of intrinsically disordered proteins, which are characterized by a low level of secondary structure. This makes MECP2 well suited to interact with a large variety of different types of macromolecules, such as other proteins, DNA and RNA. Central to the primary structure the MECP2 protein is its DBD that includes a MBD (methyl-DNA-binding domain), but also contains several other structures, such as AT hook motifs.

DNMTs are chromatin modifying enzymes that catalyze in an one-step reaction the transfer of a methyl group from S-adenosyl-L-methionine (SAM) to cytosines of genomic DNA. Since **the DNA methylation pattern of somatic cells represents an epigenetic program of global repression of the genome** and specific settings of imprinted genes (Sect. 7.3), it is important to maintain the DNA methylome during replication. This is the main responsibility of DNMT1 in collaboration with its partner UHRF1 (ubiquitin-like plant homeodomain and RING finger domain 1) that preferentially recognizes hemi-methylated CpGs. In contrast, in the absence of functional DNMT1/UHRF1 complex, successive cycles of DNA replication lead to passive loss of 5mC, such as the global erasure of 5mC in the maternal genome during pre-implantation (Sect. 11.1). In particular, during the development of primordial germ cells (PGCs) in embryogenesis, genomic DNA is widely demethylated. This creates pluripotent states in early embryos and erases most of the parental-origin-specific imprints in developing PGCs. With the exception of imprinted genomic regions (Sect. 7.3), DNMT3A and DNMT3B perform **de novo DNA methylation during early embryogenesis**, i.e., together with DNMT1 they act as writers of DNA methylation (Fig. 7.2, left).

Active demethylation of genomic DNA is a multistep process that involves the methylcytosine dioxygenase enzymes TET (ten-eleven translocation) 1–3, which convert 5mC to 5hmC (Fig. 7.2). 5hmC is found in most cell types but only in levels of 1–5% compared to 5mC rates. However, adult neurons are an exception, since their 5hmC level is 15–40% of that of 5mC. In two further oxidation steps, TETs convert 5hmC into 5fC and to 5caC. 5fC and 5caC are significantly less prevalent (0.06–0.6% and 0.01% of 5mC rates, respectively) than 5hmC, i.e., TETs tend to preferentially halt at the 5hmC stage. Oxidized cytosines are deaminated to 5-hydroxyuracil (5hmU) so that they create a 5hmU:G mismatch, which is recognized and removed by the enzyme TDG (thymine-DNA glycosylase) (Fig. 7.2, top). The abasic site is then repaired by the base excision repair (BER) machinery, i.e.,

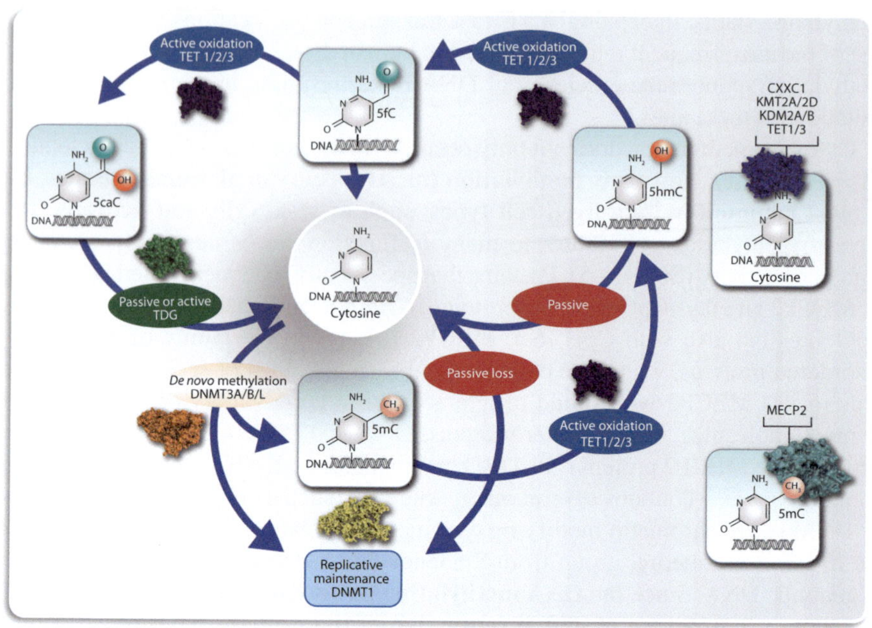

Fig. 7.2 Writing, erasing and reading cytosine methylations. DNMT1, as well as DNMT3A and DNMT3B, catalyze the methylation of cytosines at position 5, i.e., they act as "writer"-type chromatin modifying enzymes (**left**). The dioxygenase enzymes TET1, TET2 and TET3 oxidize 5mC to 5hmC and further to 5fC and 5caC. This leads via the action of the DNA glycosylase TDG to the loss of DNA methylation, i.e., both types of enzymes function as "erasers" (**center**). Different sets of proteins either specifically recognize unmethylated cytosines or 5mC, i.e., they are "readers" (**right**)

by a regular DNA repair process. This results in the overall demethylation of the respective cytosine. Thus, **the oxidative modification of 5mC via the TET/TDG pathway allows a dynamic regulation of DNA methylation patterns**.

DNA-binding proteins that specifically recognize either unmethylated or methylated genomic DNA (Fig. 7.2, right), such as CXXC1 (CXXC finger protein 1) or MECP2 (Sect. 11.4), then read the information stored in DNA methylation patterns and translate them into biological actions.

7.2   The DNA Methylome

Maps of genome-wide DNA methylation, histone modifications, patterns of chromatin accessibility, transcription factor binding as well as RNA expression from multiple tissues are provided by the Big Biology projects, such as *ENCODE* and *Roadmap Epigenomics* (Sect. 6.1). **A key example of the epigenome is the DNA methylome**, i.e., the genome-wide map of 5mC patterns and its oxidized modifications.

Despite an overall consistency in tissue-specific DNA methylomes between human individuals, variations in these patterns exist from person to person. This applies to each of the approximately 400 different human tissues and cell types of each human individual. Although unrelated human individuals already differ among each other in approximately 4–5 million SNPs out of the 3.05 Gb of their haploid genome (Sect. 1.2), the potential number of variations in their epigenome is far larger. In response to cellular perturbations by diet, microbe encounter, cellular stress or other environmental influences epigenomes vary a lot over time. Thus, **individuals vary far more on the level of their epigenomes than on the level of their genomes** (Fig. 7.3). This suggests that phenotypic differences between individuals (as well as their predisposition for diseases) are rather based on the epigenome than on the genome (Sect. 12.1).

DNA methylation is often associated with transcriptional silencing of repetitive DNA, such as SINEs, LINEs and LTRs (Box 1.3), and genes that are not needed in a specific cell type. LINEs and LTRs carry strong promoters that must be constitutively silenced via placing them into constitutive heterochromatin in order to prevent their activity (Fig. 7.4a). Therefore, these genomic regions are generally hypermethylated.

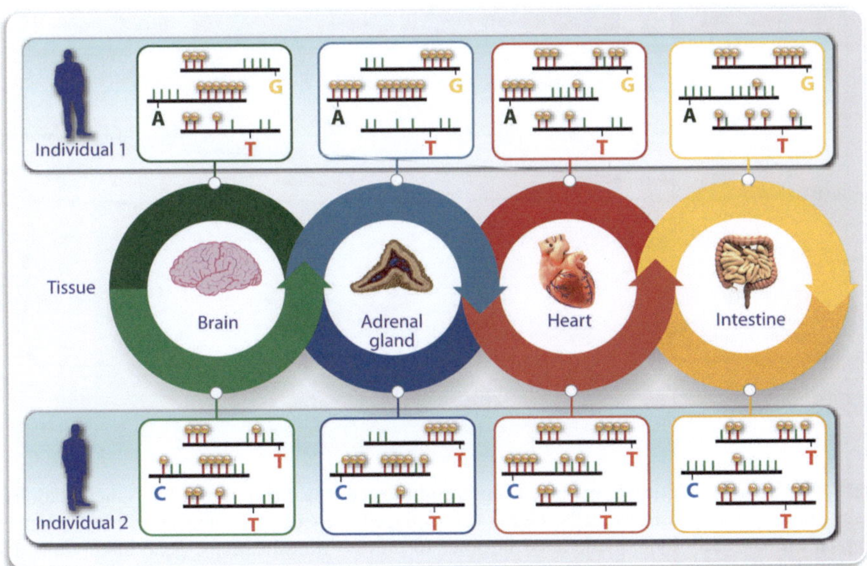

Fig. 7.3 Individuals show epigenetic heterogeneity. Tissue- and cell type-specific DNA methylations are displayed by clusters of methylated CpGs (Sect. 7.1) that vary from tissue to tissue of the same individual. Filled circles illustrate methylated CpGs and lack of a circle unmethylated CpGs. SNPs are monitored by the corresponding base

The silencing of the repetitive DNA happens primarily during early embryogenesis (Sect. 11.1), while in adult tissues de novo silencing is initiated by proteins like MECP2. The latter proteins bind symmetrically methylated CpGs, but they have no sequence specificity, i.e., **they are not classical transcription factors but act as readers** (Fig. 7.2, right) and adaptors for the recruitment of chromatin modifying enzymes, such as HDACs and KMTs (lysine methyltransferases), to methylated genomic DNA.

The DNA methylome is bimodal, i.e., it occurs in two major modes. These are a low methylation level at CpG-rich promoters and binding sites for methylation-sensitive transcription factors, such as CTCF (Sect. 7.3), while the remaining CpGs are by default methylated. Methylated genomic DNA is transcriptionally repressed, i.e., in most cases there is an inverse correlation between DNA methylation of regulatory genomic regions, such as promoters and enhancers, and the expression of the genes that they are controlling. However, at their gene bodies, highly expressed genes show significant levels of DNA methylation, i.e., some methylated CpGs downstream

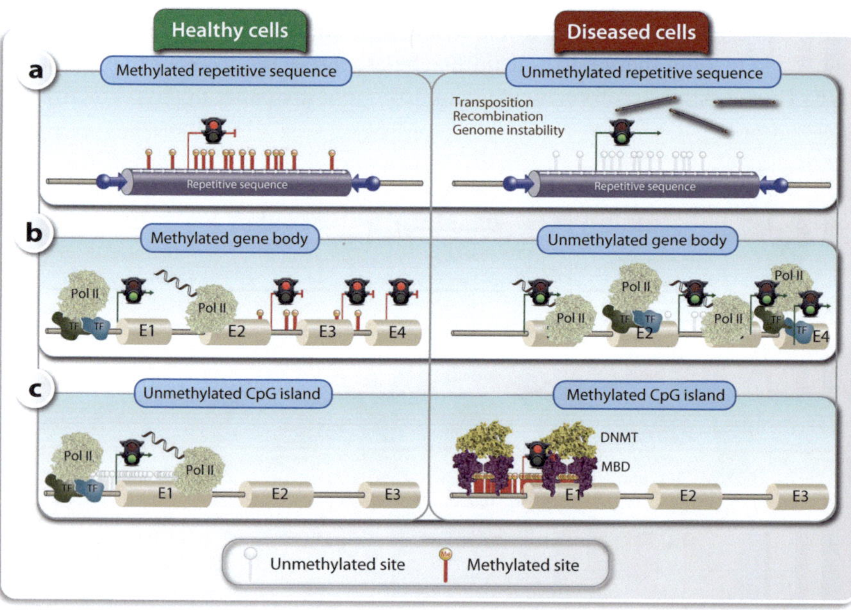

Fig. 7.4 DNA methylation in different regions of the genome. Scenarios of DNA methylation of healthy cells (**left**) or diseased cells (**right**) are displayed. Repetitive sequences within the human genome are normally hypermethylated in order to prevent translocations, gene disruptions and general chromosomal instability through the reactivation of retrotransposons (**a**). This pattern is altered in disease. Methylation of the transcribed region of a gene facilitates transcription (traffic lights) by the prevention of transcription initiations (**b**). Gene body tends to get demethylated in disease so that transcription may be initiated at several incorrect sites. CpG islands close to TSS regions are normally unmethylated (**c**). This allows transcription, while hypermethylation causes transcriptional inactivation

of TSS regions positively correlate with gene expression (Fig. 7.4b). Genes driven by CpG-rich promoters are silenced when methylated (Fig. 7.4c), while genes without CpG islands close to their TSS regions are regulated by other mechanisms than DNA methylation, such as transcription factors binding to enhancers.

In general, DNA methylation and histone modifications have different roles in gene silencing. **While most DNA methylation loci represent very stable silencing marks that are seldom reversed, histone modifications mostly lead to labile and reversible transcriptional repression**. For example, genes for pluripotency transcription factors in embryogenesis, such as *OCT4* and *NANOG*, need to be permanently inactivated in later developmental stages, in order to prevent possible tumorigenesis. This happens via H3K9 methylation at unmethylated CpGs on TSS regions of these genes, the attraction of HP1 (heterochromatin protein 1), de novo DNA methylation via DNMT3A and DNMT3B and finally transcriptional silencing for the rest of the life of the individual. In contrast, when in differentiated cells these pluripotency genes are silenced only by histone modification, these cells can be rather easily converted to iPS (induced pluripotent stem) cells (Sect. 11.2). Nevertheless, **the methylation status of some 20 % of all CpGs within the human genome is dynamically modified**. Differential DNA methylation is established by de novo methylation combined with active demethylation of CpG islands. During early embryogenesis, i.e., in the pre-implantation phase, most CpGs are unmethylated (Sect. 11.1). After implantation DNMT3A and DNMT3B de novo methylate those CpGs that had not been packed with H3K4me3-marked nucleosomes. In contrast, H3K4me3-marked CpGs on TSS regions of CpG-rich promoters stay unmethylated.

Methylation and demethylation of CpGs modulate the DNA-binding affinity of transcription factors, i.e., DNA methylation is a signal being differentially recognized by specific protein domains. Interestingly, a third of the 1600 human transcription factors are positively affected by methylation of their DNA binding sites, half of all do not bind DNA when it is unmethylated and only a fourth of all are negatively influenced by DNA methylation. A well-known example of the latter is CTCF in the context of genomic imprinting (Sect. 7.3). Thus, **there are different forms of gene silencing ranging from flexible repressor-based mechanisms to a highly stable inactive state being maintained by DNA methylation**.

7.3   CTCF and Genomic Imprinting

The mechanistic understanding of the functional implications of the DNA methylome is largely based on the specific interaction of transcription factors and other DNA-binding proteins with methylated and unmethylated genomic DNA. These "readers" (Fig. 7.2) translate the methylation signal into activities of biological processes. For example, in the process of alternative splicing, MECP2 binding to methylated genomic DNA of an exon leads to its inclusion into mRNA, while in the absence of MECP2 binding the exon is excluded (Fig. 7.5, right). The central point in this translation process is that methylation and demethylation of CpGs modulate the

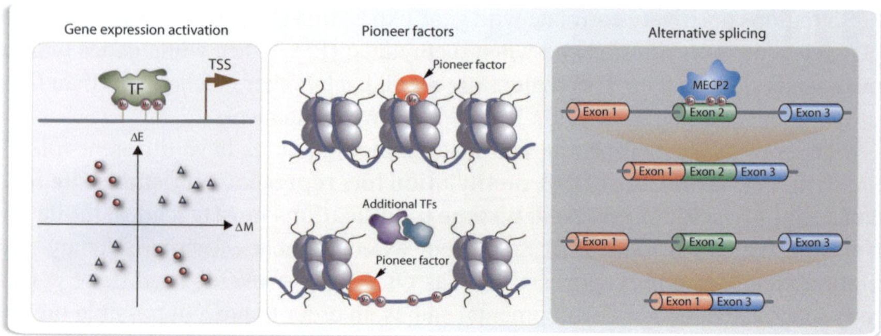

Fig. 7.5 Regulating biological processes by DNA methylation. Many transcription factors (TFs) bind to methylated DNA (**left**). Genome-wide profiling of changes in gene expression (ΔE) and methylation levels (ΔM) demonstrated that the methylation of CpGs either positively (blue triangles) or negatively (red dots) correlate with gene expression. Transcription factors that bind methylated genomic DNA when wrapped around nucleosomes have a high likelihood being pioneer factors (**center**). Methylation-sensitive transcription factors binding to exons, such as MECP2, can regulate alternative splicing (**right**). When the exon is not methylated, MECP2 does not bind and the exon is excluded

DNA-binding affinity of transcription factors, i.e., DNA methylation is a signal that is differentially recognized by specific protein domains. A combined epigenomics-proteomics approach demonstrated that nearly half of the 1600 human transcription factors have a DNA methylation preference: 34% of these transcription factors are positively affected by methylation, such as several homeodomain transcription factors, half of which do not bind DNA when is unmethylated and only 23% of them are negatively influenced.

Transcription factors recognize methylated DNA also without a MBD, such as ZBTB33 that binds a specific methylated sequence with its C2H2 zinc finger domain. Moreover, the pioneer factor CEBPα, the zinc finger protein (ZFP) 57 and its co-factor TRIM28 (tripartite motif containing 28) interact with specific methylated sequences. Some transcription factors, such as OCT4, SOX2 and Krüppel-like factor 4 (KLF4), specifically bind methylated DNA even when it is wrapped around nucleosomes (Fig. 7.5, center). They act as pioneer factors (Sect. 11.2) by recruiting chromatin modifiers that help to open the heterochromatic DNA, change its status to euchromatin and initiate transcription. Like 5mC, also 5hmC is a signal that influences transcriptional activity by attracting or repelling specific transcription factors. For example, MECP2 binds to hemi- or fully 5hmC-marked CpGs, i.e., it has a dual binding capacity for both 5mC and 5hmC.

There are a number of chromatin-associated proteins that have a strong preference for unmethylated CpGs. Some of them, such as TET1 and TET3, carry a zinc finger CXXC domain. Other examples are CFP1 and the histone demethylases (KDMs) KDM2A and KDM2B (Fig. 7.2, right). CPF1 binds unmethylated CpGs and recruits KMTs that leave H3K4me3 marks at active promoters. DNMT3A and DNMT3B recognize unmethylated CpGs but are allosterically inhibited by H3K4 methylation.

In this way, genomic DNA at these regions is kept unmethylated and the chromatin active. In parallel, the histone variant H2A.Z (Box 2.2) is strongly enriched at unmethylated, active promoters. The H3K4-KMTs KMT2A and KMT2D also carry CXXC domains, bind to unmethylated TSS regions of developmental genes and protect them from DNA methylation. This is another example how the interplay of histone modification and histone variants keeps the DNA methylation status low at selected genomic regions.

Insulators are genomic loci that separate genes located in one chromatin region from promiscuous regulation by transcription factors binding to enhancers of neighboring chromatin regions. The methylation-sensitive transcription factor CTCF is the main protein binding to insulator regions. In complex with other proteins, such as cohesin, CTCF mediates the formation of architectural loops, such as TADs, as well as of regulatory loops (Fig. 2.6). In addition to the prevention of cross-border enhancer activity, insulators can act as boundary elements that inhibit spreading of heterochromatin from silenced genomic regions to transcriptionally active parts of the genome. This means that these boundary elements "insulate" closed from open chromatin, i.e., inactive from active genes. Thus, **CTCF-bound insulators are epigenetic structures that are important for both specific gene regulation as well as chromatin architecture**.

The transcription factor CTCF has a DBD formed by eleven zinc fingers (Fig. 7.6a). The combinatorial use of these zinc fingers creates a conformation that allows CTCF to recognize not only a large variety of DNA sequences but also numerous coregulatory proteins. However, the central 4–5 zinc fingers of CTCF bind to a consensus core sequence of some 12 base pairs in length. This unique structural feature provides CTCF with a versatile role in genome regulation, such as binding to a large variety of insulator regions. **This can result in enhancer activity blocking, inhibition of heterochromatin spreading and inter- and intrachromosomal organization**.

CTCF is ubiquitously expressed in basically all human tissues, but the levels of its expression and nuclear distribution vary in a cell type- and species-specific manner (Fig. 7.6b). The protein is evolutionarily very conserved both in its protein structure as well as in its DNA-binding pattern (Fig. 7.6c). Genome-wide, there are approximately 30,000 CTCF-binding sites, some 15% of which are involved in the formation of TADs and only a few hundred control imprinting. All CTCF-binding sites are sensitive to methylation, i.e., CTCF binding to methylated sites is drastically reduced.

DNA is very flexible in forming any type of loops. Nucleosomes, around which the genomic DNA is wrapped some 2-times per 200 base pairs, show the smallest scale of DNA looping. The next level is represented by loops between enhancer regions and TSS regions of several kb in size that bring transcription factors and the basal transcriptional machinery into close vicinity. A further level of DNA looping in the scale of several hundred kb is mediated by CTCF and organizes the genome in a few thousand TADs (Sect. 2.4). These domains are conserved between cell types and species indicating that this organization is an evolutionary feature. Additionally, the boundaries of these domains are enriched for CTCF, but also with other factors,

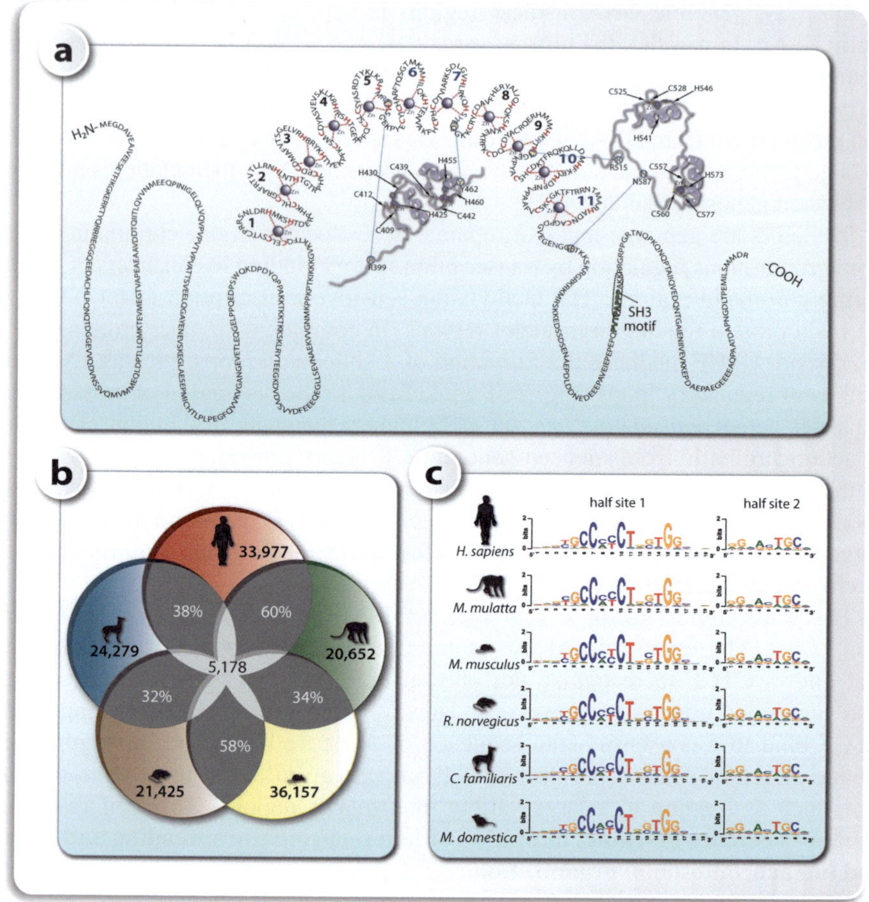

Fig. 7.6 The genome regulator CTCF. CTCF is an unusual transcription factor containing 11 DNA-interacting zinc finger domains (**a**). Venn diagram of interspecies conservation of CTCF sites (**b**). Canonical CTCF motifs obtained by de novo motif discovery (**c**)

such as housekeeping genes and proteins found at active promoters and gene bodies. This suggests that topological domains are generated, in part, by transcriptional activity. Thus, **the interactions of CTCF and its partner protein cohesin together with lamins of the nucleoskeleton are important for the position of genes within subnuclear compartments** (Fig. 7.7).

Higher-order chromatin structures, such as DNA loops that are stabilized by CTCF binding, represent another form of epigenetic memory (Sect. 11.5), which can be modulated by DNA methylation. Interestingly, only a small subset of unmethylated CTCF-binding sites keep CTCF proteins bound throughout the cell cycle, in order to protect these sites against de novo methylation. Thus, only those higher-order

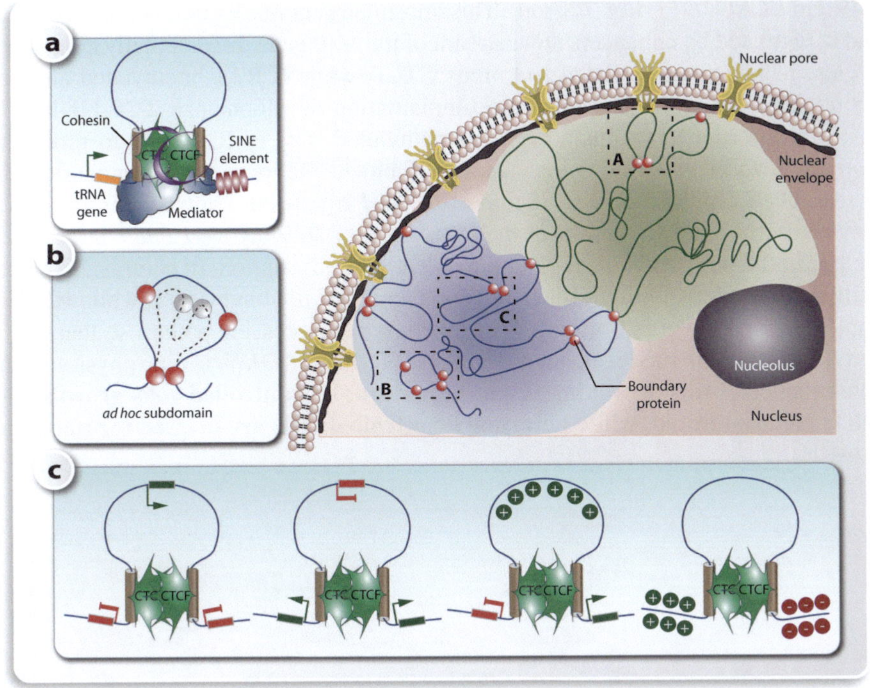

Fig. 7.7 Topological domains in the genome. Two chromosomes (green and blue lines) and their respective chromosome territories (green and blue areas) are shown. Proteins of the chromatin boundary (red circles), such as CTCF and cohesin, divide the genome into distinct domains. This topological organization implies interactions within and between chromosomes and between chromatin and the lamina of the nucleus. Different examples of CTCF-mediated looping are shown, such as regulatory loops (**a**), architectural loops and their subdomains (**b**) and different configurations CTCF-mediated loops separating active genes (green), inactive genes (red), euchromatin (+) and heterochromatin (−) from each other (**c**)

chromatin structures that are mediated by unmethylated CTCF sites can be inherited through mitosis, i.e., **CTCF-mediated chromatin structures represent a heritable component of phenotype-specific epigenetic programs**.

The DNA-methylation sensitive binding of CTCF to **imprinting control regions** (ICRs), which are a special subset of insulators that control the monoallelic expression of imprinted genes. This provides a mechanistic explanation of the epigenetic process of genomic imprinting occurring in most mammals and even in in some plants. In humans there are more than 100 maternally and paternally controlled genes (www.geneimprint.com/site/genes-by-species.Homo+ sapiens). Most imprinted genes occur in clusters, a key example of which is the chromosome 11p15 region that contains the protein-coding genes *IGF2* (insulin-like growth factor 2), *KCNQ1* (potassium voltage-gated channel subfamily Q member 1) and *CDKN1C* (cyclin-dependent kinase inhibitor 1C) as well as the ncRNA genes

H19 and *KCNQ1OT1* (Fig. 7.8, top). This imprinted genomic locus contains two ICRs and is regulated by enhancers downstream of the *H19* gene. In maternally controlled alleles, ICR1 is unmethylated and binds CTCF, while ICR2 is methylated and not bound (Fig. 7.8, center). During postimplantation development CTCF binding is essential in order to maintain the hypomethylated state of ICR1 and to protect it from de novo methylation in oocytes. CTCF blocks the long-range communication of the enhancers with the TSS region of the *IGF2* gene but allows the initiation of *H19* transcription. This results in the expression of *H19*, *KCNQ1 and CDKN1C* as well as in the repression of *IGF2* and *KCNQOT1* transcription. In contrast, in paternally controlled alleles, ICR1 is methylated and does not bind CTCF, while ICR2 is unmethylated (Fig. 7.8, bottom). This reverses the expression pattern so that *IGF2* and *KCNQOT1* are produced but not *H19*, *KCNQ1* and *CDKN1C*. The physiological consequence of this imprinting is that **in maternally controlled cells growth and cell cycle are limited, while paternally controlled cells are primed for maximal growth**.

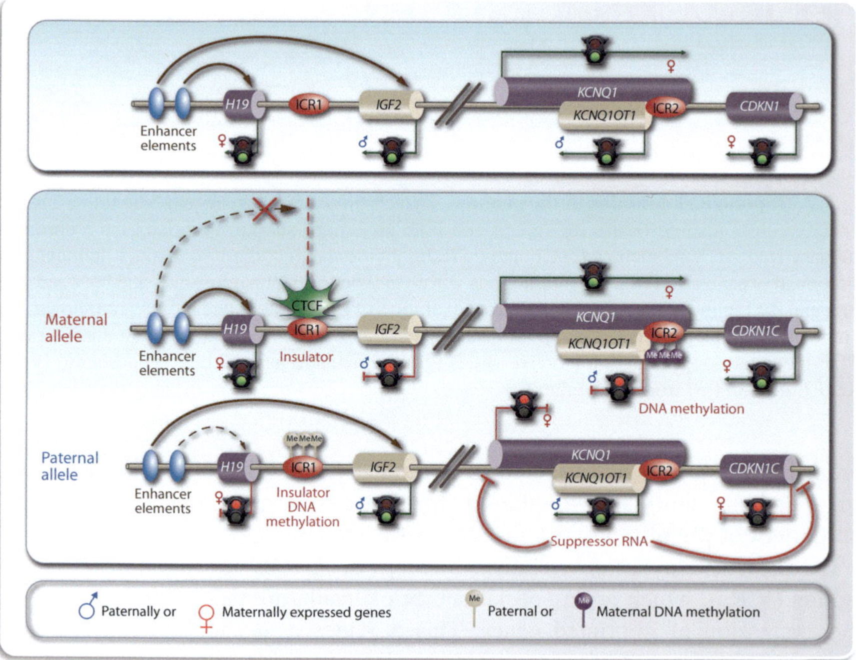

Fig. 7.8 Control mechanisms of the 11p15 imprinted cluster. General structure of the 11p15 cluster (**top**) and of scenarios of maternally (**center**) and paternally (**bottom**) controlled alleles. *IGF2* encodes for a growth factor, *H19* for a long ncRNA limiting body weight, *KCNQ1* for a potassium channel, *KCNQ1OT1* for an antisense transcript of *KCNQ1* that interacts with various chromatin components and *CDKN1C* for a cell cycle inhibitor

Another well studied example of **imprinting is the inactivation of one X chromosome in female cells**. The inactive X chromosome is observed as Barr body in female interphase cells. The epigenetic process behind the XCI mechanism is the long ncRNA *Xist* (X inactive specific transcript) (Sect. 10.3), which is exclusively expressed from the X inactivation center of the inactive X chromosome. The action of *Xist* represents a special form of imprinting that affects a whole chromosome.

Imprinted genes have also an important role in adaptation to feeding, social behavior and metabolism, i.e., postnatal processes that are very responsive to environmental influences. In this way, genomic imprinting is an epigenetic mechanism regulating gene dosage. Imprinting may have evolved in response to intra- and extracellular signals, in order to modulate the expression levels of these genes as required by various conditions.

(Clinical) conclusion: DNA methylation is the clinically best understood epigenetic mark. In health, tissue-specific DNA methylation patterns are essential for keeping terminally differentiated cells in their status. In contrast, aberrant DNA methylation is a key marker for many diseases, such as cancer, imprinting disorders and premature aging syndromes.

Additional Reading

John, R.M., Higgs, M.J. and Isles, A.R. (2023). Imprinted genes and the manipulation of parenting in mammals. Nat Rev Genet *24*, 783–796.

Luo, C., Hajkova, P. and Ecker, J.R. (2018). Dynamic DNA methylation: in the right place at the right time. Science *361*, 1336–1340.

Monk, D., Mackay, D.J.G., Eggermann, T., Maher, E.R. and Riccio, A. (2019). Genomic imprinting disorders: lessons on how genome, epigenome and environment interact. Nat Rev Genet *20*, 235–248.

Monteagudo-Sanchez, A., Noordermeer, D. and Greenberg, M.V.C. (2024). The impact of DNA methylation on CTCF-mediated 3D genome organization. Nat Struct Mol Biol *31*, 404–412.

Chapter 8
Histone Modifications

Abstract   In this chapter, we will present posttranslational modifications as a general mechanism for instructing proteins about their function. Moreover, they allow proteins to "memorize" their encounters, such as contacts with other proteins. Histone proteins are the key examples of this **protein communication system**. In particular, lysine acetylations and methylations of histone tails have a large functional impact. Genome-wide profiling of the large set of posttranscriptional histone modifications provides the basis of the histone code. This code leads to the understanding, how the epigenome directs transcriptional regulation and stores information. Histone modifying enzymes, such as HATs, HDACs, KMT and KDMs, either add or remove posttranslational modifications to histone proteins and in this way change the functional profile of the epigenome.

Keywords   Posttranslational histone modification · Histone code · Chromatin modifying enzymes · HATs · HDACs · KMTs · KDMs · Writers · Erasers · Readers

8.1   Histones and Their Modifications

Posttranslational modifications of histones are frequent and important epigenetic signals that control many biological processes, such as cellular differentiation in the context of development (Chap. 11). **Acetylations and methylations of lysines at histone tails are understood best and are the most important epigenetic marks affecting histones**. However, there are also a number of other acylations, such as formylation, propionylation, malonylation, crotonylation, butyrylation, succinylation, glutarylation and myristoylation, the functional impact of which is far less understood. In addition, there are phosphorylations at tyrosines, serines,

C. Carlberg, *Gene Regulation and Epigenetics*,
https://doi.org/10.1007/978-3-031-68730-3_8

histidines and threonines, ADP ribosylations at lysines and glutamates, citrullinations of arginines, hydroxylations of tyrosines, sumoylations and ubiquitinations of lysines as well as O-GlcNAcylation of serines and threonines. Basically all of these posttranslational modifications are found also with many other proteins, i.e., they are not specific for histone proteins. For example, there are 8000 proteins known, to which the sugar β-D-N-acetylglucosamine is added by the enzyme O-linked N-acetyl-glucosaminyltransferase and removed by the O-linked N-acetyl β-D-glucosaminidase.

Covalent modifications of histone proteins alter the physio-chemical properties of the nucleosome and are recognized by specific proteins, referred to as "readers" (compare also DNA methylation readers, Sect. 7.2). Basically, all covalent histone modifications are reversible via the action of specific enzymes. In general, chromatin acetylation is associated with transcriptional activation and controlled by two classes of antagonizing chromatin modifying enzymes, HATs and HDACs. When a HAT adds an acetyl group to the amino group in the side chain of a lysine, the positive charge of this amino acid is neutralized (Fig. 8.1). In reverse, an HDAC can remove the acetyl group and restore the positive charge of the lysine residue. Thus, **chromatin modifying enzymes determine through the addition or removal of a rather small acetyl group the charge of the nucleosome core, which has major impact on the attraction between nucleosomes and the density of chromatin packing**. In analogy, for histone methylation at lysines there are two classes of enzymes with opposite functions, KMTs and KDMs. Since histone methylation can be a repressive as well as an active marker, the exact position in the histone tail and its degree of methylation (mono-, di- or trimethylation) is critical.

Lysine (K) is the most frequently modified amino acid in proteins, since it can accommodate a number of different modifications, such as several types of acylations and methylation and reactions with ubiquitin and ubiquitin-like modifiers. These modifications occur in a mutually exclusive manner, so that specific lysine residues, such as H3K27, can serve as hubs for the integration of different signaling pathways (Fig. 8.2). Methylation is a special type of posttranslational modification. Since a single methyl group is small, it contributes only in a minor way to the steric properties of amino acids. Moreover, the methylation of lysines and arginines does not affect the charge of these residues, i.e., also in their methylated form they are positively charged. Lysines can be methylated up to 3-times and arginines up to 2-times, respectively. **Histone methylations are more stable modifications than phosphorylations or acetylations, i.e., their turnover is lower and they mark more stable epigenetic states**.

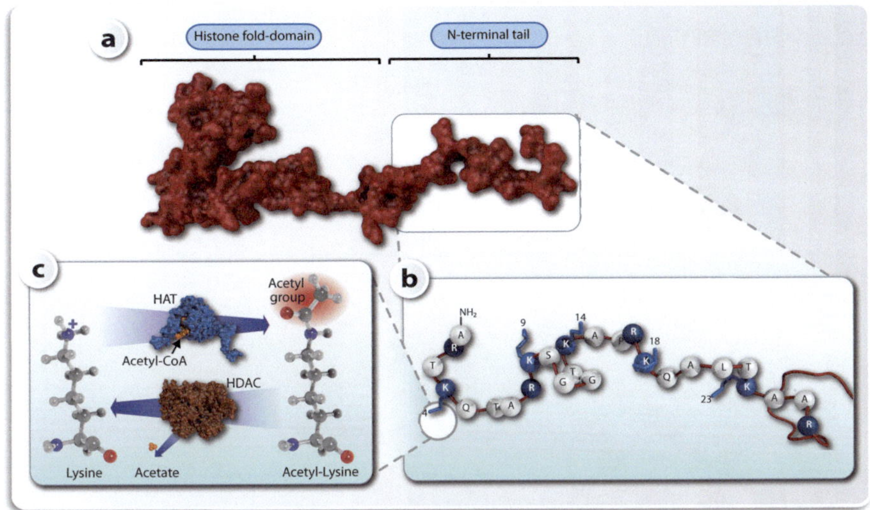

Fig. 8.1 Histone acetylation. Acetylation is shown as an example of a posttranslational modification of histone proteins. A Connolly surface model with secondary structures of histone H3 (**a**) is displayed in combination with a zoom into its amino-terminal tail. The positively charged amino acids lysine (K) and arginine (R) are indicated in blue (**b**). The activity of HATs removes the positive charge, while HDACs can reverse this process (**c**)

8.2 The Histone Code

The integrated data of maps of genome-wide histone modifications with patterns of chromatin accessibility, transcription factor binding as well as RNA expression from multiple tissues identified novel relationships between histone modifications and related chromatin structures. Reversible posttranslational modifications like phosphorylation, acetylation and methylation, of key amino acid residues, such as lysine and arginine, within proteins are the major mechanisms of communication and information storage in the control of signaling networks in cells. This means that **many proteins "remember" their functional tasks via their specific pattern of posttranslational modifications**.

Key examples of such information-processing circuits via posttranslational modifications are the nucleosome-forming core histones H2A, H2B, H3 and H4 as well as the linker histone H1 (Fig. 8.3). The tails and globular domains of these histone proteins provide over 130 amino acid residues for posttranslational modifications, the information content of which is summarized as the histone code (Box 8.1). The theoretical number of possible combinations of signals forming the histone code is

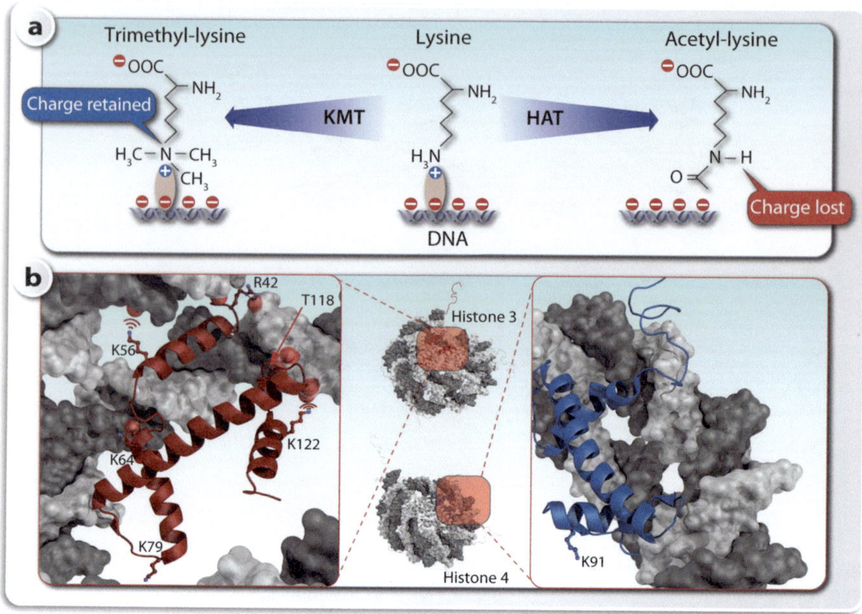

Fig. 8.2 Nucleosome stability through histone modifications. Unmodified lysine residues are positively charged and can form a salt bridge with negatively charged genomic DNA (both at physiological pH). The acetylation of lysines by HATs introduces a bulkier side chain and in parallel removes the positive charge (**a, right**). This decreases the affinity between DNA and the nucleosome and may destabilize the latter. The methylation of lysines by KMTs does not change the charge but, dependent on the number of added methyl groups, introduces various degrees of bulkiness. Crystal structures of the nucleosome are shown with highlighted key amino acids (**b**)

very large. In analogy to the alphabets of human languages, posttranslational modifications comprised in the histone code exemplify "letters" that may be combined to a large number of "words" with different meanings (Table 8.1). Thus, **histone modifications represent a text rich in information about the local chromatin status**. This more fine-grained distinction suggests that the epigenome has far more differential functions than "on" and "off", such as accessible and non-accessible chromatin.

Box 8.1: The Histone Code Model The model suggests that histone modifications modulate the structure of the nucleosome. This provides a platform for the recruitment of chromatin modifying enzymes that specifically recognize the respective histone modifications (readers). Moreover, multiple histone modifications act in a combinatorial fashion to specify distinct chromatin states. This allows a large number of posttranslational histone modifications to generate a very specific chromatin structure. This determines a specific expression level

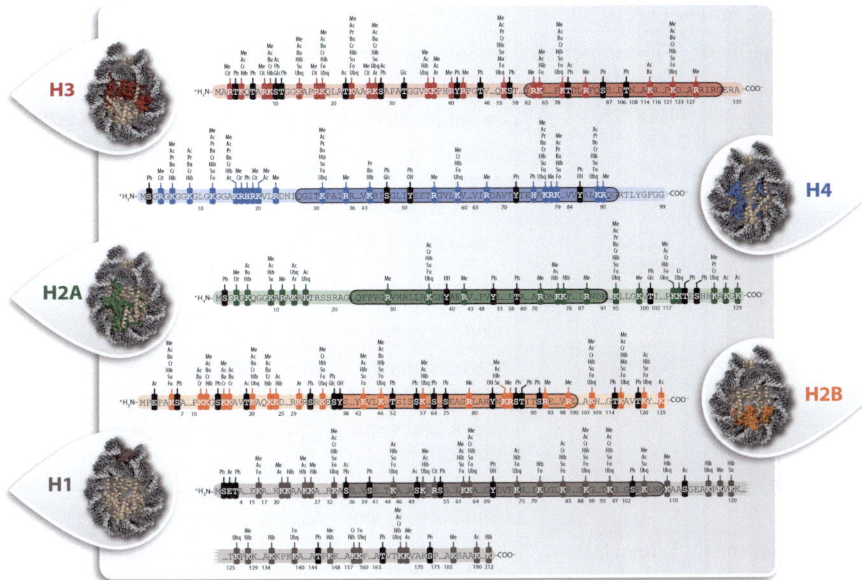

Fig. 8.3 Posttranslational modifications of histone proteins. All presently known posttranslational modifications of the nucleosome forming core histones H2A, H2B, H3 and H4 and the linker histone H1 are indicated. Amino acids that can be modified (K, lysine, R, arginine, S, serine, T, threonine, Y, tyrosine, H, histidine, E, glutamate) are highlighted, most of them can carry different modifications, but they do not occur in parallel. Me = methylation (K, R), Ac = acetylation (K, S, T), Pr = propionylation (K), B = butyrylation (K), Cr = crotonylation (K), Hib = 2-hydroxyisobutyrylation (K), Ma = malonylation (K), Su = succinylation (K), Fo = formylation (K), Ub = ubiquitination (K), Cit = citrullination (R), P = phosphorylation (S, T, Y, H), OH = hydroxylation (Y), Glc = glycation (S, T), Ar = ADP-ribosylation (K, E)

for each class of genes. The integration of histone modification maps with patterns of chromatin accessibility, transcription factor binding as well as RNA expression from multiple tissues identified novel relationships between histone modifications and related chromatin structures. This leads to the development of new hypotheses regarding the regulatory functions of chromatin features that are all part of the histone code model (Table 8.1). Some important elements of the model are:

- **Acetylation and deacetylation of histone tails represent major regulatory mechanisms during gene activation and repression**. Actively transcribed regions of the genome tend to be hyperacetylated, whereas inactive regions are hypoacetylated. However, histone hyperacetylation has been associated with histone deposition during replication and repair. Importantly, in case of histone acetylation more the overall degree of acetylation rather than any specific residue is critical.

Table 8.1 The histone code

Histone	Epi-mark	Most conserved co-marks	Putative function	Biological inference
H2	A.Z	H3K4me2/3	Poised promoter	Negatively associated with gene activation in ES cells and during their differentiation
	BK5me1		Active enhancer	
H3	K4me1	K27me3, K4me2	Poised enhancer	
	K4me2	K4me3	Active or poised regulatory regions	
	K4me3	K4me2	Active or poised regulatory regions	Poised enhancers regulate as many genes as bivalent promoters do
	K9me1		Active genes and enhancers	
	K9me3		Active or poised regulatory regions	Negatively correlated with sequence conservation
	K27ac	K4me1/2/3	Active enhancer (me1/me2) or promoter (me2/me3)	H3K27ac marks promoters as well
	K27me3	K4me1/2/3	Poised enhancer (me1/me2) or bivalent promoter (me2/me3)	Poised enhancers regulate as many genes as bivalent promoters do
	K27me1		Active enhancer	
	K36me3	K27ac, K4me1	Active enhancer	Not correlated with H3K27me3, may be a neglected mark of active enhancers
	K36me1		Active enhancer	
H4	K20me1		Active genes	
	K20me3		Active or poised regulatory regions	
	C^m		Repressed region	C^m either only mildly influences gene regulation or influences it in a way that is independent from histone modifications

The non-exclusive list of the relation of posttranslational modifications and their functional impact are summarized as the histone code model. Most of the features are confirmed by genome-wide analysis

- In contrast to acetylation, **there is a clear functional distinction between histone methylation marks**, both concerning the exact histone residues as well as their degree of modification, such as mono-, di- or trimethylation. For example, H3K9me3 and H4K20me3 are enriched near boundaries of large heterochromatic domains, while H3K9me1 and H4K20me1 are found primarily in active genes.
- **H3K4me3 is detected specifically at active promoters, while H3K27me3 is correlated with gene repression over larger genomic regions**. Both modifications are usually located in different chromatin domains, but they co-exist in a subset of genomic regions that are termed bivalent domains. These regions seem to have crucial roles, e.g., in ES cell differentiation (Sect. 11.3), by providing the potential for both transcriptional activation and repression. Moreover, their dysregulation can cause different types of diseases.
- **H3K36me3 marks in gene bodies correlate with levels of gene transcription**, since KMTs deposit this mark when interacting with elongating Pol II, i.e., expressed exons have a strong enrichment for this histone mark.
- **Histone modification profiles allow the identification of distal enhancer regions**, as they show relative H3K4me1 enrichment and H3K4me3 depletion. Interestingly, chromatin patterns at enhancer regions seem to be far more variable and cell specific than those at core promoter or insulator regions. Enhancer regions also show enrichment not only for H3K27 acetylation, but also for H2BK5me1, H3K4me2, H3K9me1, H3K27me1 and H3K36me1, suggesting the redundancy of these histone marks. Each of the modifications is detected at a rate of only 20–40% of all potential enhancers, i.e., none of them is associated with all enhancer regions.
- **Early replicating genes** are marked by H3K4me1, 2 & 3, H3K36me3, H4K20me1 as well as H3K9 and H3K27 acetylation. In contrast, **late replicating genes** often correlate with the marks H3K9me2 and H3K9me3. Moreover, boundaries between replicating zones show a pattern of histone signature modification, such as H3K4me1, 2 & 3, H3K27ac and H3K36me3. This suggests that **histone modifications serve as boundary elements**, comparable to insulators that block spreading of late-replicating heterochromatin.

Posttranslational modifications of histones either directly affect chromatin density and accessibility or serve as binding sites for effector proteins, such as chromatin modifying (Sect. 8.3) or remodeling complexes (Sect. 9.2). This ultimately modulates the initiation and elongation of transcription, i.e., histone modifications have an impact on the transcriptome (Fig. 8.4). In addition, histone marks are able to store information. Multiple histone modifications act in a combined way to generate a very specific chromatin structure that determines a specific expression level for each class of genes (Sect. 9.4).

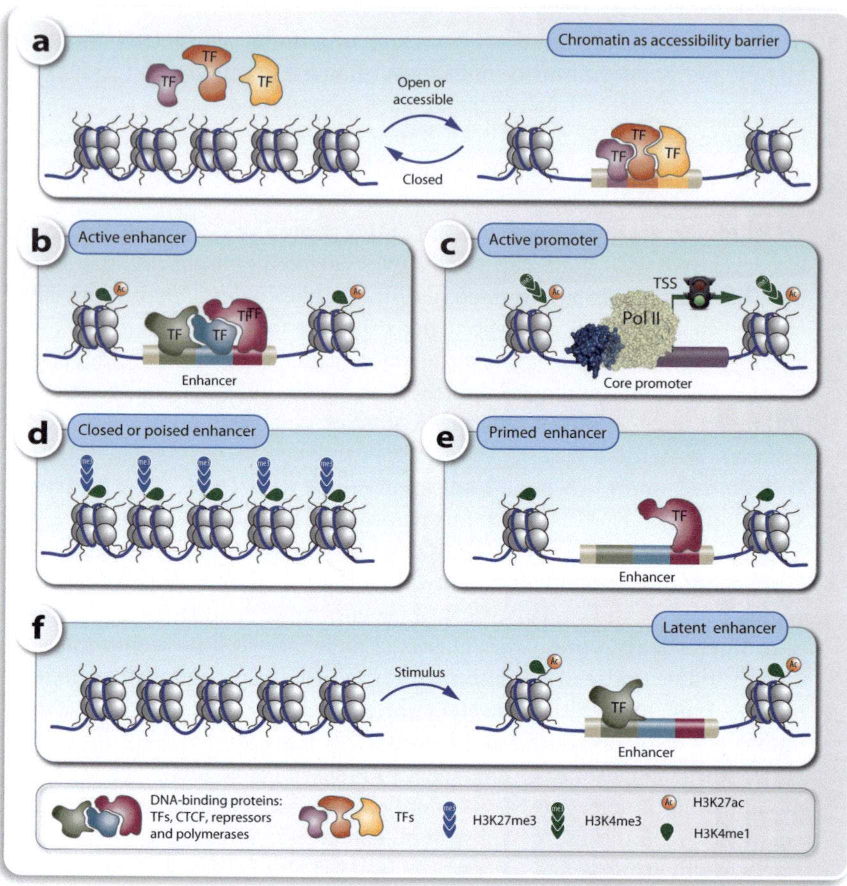

Fig. 8.4 Chromatin accessibility and histone marks. Chromatin restricts the access of transcription factors, Pol II and other nuclear proteins to the genome (**a**). Enhancers are often marked by H3K27ac and H3K4me1 modifications (**b**), while H3K27ac and H3K4me3 mark active promoters (**c**). Closed enhancers are termed poised when they carry both active histone marks (e.g., H3K4me1) as well as repressive marks (e.g., H3K27me3) (**d**). Enhancers are considered primed when they are pre-marked by H3K4me1 (**e**), while they are termed latent when they do not show any specific histone mark, but can become accessible through a stimulus (mostly an extracellular signal activating a signal transduction cascade) so that flanking nucleosomes acquire H3K4me1 and H3K27ac marks (**f**)

For some common histone marks, correlations between their presence and the activity of different genomic elements are well established, such as H3K9me3 and H3K27me3 at inactive or poised promoters (Fig. 8.4d), H3K27ac and H3K4me3 on active enhancers and promoters as well as H3K36me3 in transcribed gene bodies, respectively (Table 8.1). **Poised genes, promoters and enhancers** are found within facultative heterochromatin, i.e., the respective genomic regions are in a waiting position for incoming signals.

The integration of epigenome-wide datasets allows the identification of novel relationships between histone modifications and related chromatin structures. Chromatin states, as marked by histone modifications, characterize genomic elements, such as enhancers, promoters, insulators and gene bodies (Fig. 8.5). Thus, **chromatin represents a cell type-specific filter of genomic sequence information that determines, based on the histone code, which genes are transcribed into RNA**. Histone modification profiles allow the identification of distal enhancer regions, as they show relative H3K4me1 enrichment and H3K4me3 depletion. However, chromatin patterns at enhancer regions are variable, as they show enrichment not only for H3K27ac, but also for H2BK5me1, H3K4me2, H3K9me1, H3K27me1 and H3K36me1, suggesting the redundancy of these histone marks (Box 8.1).

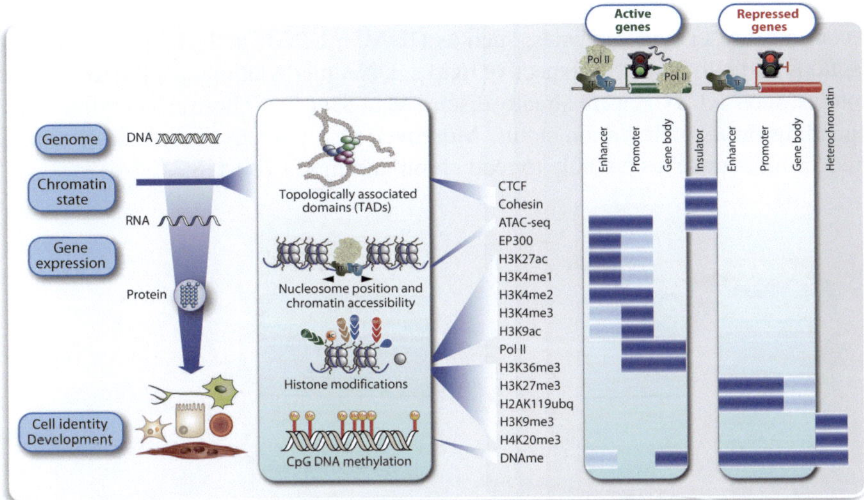

Fig. 8.5   Impact of the epigenome on gene expression. Chromatin acts as a filter for the genome concerning gene expression and, in this way, determines cell identity (**bottom left**). Epigenomic regulation happens at various scales of chromatin states (**center**), such as topological organization, chromatin accessibility, histone modifications and DNA methylation. Key histone modifications and binding of nuclear proteins that are characteristic for these chromatin states are indicated and distinguished between active and repressed genes (**right**). CTCF and cohesion are involved in chromatin organization, the HAT EP300 (also called KAT3B, lysine acetyltransferase 3B) marks enhancers and both Pol II and H3K36me3 indicate actively transcribed genes. For protein binding or histone modification lighter shades mark a lower or variable degree of modifications, while for DNA methylation it indicates that the genomic region can be regulated by methylation

8.3 Chromatin Modifying Enzymes

An average human cell has only some 100,000 open loci within its chromatin, i.e., more than 90% of the genome is buried in heterochromatin and therefore not accessible to transcription factors and Pol II. However, **many of these accessible chromatin regions are not static as they are dynamically controlled by chromatin modifying and remodeling proteins** (Fig. 8.6). These enzymes catalyze the methylation of genomic DNA (Sect. 7.1), the posttranslational modification of histone proteins (Sect. 8.1) or the positioning of nucleosomes (Sect. 9.1). The human genome expresses in a tissue-specific fashion hundreds of these chromatin modifying and remodeling enzymes that recognize (read), add (write) and remove (erase) chromatin marks. Writer-type enzymes, such as HATs, KMTs and DNMTs, add acetyl or methyl groups to histone proteins or cytosines of genomic DNA, respectively. In contrast, eraser-type enzymes, such as HDACs, KDMs and TETs, reverse these reactions and eliminate the respective marks. DNA methylation-specific reader-type proteins, such as CTCF, were already discussed in Sect. 7.3. These proteins bind DNA depending on its methylation status. Moreover, also components of the chromatin remodeling complexes are able to read chromatin marks (Sect. 8.2).

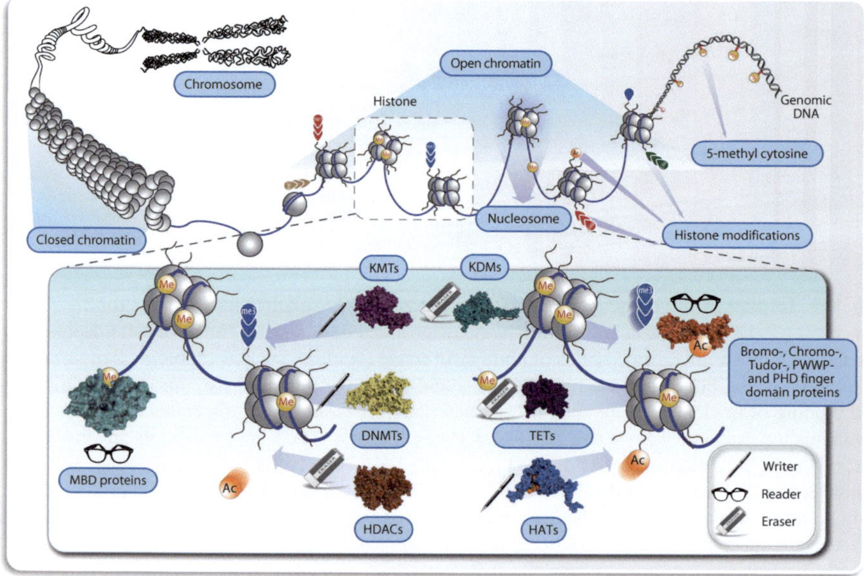

Fig. 8.6 Central role of chromatin. Covalent modifications of histones and genomic DNA, such as methylations, control the accessibility of chromatin to transcription factors and other regulatory proteins (**top**). These chromatin marks are introduced by writers, interpreted by readers and can be removed by erasers (**bottom**). The interplay between these nuclear proteins is essential for controlling gene expression

Acetyltransferases, which specifically acetylate lysines are termed KATs, are found in the nucleus and the cytoplasm. KATs that have histone proteins as substrates are mostly termed HATs. Cytoplasmic HATs acetylate histones H3 and H4 post-translationally, which is important for being deposited onto chromatin during DNA replication and repair. KATs/HATs use acetyl-CoA as an essential cofactor to donate an acetyl group to the target lysine residue, but they can also use different acyl-CoAs as substrates for histone lysine acylation. The human genome encodes for 22 HATs, of which 11 are KATs. CREBBP (CREB-binding protein, also called KAT3A) and EP300 (E1A-binding protein p300, also called KAT3B) do not only acetylate histones, but also modify GTFs, such as TFIIE, signal-dependent transcription factors, such as p53, and architectural proteins, such as HMGA1 (high mobility group AT-hook 1).

KATs are found at genomic regions that show high levels of histone acetylation, Pol II binding and gene expression. For example, KAT3A and KAT3B associate both with enhancer and TSS regions, whereas the binding of KAT2B (PCAF), KAT5 (TIP60) and KAT8 (MYST1) is enriched in both TSS and transcribed regions of active genes. In this context, KAT3A and KAT3B interact with the activation domains of numerous activated transcription factors, such as ligand-activated nuclear receptors or phosphorylated p53. The human genome also encodes for 18 members of the HDAC family taking care on the acetylation status of chromatin. The Zn^{2+}-dependent HDACs 1–11 act predominately in the nucleus and the cytoplasm, while NAD^+ (nicotinamide adenine dinucleotide)-dependent sirtuins (SIRTs) 1–7 are found in addition also in mitochondria.

There are 66 human genes encoding KMTs and 20 for KDMs. Both types of enzymes use both histone and non-histone proteins as substrates, i.e., these enzymes control the methylation status of chromatin and other proteins. KDMs are either flavin adenine dinucleotide (FAD)-dependent monoamine oxidases or Fe(II) and α-ketoglutarate-dependent dioxygenases, i.c., like SIRTs they sense via these metabolites the energy status of cells (Sect. 16.2).

The effects of chromatin modifying enzymes, such as HATs and HDACs, are primarily local and may cover only a few nucleosomes up and downstream of the starting point of their action. The same applies to KMTs and chromatin remodeling enzymes, such as the SWI/SNF (switching/sucrose non-fermenting) complex (Sect. 9.2). In case there is more HAT activity, chromatin is locally acetylated, the attraction between nucleosomes and genomic DNA decreases and the latter gets accessible for activating transcription factors, GTFs and Pol II. In this euchromatin state chromatin remodeling enzymes fine-tune the position of the nucleosomes, in order to obtain full accessibility of the respective binding sites. In the opposite case, when HDACs are more active, acetyl groups get removed and the packing of chromatin locally increases. KMTs then methylate the same or neighboring amino acid residues in the histone tails that attract heterochromatin proteins, such as HP1, and further stabilize the local heterochromatin state.

Specific histone marks are specifically recognized (read) by a large number of chromatin modifying enzymes via a small set of common recognition domains (Fig. 8.7). Repressive proteins of the Polycomb family use chromatin organization modifier domains (chromodomains) in order to interact with methylated chromatin. Some 10 different HATs have a PHD (plant homeodomain) finger, which is a specific reading motif for H3K4me2 and H3K4me3 marks. There are even 46 human proteins like HATs, HAT-associated proteins, KMTs, helicases, ATP-dependent chromatin remodeling proteins, transcriptional coactivators and nuclear scaffolding proteins that carry a bromodomain in order to recognize acetylated lysines. Chromodomains are far more specific for a given chromatin modification than bromodomain proteins, i.e., chromodomain-containing nuclear proteins recognize their genomic targets with far higher accuracy. In addition, a smaller set of proteins use a YEATS domain for sensing acetylated and crotonylated lysines. Finally, proteins with a Tudor domain recognize methylated lysines and arginines.

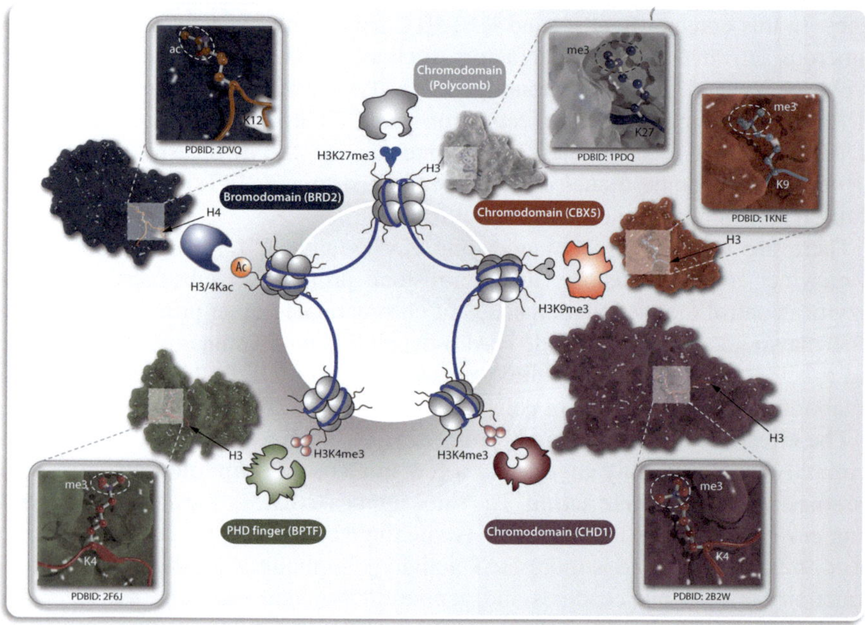

Fig. 8.7 Histone reading proteins. There are three major types of domains, such as chromodomains, PHD fingers and bromodomains, by which histone modifications are recognized (read). While chromodomains and PHD fingers bind to specific histone methylations, bromodomains are rather unspecific and recognize all forms of acetylated histones. BPTF = bromodomain PHD finger transcription factor, BRD2 = bromodomain containing 2, CBX5 = chromobox 5, also called HP1

8.4   Gene Regulation Via Chromatin Modifying Enzymes

Cells are constantly exposed to a multitude of signals, such as the extracellular matrix, cytokines, peptide hormones and other active compounds, the majority of which are transmitted by receptors at the membrane. These extracellular signals induce via membrane proteins intracellular signal transduction cascades. **These pathways often terminate at nuclear proteins, such as transcription factors, chromatin modifying and remodeling enzymes, i.e., they modulate the epigenome and transcriptome**. Most of the signals vary over time and usually have an "on" or "off" character, but the resulting changes in the transcriptome rather resemble a waveform (Fig. 8.8, left). For example, a signal can either directly activate chromatin modifying enzymes, which then write or erase histone marks, or act indirectly via the activation of chromatin remodelers that alter the nucleosome composition. In this way, chromatin-associated proteins act as signal converters and integrators.

Since the methylation of histones has a longer half-life than its acetylation or phosphorylation, signals can be stored within the epigenomic landscape for shorter and longer time periods. Accordingly, **the histone methylome is suited for a long-term epigenetic memory** (Sect. 11.5). Histone modifications act in combination with DNA methylation and transcription factor activity (Fig. 8.8, right), which increases the diversity of their outputs. Some of the information stored in the histone modification pattern can even be maintained throughout DNA replication, i.e., a part of the histone marks can be inherited.

Genome-wide histone modification maps correspond to different genomic features, such as promoters, enhancers and gene bodies, or activation states, such as actively transcribed, poised or silenced and often exist in combinations. The three main modes for the activity of genes are distinguished:

- **Active genes** are expressed genes that are associated with histone acetylation, H3K4me1, 2 & 3 marks and H2A.Z occurrence in their TSS regions as well as a number of different marks (H2BK5me1, H3K9me1, H3K27me1, H3K36me3, H3K79me1, 2 & 3 and H4K20me1) in their gene bodies. Highest levels of both HATs and HDACs are found associated within these genes. Their presence correlates positively with mRNA expression and Pol II levels. The antagonizing activities of the chromatin modifying enzymes are the basis for precise fine-tuning of gene expression via the homeostasis of active chromatin loci and apply both for TSS and enhancer regions. Thus, active genes are found in euchromatin.

- **Poised genes** are not expressed and do not associate with significant histone acetylation, but they show H3K4 methylation marks and occurrence of the histone variant H2A.Z. As long these genes wait for their activation, only low level of HATs or HDACs are associated with them, they are found in facultative heterochromatin. A cycle of transient acetylation and deacetylation keeps these genes primed but inactive and maintains their promoter regions in a poised state waiting for future activation via external signals.

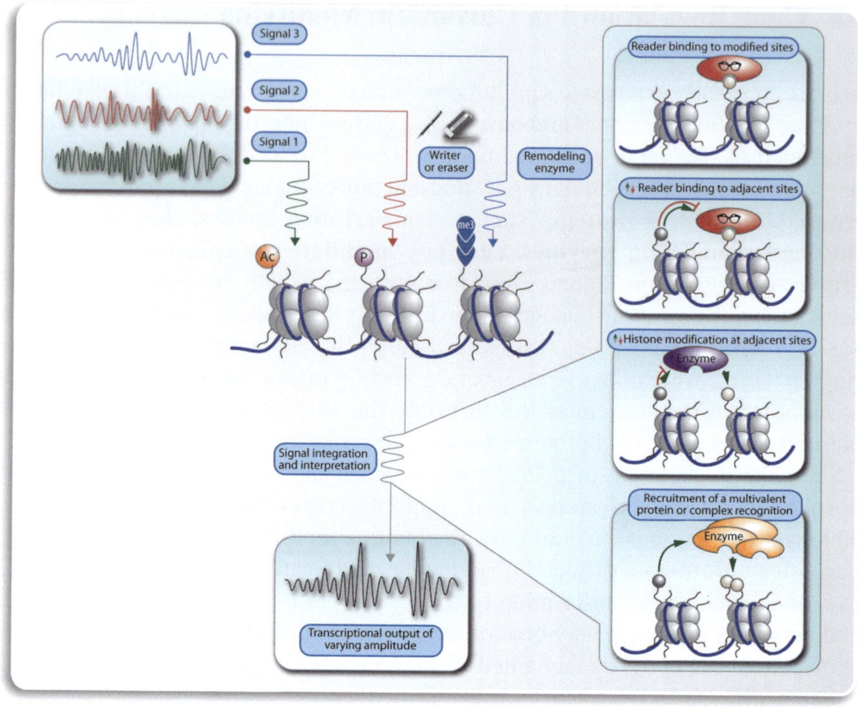

Fig. 8.8 Signal storage and interpretation via chromatin modifying enzymes and readers. Signals deriving from various, mostly membrane-based signal transduction cascades are integrated on chromatin through modifications at histone tails. Multiple inputs occurring over time can be stored. These inputs can affect chromatin directly or are transmitted via chromatin modifying enzymes, such as the writers HATs and KMTs, as well as the erasers HDACs and KDMs and chromatin remodeling proteins (**left**). There are a number of mechanisms how this dynamic epigenetic landscape is constantly interpreted by reader proteins, such as changing the ability of a reader protein to recognize a neighboring mark, recruiting enzymes that modify additional sites and creating a combinatorial display for recognition in various binding events (**right**). The net result of the signal integration can be observed as transcriptional output, i.e., as a change of the transcriptome (**bottom left**)

- **Silent genes** either carry H3K27me3 marks together with Polycomb proteins or do not have at all any known chromatin marker. At these genes neither HATs nor HDACs are found and they are located within heterochromatin.

Taken together, **chromatin modifying enzymes maintain the epigenome and in this way control gene expression**. Thus, these nuclear enzymes have central importance during embryogenesis as well as in cell fate decisions in health and disease.

> **(Clinical) conclusion**: There more than 130 ways to posttranslationally modify histone proteins. This tremendous capacity to store information represents the second column of the epigenome. Histone acetylations are primarily used for short storage of epigenetic information, such as transient changes in gene expression. In contrast, signals based on histone methylation last longer and interfere with long-term epigenetic memory stored on the level DNA methylation (Sect. 11.5).

Additional Reading

Blackledge, N.P. and Klose, R.J. (2021). The molecular principles of gene regulation by Polycomb repressive complexes. Nat Rev Mol Cell Biol 22, 815–833.

Dupas, T., Lauzier, B. and McGraw, S. (2023). O-GlcNAcylation: the sweet side of epigenetics. Epigenetics Chromatin 16, 49.

Janssen, S.M. and Lorincz, M.C. (2022). Interplay between chromatin marks in development and disease. Nat Rev Genet 23, 137–153.

Martire, S. and Banaszynski, L.A. (2020). The roles of histone variants in fine-tuning chromatin organization and function. Nat Rev Mol Cell Biol 21, 522–541.

Morgan, M.A.J. and Shilatifard, A. (2020). Reevaluating the roles of histone-modifying enzymes and their associated chromatin modifications in transcriptional regulation. Nat Genet 52, 1271–1281.

Sheikh, B.N. and Akhtar, A. (2019). The many lives of KATs - detectors, integrators and modulators of the cellular environment. Nat Rev Genet 20, 7–23.

Chapter 9
Chromatin Remodeling and Organization

Abstract In this chapter, we will learn that nucleosome positioning around TSS regions has an important role on coordinated gene activation of promoter regions. In this context, we will discuss that chromatin modification and remodeling machineries allow the transition from a repressed state to an active state. Chromatin remodeling factors are multiprotein complexes that use the energy of ATP hydrolysis, in order to remodel or remove nucleosomes regulating the exposure of genomic DNA to transcription factors. Global changes in gene expression correlate with spatial chromatin reorganizations that play a significant role during development. Thus, transcriptionally active genes are involved in the process that directs the architecture of the nucleus in differentiated tissues and cell types. In the interchromatin compartment of the nucleus there are subnuclear structures, such as transcription factories, that contain high concentrations of Pol II. Transcription factories function as an attractor for commonly regulated genes with shared nuclear positions. This suggests that the transcriptional status of a gene is based on the position in the sphere of the nucleus.

Keywords Nucleosome positioning · Chromatin remodeling · ATP-dependent remodeling complex · Transcriptional dynamics · 3D chromatin organization · Chromosome territory · Transcription factory

9.1 Nucleosome Positioning at Promoters

Constitutively active genes like housekeeping genes have open chromatin at the genomic regions containing their critical TFBSs (Fig. 9.1a). Although there is always a dynamic competition between nucleosomes and transcription factors at regulatory regions, housekeeping genes use mechanisms that favor the binding of transcription factors over that of nucleosomes, such as the recruitment of appropriate chromatin modifying enzymes (Sect. 8.3). Thus, constitutively active genes typically have a nucleosome-depleted region upstream of their TSS, within which key TFBSs reside.

C. Carlberg, *Gene Regulation and Epigenetics*,
https://doi.org/10.1007/978-3-031-68730-3_9

Experimentally these regions are often detected as DNase hypersensitivity regions and were traditionally considered to be nucleosome-free. However, in reality there is a gradient of depletion, so that the term "nucleosome-depleted region" is more appropriate.

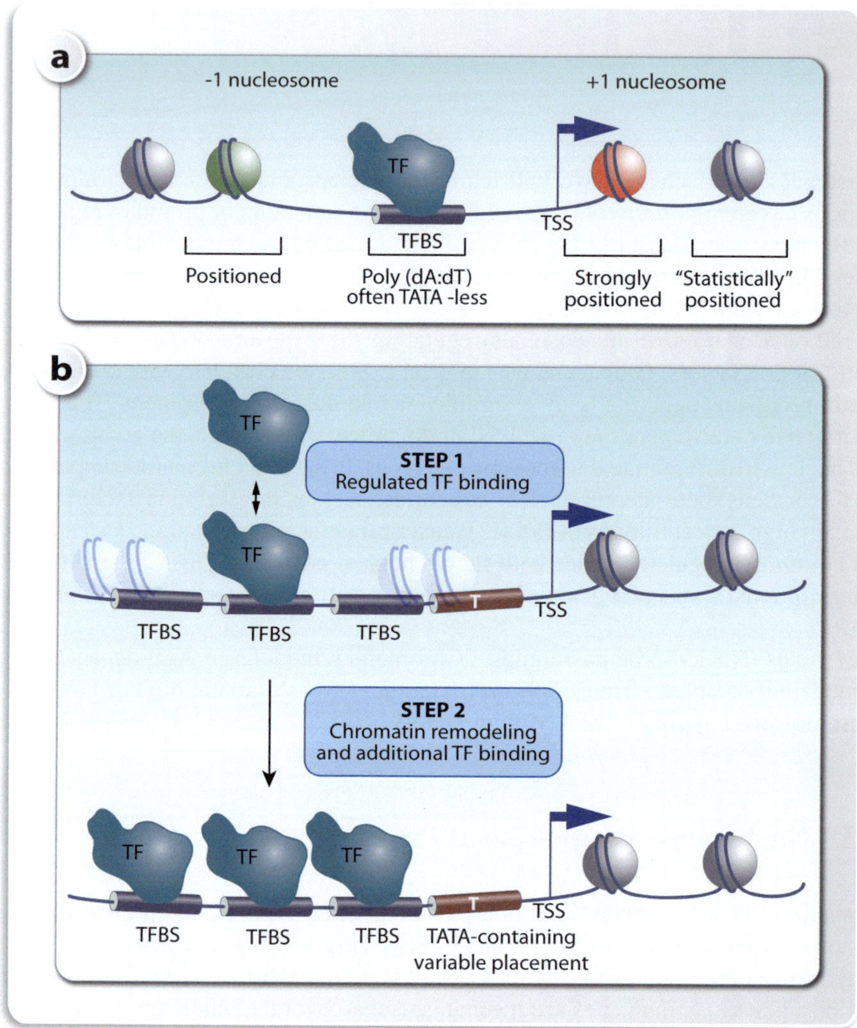

Fig. 9.1 Properties of open and closed promoters. A common feature of constitutively active genes is an open TSS region, i.e., a depleted proximal nucleosome neighboring to the TSS (**a**). In contrast, a common feature of highly regulated genes is to have in their repressed state a closed core promoter, i.e., a nucleosome next to the TSS (not shown). On covered promoters, nucleosome positioning sequences of varying strength and locations help to define nucleosome positions and promoter architecture (**b**)

Robust transcriptional activity, such as of housekeeping genes, requires nucleosome depletion, whereas transcriptional regulation of other genes involves nucleosomes repositioning (Fig. 9.1b). Genome-wide studies indicated that often a 200 base pairs nucleosome-depleted region upstream from the TSS is flanked on either side by well-positioned nucleosomes. The +1 nucleosome plays a central role in determining the activity of Pol II. At active genes the +1 nucleosome is found approximately 40 base pairs downstream of the TSS, while at inactive genes the nucleosome it is only 10 base pairs downstream. A common finding from Pol II ChIP-seq studies is a clear enrichment of Pol II at TSS regions compared with the gene body. Thus, **Pol II is frequently stalled at the +1 nucleosome**. This stalling is also referred to as "poising" (Sect. 3.2), when transcription is blocked until a signal for activation or release is received, or as "pausing", when Pol II is slowed down immediately downstream of the TSS. Therefore, the +1 nucleosome either physically blocks the progression of Pol II or regulates the presence and/or activity of proteins that support Pol II to overcome the stalling. For example, the +1 nucleosome shows high levels of H3K4me3 that is bound by the TAF3 subunit of the basal transcriptional machinery. Even though the H3K4me3 mark is generally associated with active promoters, it is also present on promoters with non-elongating Pol II. Thus, **H3K4me3 is not exclusively a marker for active promoters but represents also poised TSS regions**.

For some genes, such as those being important during embryogenesis, poising is a strategy for rapidly starting transcription in response to a stimulus (Sect. 11.3). However, for genes with broad core promoters poising or other kind of stalling may only reflect open chromatin. Moreover, the least efficient phase in transcription is early elongation, so that accumulation of Pol II not much downstream of the TSS can also be a kinetic effect. Nevertheless, in poised genes elongation is actively regulated, in order to release Pol II for achieving transcriptional bursts, i.e., rapid increases of transcribed mRNA. Genes with active Pol II show phasing of nucleosomes within their coding region, i.e., their accurate positioning in relation to the +1 nucleosome. This region serves as a boundary for positioning nucleosomes after Pol II pauses. For example, the positioning of nucleosomes at exons can function as "speed bumps" that enhance splicing by slowing down Pol II. **The increased Pol II occupancy at the gene bodies then provides time to recruit the splicing machinery during transcription and results in improved recognition of splicing signals.**

9.2 Chromatin Remodeling

A cell's phenotype depends on its gene expression pattern, which basically is influenced how genomic DNA is packed into chromatin. Nucleosomes often block the access of transcription factors to their genomic binding loci, since the packing of genomic DNA around histone octamers hides one side of the DNA. Within a stretch of 200 base pairs of genomic DNA 147 base pairs contacting a histone octamer, i.e., some 75% are not easily accessible to regular transcription factors. Binding sites that are located close to the center of these 147 base pairs are generally inaccessible to transcription factors. Sites closer to the edge of the nucleosome-covered sequence are a bit better reachable, but **only within the 50 base pairs between two neighboring nucleosomes genomic DNA is fully accessible**. This is sufficient space for the binding of transcription factors (Chap. 3). Thus, in some cases only minor shifts in the position of the nucleosomes are necessary in order to get TFBSs accessible, whereas in other cases a whole nucleosome needs to be depleted.

Since nucleosomes have a rather strong electrostatic attraction for genomic DNA, the catalysis of the sliding, removal or exchange of individual subunits, or even the eviction of whole nucleosomes has to dissolve all histone-DNA contacts and requires the investment of energy in the form of ATP. Thus, chromatin remodelers are multiprotein complexes that use the energy of ATP hydrolysis in order to affect nucleosomes in at least four ways (Fig. 9.2). These are:

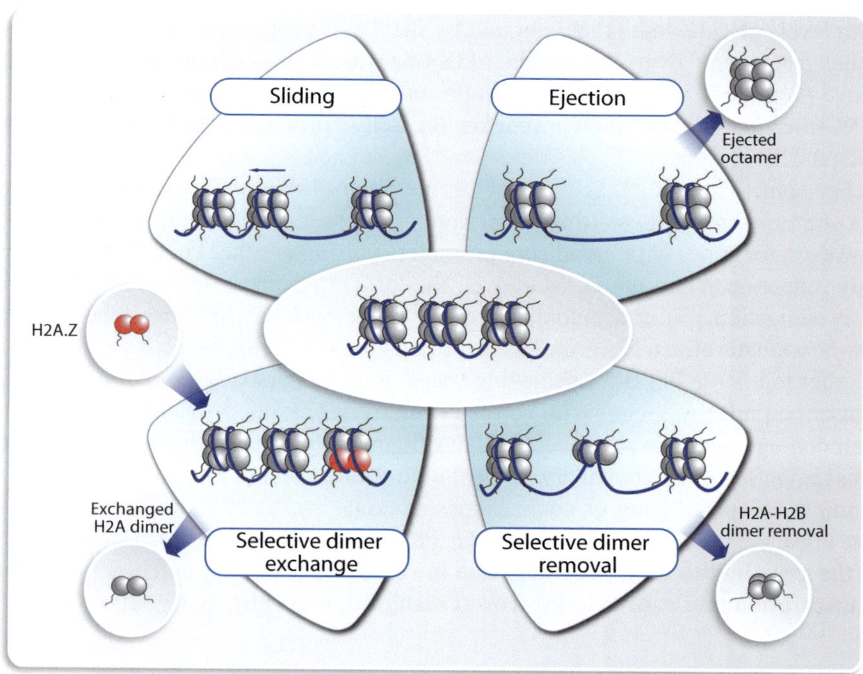

Fig. 9.2 Mobility and stability of nucleosomes. Chromatin remodelers enable access to genomic DNA through sliding, ejection, H2A-H2B dimer removal or selective dimer exchange from nucleosomes. ATP-dependent remodeling complexes as well as thermal motion influence the mobility of nucleosomes. The stability of nucleosomes is affected by its detailed octamer composition and the pattern of histone modifications. For example, the incorporation of histone variants into nucleosomes alters the interactions with histone and non-histone proteins

- the movement (**sliding**) of the histone octamer to a new position within the same chromatin region
- the complete displacement (**ejection**) of the histone octamer, e.g., from TSS regions of heavily expressed genes
- the **exchange** of regular histones by their variant forms, such as H2A by H2A.Z (Box 2.2)
- the **removal** of H2A-H2B dimers from the histone octamer.

ATP-dependent remodeling is crucial for both the assembly of chromatin structures and their dissolution. About 30 human genes encode for subunits of four different chromatin remodeling complexes (Fig. 9.3). Although these different complexes share common properties, they are also specialized for particular tasks. The ATPases in the core of the chromatin remodeling complexes are genetically non-redundant, but they all increase nucleosome mobility with different efficiencies and outcomes. The investment of energy for this process is necessary, since both sliding and ejection of nucleosome from genomic DNA has to dissolve all histone-DNA contacts requiring approximately 12–14 kcal/mol. The remodeling processes involve the dissociation of genomic DNA at the edge of the nucleosome and form a DNA bulge on the histone octamer surface. The DNA loop then propagates across the surface of the nucleosome in a wave-like manner, resulting in the repositioning of DNA without changes in the total number of histone-DNA contacts.

Remodeling complexes contain proteins with bromodomains, chromodomains and PHD domains that can read histone marks (Sect. 9.3). For example, ISWI (imitation SWI) remodelers use a PHD domain for targeting H3K4me3 marked histones, in CHD (chromodomain helicase DNA binding) remodelers chromodomains bind to methylated histones, in SWI/SNF remodelers a bromodomain is used for recognizing acetylated histone H3, while INO80 (INO80 complex subunit) remodelers employ a bromodomain for binding to acetylated H4.

The action of the SWI/SNF complex is mostly associated with transcriptional activation. Interestingly, the activity of many chromatin remodelers is affected by the presence of histone variants that they themselves introduce into the chromatin, *i.e.*, they control each other's action through the exchange of histones (Fig. 9.4). The histone variants MacroH2A and H2A.Bbd (Box 2.2) reduce the efficiency of the SWI/SNF complex, whereas H2A.Z stimulates remodeling by ISWI complexes. The INO80 complex removes H2A.Z from inappropriate locations. In general, H2A.Z resides at open TSS regions and positively regulates gene transcription. The unique amino-terminal tail of this histone variant becomes acetylated when a gene is active.

9.3 Organization of Chromatin in the Nucleus

Within the interphase nucleus the position of a gene, such as being located in the center or at the borders, is important for its expression. This leads to the question, whether the nuclear location is an independent and functionally important epigenetic

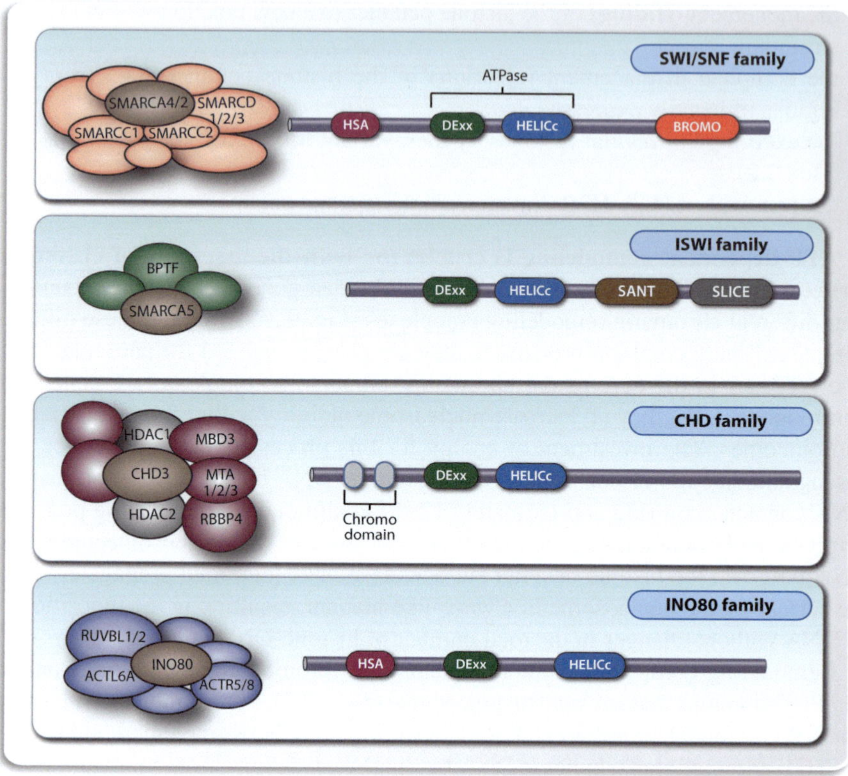

Fig. 9.3 Chromatin remodeling complexes. Chromatin remodelers are divided into four main families on the basis of the sequence and structure of the ATPase subunit: SWI/SNF, ISWI, CHD and INO80 complexes. Most of these names derive from the nomenclature in yeast, where these complexes were first discovered and characterized. Remodeling complexes contain proteins with bromodomains and chromodomains that can read histone marks. Some histone modifications promote ATP-dependent chromatin remodeling by creating binding sites for remodelers. For example, acetylation of nucleosomes promotes the recruitment of SWI/SNF remodelers through their acetyl group binding bromodomain and increases remodeling efficiency. In contrast, the activities of ISWI and CHD complexes are inhibited by histone acetylation. In general, ISWI complexes remodel nucleosomes that lack acetylation, such as at H4K16, i.e., their activity is focused on transcriptionally inactive regions. Furthermore, CHD complexes contain HDACs 1 and 2, i.e., they also have HDAC activity. ACTL6A = actin like 6A, ACTR = actin related protein, MTA = metastasis associated, RBBP4 = RB binding protein 4, chromatin remodeling factor, RUVBL = RuvB like AAA ATPase, SMARC = SWI/SNF related, matrix associated, actin dependent regulator of chromatin

parameter or whether it is only the consequence of the action of transcription factors in context with chromatin. Since each tissue is characterized by its own selection of active and inactive genes, different **cell types can be distinguished by individual patterns of active and inactive chromatin regions**. The clustering of heterochromatin marked by H3K27me3 at the nuclear periphery creates silencing foci, referred

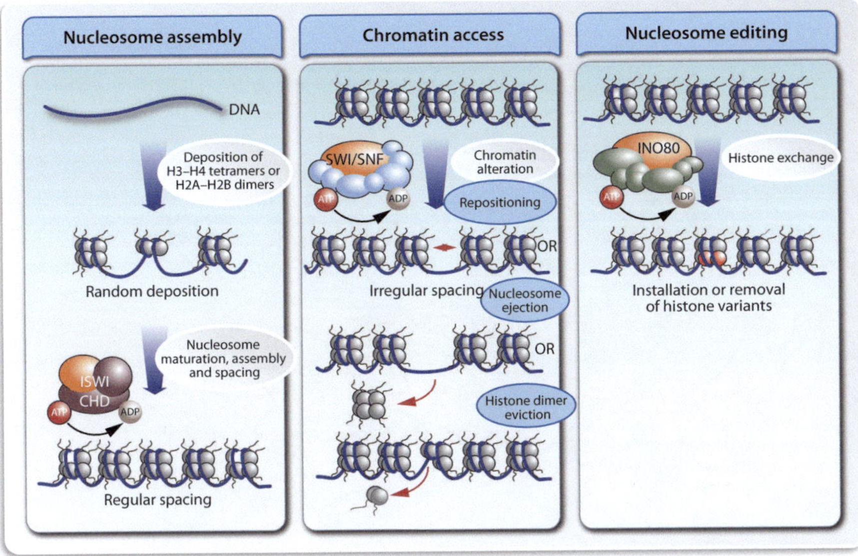

Fig. 9.4   Function of human chromatin remodeling enzyme complexes. ISWI and CHD remodelers are involved in the deposition of histones, the maturation of nucleosomes and their spacing (**left**). SWI/SNF remodelers alter chromatin by repositioning nucleosomes, ejecting octamers or evicting histone dimers (**center**). INO80 remodelers change nucleosome composition by exchanging core and variant histones, such as installing H2A.Z variants (**right**). ISWI complexes create nucleosome arrays of uniform spacing by sliding the nucleosomes until the linker DNA reaches the same fixed distance. In contrast, SWI/SNF complexes disorganize the nucleosome position that makes TFBSs accessible. Most chromatin remodeling complexes can eject nucleosomes, but ISWI complexes lack this activity

to as Polycomb bodies (Fig. 9.5). These are complexes of members of the Polycomb family, such as the components of PRC1 and PRC2. PRCs act as transcriptional repressors that are essential for maintaining tissue-specific gene expression programs, i.e., they ensure the long-term repression of specific target genes.

The nuclear lamina also binds and silences large regions of heterochromatin organized in LADs (Sect. 2.4), i.e., interactions of genomic regions with lamin proteins are central in reducing gene expression. Thus, in association with the nuclear periphery silenced genes and gene-poor chromosome territories are found. However, **the position of chromatin and with that the position of genes, is not fixed** but there are dynamic changes in the contacts between the nucleoskeleton and genomic DNA involving single genes or small gene clusters. These changes are most pronounced during development. For example, ES cells possess dispersed chromatin with limited compaction. However, during differentiation the cells show changes in their chromatin structure that include larger compaction of genomic domains. In the same way, embryonic development proceeds from a single cell with dispersed chromatin to differentiated cells with nuclei that show compact chromatin domains being located

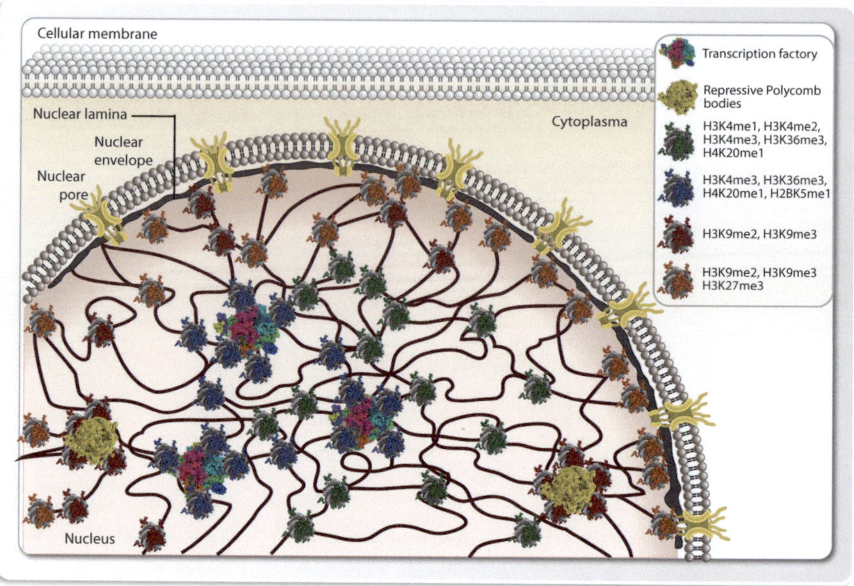

Fig. 9.5 Chromatin modification signatures associate with relative position features in the nucleus. Histone modifications correlate with the position within the nucleus: chromatin modifications that are generally associated with active transcription (green nucleosomes) are often found in the center of the nucleus, whereas chromatin with generally repressive modifications (orange nucleosomes) is associated with the nucleoskeleton. Regions with active modifications (blue nucleosomes) may participate in transcription factories (purple Pol II in the center). Blocks of histone H3K27me3 (dark red nucleosomes) may be components of Polycomb bodies (yellow)

in the periphery. This indicates that the physical relocation of a gene from the nuclear periphery to the center unlocks it, in order to be expressed in a future developmental stage.

Although chromosomes are microscopically not visible in an interphase nucleus, they occupy specific chromosome territories. This represents another level of chromatin architecture. In their territories, chromosomes fold so that active and inactive TADs and LADs are found in distinct nuclear compartments. Active regions are preferentially located in the nuclear interior, whereas inactive TADs accumulate at the periphery. In addition, TADs that are heavily bound by tissue-specific transcription factors are in different neighborhoods than those interacting with repressive PRCs. To some extent, chromosome territories intermingle, which could explain interchromosomal interactions. Nevertheless, interactions between loci on the same chromosome are much more frequent than contacts between different chromosomes. Since the volumes of chromosome territories depend on the linear density of active genes on each chromosome, **chromatin with higher transcriptional activity occupies larger volumes in the nucleus than silent chromatin.**

The two main experimental approaches for studying 3D genome organization are FISH (fluorescence in situ hybridization) and 3C-based methods (Sect. 6.2). FISH is a molecular cytogenetic method using fluorescent oligonucleotides that detect genomic regions with a high degree of sequence complementarity. It is often used for the diagnostic of genetic diseases, but is also applied for defining spatial–temporal patterns of gene expression within cells and tissues. Moreover, FISH live-cell imaging provided dynamic views of chromatin domains, while 5C and Hi-C map the whole genome in kb resolution for chromatin loops, such as TADs and LADs. The latter methods are performed with large cell populations and thus provide a probabilistic view of chromosome folding and nuclear organization. In contrast, in FISH multiple individual cells are analyzed, i.e., both methods are complementary to each other and may be used for cross-wise validation. However, there is significant cell-to-cell and time-dependent variation in chromatin folding, so that the spatial distance between two genomic loci can show a rather wide distribution. **3C-based methods detect only events in cells, in which two loci are in close proximity, while FISH can determine the spatial distance between the loci in any cell**. Thus, both methods investigate different subpopulations of cells and may lead to apparent inconsistencies. The *4D Nucleome* project (https://commonfund.nih.gov/4Dnucleome/index) aims to overcome this problem via developing and employing a range of genomic, imaging and modeling methods to study 3D genome organization.

Global analyses of chromatin contacts in human cells, such as performed by Hi-C, indicate that individual TADs are separated by each other by boundary regions that are enriched both for enhancer markers, such as H3K4me1, and signs for repression, such as H3K9me3 (Fig. 9.6). Often TAD boundaries are identical with insulators and bind CTCF, i.e., they separate functionally distinct regions of the genome from each other (Sect. 7.3). In this way, CTCF is not only involved in smaller-scale DNA looping resulting in enhancer-promoter contacts, but also in larger-scale loop formation. Thus, **TADs are the units of chromosomal organization and segregate the human genome into at least 2000 domains containing coregulated genes**. Based on higher resolution Hi-C maps the number of TADs may be 5-times larger and their average size is accordingly lower.

The interchromosomal space located between chromosome territories contains a variety of nuclear substructures that are referred to as "speckles", "foci", "spots" and "bodies". The composition and number of these substructures depends on the cell type. The key example for this spatial organization is the activity of RNA polymerase I in the nucleolus, in which ribosomal genes are concentrated. **RNA polymerase I and its associated partner proteins are found in 200–500 nm diameter complexes in centers within the nucleolus that are termed "factories"**. In these factories rRNA transcripts move across the surface and extrude nascent transcripts into the surrounding component of the nucleolus.

Interestingly, **actively transcribing Pol II is also distributed non-uniformly within interchromosomal spaces and is concentrated in transcription factories** (Fig. 9.7). These dynamic foci of Pol II, transcription factors, coactivators and chromatin modifying enzymes are also referred as hubs, clusters or condensates.

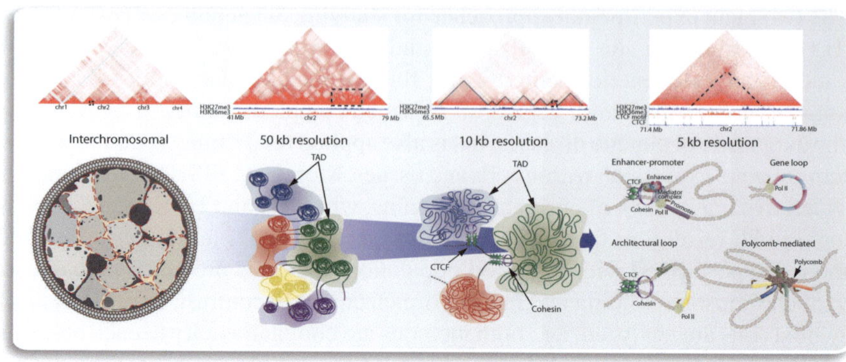

Fig. 9.6 Hierarchy of chromatin architecture. Hi-C data of four levels of resolution (**top**) are schematically interpreted (**bottom**) as interchromosomal interactions between intermingled chromosome territories (**left**), TADs with interdomain interaction (**center left**), TADs (**center right**) and enhancer-promoter loops (**right**). Within Hi-C maps, chromatin loops and TADs are typically recognizable as neighboring triangles. These indicate that regions within the same TAD interact with each other more often than with regions of neighboring TADs. The distinction into TADs correlates with many features of the linear genome, such as patterns of histone modifications or gene expression as well as association with nuclear lamina

In contrast to macromolecular complexes of fixed stoichiometries, such as ribosomes, within condensates molecules associate at variable stoichiometries, i.e., they are spatially enriched relative to the surrounding cellular environment. The number of transcription factories/condensates per nucleus varies from a few hundreds to several thousands and differs between cell types and their differentiation state. The size of these factories ranges between 45 and 100 nm in diameter as determined by electron microscopy. They include, based on the number of nascent RNA transcripts, up to eight Pol II molecules. The condensates do not only contain proteins but they also facilitate enhancer-promoter contacts that may have a linear distance of hundreds of kb.

The model of Pol II being immobilized at preassembled transcription factories implies the idea that **gene loci move to the RNA polymerase being already present in a factory rather than the whole transcriptional machinery would be recruited to the chromatin template and moved along it**. This may happen by a controlled and directed motion of chromatin fibers and may promote the assembly of transcription factories. Accordingly, during transcriptional elongation distinct genes are brought into close vicinity and pulled through the relatively immobile Pol II complexes. The spatial nuclear organization may not be absolutely essential for transcription, but it clearly enhances its efficiency. Gene transcription requires the assembly of large complexes, such as chromatin remodelers, chromatin modifying enzymes, the MED complex and the basal transcriptional machinery and involve many distinct protein–protein and protein-DNA interactions. Therefore, the efficiency of transcription is clearly enhanced, when some of these protein complexes are already concentrated in specific parts of the nuclear space. Moreover, recycling of Pol II back to TSS regions

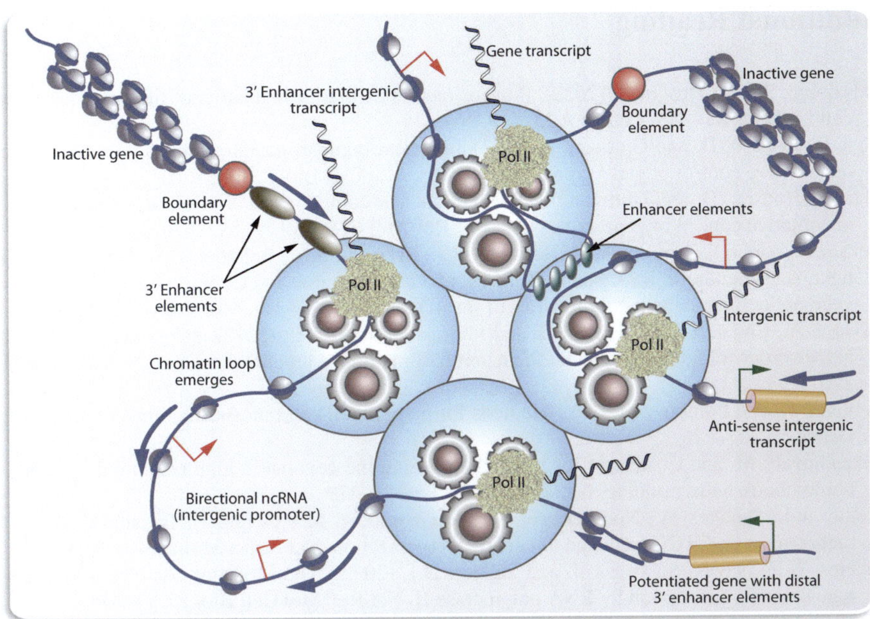

Fig. 9.7 Model of a transcription factory. Genes extend out of their chromosome territories, both in *cis* and in *trans*, in order to access a shared transcription factory. DNA binding factors are indicated by colored circles

of highly expressed genes can be facilitated, if Pol II cannot easily diffuse away from the template. Thus, **the transcription factory model is important for understanding the regulation of initiation and elongation of transcription, the genomic organization of genes, the coregulation of genes and possible instabilities of the genome**.

(Clinical) conclusion: In addition to DNA methylation and histone modification, the 3D organization of chromatin represents the third equally important level of the epigenome. Thus, the position of a gene within the nucleus is of central importance for its activity. The assembly of transcription factors, polymerases, cofactors, chromatin modifying enzymes and other nuclear proteins in transcription factories/condensates makes gene regulation more efficient and faster.

Additional Reading

Balsalobre, A. and Drouin, J. (2022). Pioneer factors as master regulators of the epigenome and cell fate. Nat Rev Mol Cell Biol *23*, 449–464.

Bhat, P., Honson, D. and Guttman, M. (2021). Nuclear compartmentalization as a mechanism of quantitative control of gene expression. Nat Rev Mol Cell Biol *22*, 653–670.

Bhat, P., Honson, D. and Guttman, M. (2021). Nuclear compartmentalization as a mechanism of quantitative control of gene expression. Nat Rev Mol Cell Biol *22*, 653–670.

Dekker, J., Alber, F., Aufmkolk, S., Beliveau, B.J., Bruneau, B.G., Belmont, A.S., Bintu, L., Boettiger, A., Calandrelli, R., Disteche, C.M., et al. (2023). Spatial and temporal organization of the genome: current state and future aims of the 4D nucleome project. Mol Cell *83*, 2624–2640.

Haws, S.A., Simandi, Z., Barnett, R.J. and Phillips-Cremins, J.E. (2022). 3D genome, on repeat: higher-order folding principles of the heterochromatinized repetitive genome. Cell *185*, 2690–2707.

Kempfer, R. and Pombo, A. (2020). Methods for mapping 3D chromosome architecture. Nat Rev Genet *21*, 207–226.

Lakadamyali, M. and Cosma, M.P. (2020). Visualizing the genome in high resolution challenges our textbook understanding. Nature methods *17*, 371–379.

Monteagudo-Sanchez, A., Noordermeer, D. and Greenberg, M.V.C. (2024). The impact of DNA methylation on CTCF-mediated 3D genome organization. Nat Struct Mol Biol *31*, 404–412.

Richter, W.F., Nayak, S., Iwasa, J. and Taatjes, D.J. (2022). The Mediator complex as a master regulator of transcription by RNA polymerase II. Nat Rev Mol Cell Biol *23*, 732–749.

Chapter 10
Regulatory Impact of ncRNA

Abstract In this chapter, we will discuss that ncRNAs are a heterogeneous group of transcripts that are not translated into proteins. The human genome is extensively transcribed also outside protein-coding regions giving rise to tens of thousands of ncRNAs. Not all of these transcripts are functional, however, many ncRNAs have regulatory specificity, i.e., some of them function similarly to proteins. For example, miRNAs are small ncRNAs that regulate posttranscriptionally the expression of most protein-coding genes. They share many similarities together with transcription factors and therefore are useful for different regulatory processes. The most effective targets of miRNAs are members of signal transduction cascades, such as receptors, kinases and transcription factors. Long ncRNAs have a number of mechanisms available to regulate biological processes, such as the initiation of the imprinting process XCI via *Xist*. A special variant of long ncRNAs are eRNAs that are produced bidirectionally at enhancer regions, when the latter interact with promoter regions.

Keywords Hidden transcriptome · Long ncRNA · miRNA · eRNA · XCI · *Xist* · Heterochromatin · Transcription factor

10.1 ncRNAs

From the evolutionary perspective, ncRNAs are an ancient version of proteins, i.e., they existed even before the occurrence of proteins and have functions similar to enzymes and structural proteins. The transcriptome-wide detection of RNA molecules, initially via DNA tiling arrays, such as oligonucleotide-based microarrays and then by NGS methods, such as RNA-seq (Box 3.2, provided the surprising result that **the proportion of the human genome being transcribed is far larger than formerly expected**. For protein-coding genes new splicing variants and additional exons and TSS regions were discovered, but also additional ncRNA molecules

C. Carlberg, *Gene Regulation and Epigenetics*,
https://doi.org/10.1007/978-3-031-68730-3_10

were found within, close to or in larger distance to protein-coding genes. These RNA molecules are either independent transcripts with own TSS regions or are processed parts of larger RNA precursors, such as spliced introns of pre-RNAs. The additional transcripts were found both in sense and in antisense orientation in relation to protein-coding genes. Some of the transcripts are remainders of the long evolution of the human genome, such as pseudogenes and integrated retrovirus genomes. Many functions of this "hidden transcriptome" are still not fully understood, but it had already been shown that they are involved in (i) regulating the expression of other genes, (ii) splicing, (iii) chromatin architecture, (iv) epigenetic regulation, (v) dysregulation in cancer and other diseases, (vi) translation and (vii) DNA repair. The present model of understanding gene regulation is still dominated by proteins, such as transcription factors and chromatin modifying enzymes, but it can be expected that in future the contribution of ncRNAs will be more and more appreciated.

From the nearly 200,000 known human RNA transcripts (Table 2.1) the majority (60%) are ncRNAs, i.e., ncRNA genes outnumber protein-coding genes. Long ncRNAs (Sect. 10.3) map to intergenic and intronic regions and many short ncRNAs derive from long ncRNAs (Table 10.1). A significant number of annotated ncRNAs originate from pseudogenes. Most pseudogenes have lost their original functional abilities but some regulate gene expression by acting as decoys for miRNAs. Open reading frames within RNA transcripts determine based on the rules of the genetic code where protein translation starts and ends. In contrast, there is no comparable genetic code for ncRNA transcripts, i.e., their function is hard to identify by computational methods alone.

The roles of ncRNA genes are quite diverse, including (i) gene regulation by miRNAs and long ncRNAs, (ii) chromatin accessibility by long ncRNAs, (iii) RNA processing in splicing like by small nucleolar RNAs (snoRNAs) and (iv) protein synthesis like by tRNAs and rRNAs. Many ncRNAs are transcribed from intergenic regions around genes, such as enhancers and promoters and therefore called eRNAs (Sect. 10.4). Thus, the gene regulatory potential of ncRNAs, in particular of miRNAs, is similar to that of transcription factors (Sect. 10.2).

Table 10.1 The complexity of ncRNAs

Short ncRNA	No	Length (nt)	Functions
miRNA	> 2500	> 100	Precursors to short regulatory RNAs (21–23 nt)
snoRNA	> 1500	> 1000	Precursors to shorter RNAs (60–300 nt) that help to chemically modify other RNAs
snRNA	> 2000	> 1000	Precursors to shorter RNAs (150 nt) that assist in RNA splicing
piRNA	~ 100	Unknown	Precursors to short (25–33 nt) RNAs that repress retro-transposition of repeat elements
tRNA	~ 600	> 100	Precursors to short transfer RNAs (73–93 nt)
Long ncRNAs			
Antisense ncRNA	> 5000	100–1000	Mostly unknown, but some are involved in gene regulation through RNA interference
Enhancer ncRNA	> 2000	> 1000	Unknown
Intergenic ncRNA	> 6000	100–10,000	Mostly unknown, but some are involved in gene regulation
Pseudogene ncRNA	> 10,000	100–10,000	Mostly unknown, but some are involved in regulation of miRNA

The number, size and function of short and long ncRNAs is indicated. nt = nucleotides

10.2 MiRNAs and Their Regulatory Potential

The stability of mRNA molecules is regulated by miRNAs for the time span that they are needed as templates for translation. Accordingly, microRNAs have sequences, which are complementary to the $3'$-UTR of their target mRNAs. Mature miRNAs are typically only 22 nucleotides long, but the length of their primary transcripts can be hundreds to thousands of nucleotides (Fig. 10.1). In the canonical pathway of miRNA biogenesis precursor RNAs are transcribed by Pol II from intergenic or intronic genomic loci. In contrast, in the non-canonical miRNA pathway, miRNAs are transcribed directly as endogenous short, hairpin RNAs or derive from mirtrons, i.e., through splicing from introns that can refold into hairpins. In both cases the primary miRNA (pri-miRNA) transcripts contain hairpin structures. These are recognized and processed by a complex of the proteins Drosha (a RNase III-type endonuclease) and DGCR8 (DiGeorge syndrome critical region gene 8). The complex generates a 70 nucleotide stem-loop structure, referred to as precursor miRNA (pre-miRNA), that is actively exported from the nucleus to the cytoplasm. There the complex of the proteins Dicer (another RNase III-type endonuclease) and TRBP (transactivation-response RNA binding protein) recognizes the pre-miRNA. Dicer cleaves this precursor and generates in this way the 22 base pairs of the mature miRNA duplex. Only one strand of this double-stranded miRNA binds to the protein AGO2 (Argonaute RISC catalytic component 2) of RISC (RNA-induced silencing complex). Nucleotides 2 to 8 of the mature miRNA are referred to as "seed sequence", which is complementary to sequences in the 3′-UTR of mRNAs. This RNA-RNA hybrid is by recognized by RISC, which inhibits target mRNA expression through its degradation. This involves the removal of the polyA tail via increasing the activity of deadenylases and blocking the initiation or elongation step of protein translation.

The copy number of an individual miRNA is in average approximately 500 molecules per cell, which is higher than the average expression of individual mRNAs. However, miRNA species differ in their concentration over a dynamic range of 4 or more orders of magnitude. For example, there are cell type-restricted miRNAs with more than 10,000 copies per cell. **The human genome encodes for more than**

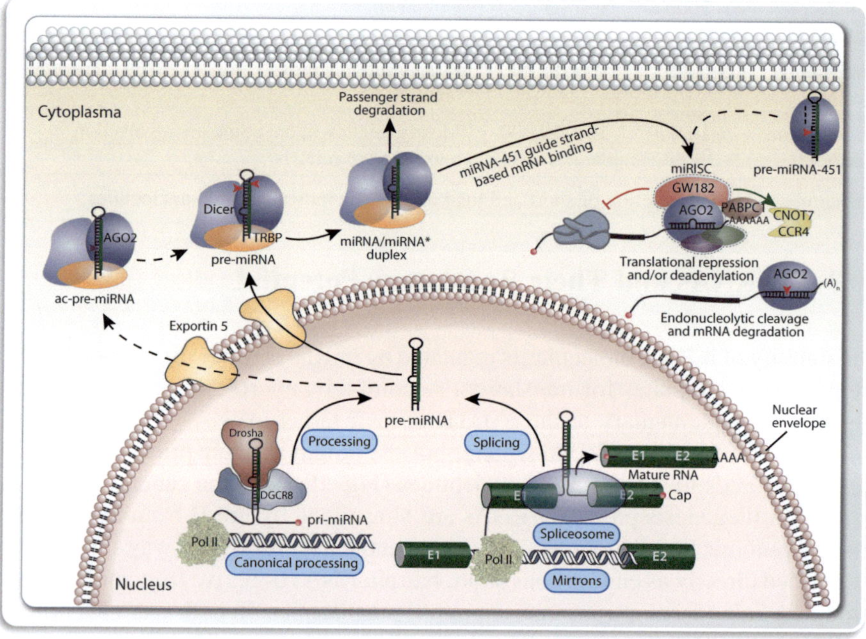

Fig. 10.1 Biogenesis of miRNA. A special feature of miRNA genes is the folding of their primary RNA transcripts (pri-miRNAs) into hairpin structures that the miRNA biogenesis machinery specifically recognizes and processes. miRNAs are encoded either by individual genes with their own TSS regions or as miRNA gene clusters that are transcribed as a single pri-miRNA. In addition, intron regions of protein-coding genes sometimes contain miRNA genes. The Drosha-DGCR8 protein complex recognizes the hairpin regions of pri-miRNAs and processes them through cleavage at the double-stranded stem region. These pre-miRNAs are exported from the nucleus to the cytoplasm, where the Dicer complex recognizes and processes them into 22 base pairs double-stranded mature miRNAs. Through binding one strand of the mature miRNA the AGO2-containing RISC specifically recognizes target mRNAs by partial base-pairing. This then blocks the translation of the mRNA target and leads to its degradation. CNOT7 = CCR4-NOT transcription complex subunit 7, PABPC1 = poly(A) binding protein cytoplasmic 1

2600 different miRNAs. The database miRBase provides information on the location and sequence of the mature miRNA sequence and also determines the miRNA nomenclature (Box 10.1).

> **Box 10.1: miRNA Nomenclature** The numbering of miRNA genes is simply sequential. The names/identifiers in the database miRBase (www.mirbase.org) and in the literature are given in the hsa-mir-121 form, where the first three letters signify the organism, as in this case "hsa" for *homo sapiens*. Then, the mature miRNA is designated as "miR-121" (with capital R) in the database and in much of the literature, whilst "mir-121" (with small form r) refers to the miRNA gene and also to the predicted stem-loop portion of the primary transcript. Distinct precursor sequences and genomic loci that express identical mature sequences get names of the form hsa-mir-121-1 or hsa-mir-121-2, respectively. Lettered suffixes denote closely related mature sequences, e.g., hsa-miR-121a and hsa-miR-121b (with capital R), are expressed from the precursors hsa-mir-121a and hsa-mir-121b (with small r), respectively.

In general, **every ncRNA has the intrinsic capacity to regulate in *cis*, since it can function while remaining connected to its own locus**. In contrast, an mRNA molecule can only act in *trans*, i.e., it needs to dissociate from its origin, be exported to the cytoplasma and gets there translated. However, **a given miRNA can regulate hundreds of mRNAs, i.e., most of its action is in *trans***. As a result, miRNAs have substantial effects on gene expression networks. However, the degree of target gene downregulation imposed by a given miRNA often is only of modest quantity. Although basically all genes can act as miRNA targets, only a subset of the interactions of miRNAs with mRNAs effectively modulates biological responses. The ideal targets of miRNAs are mRNA encoding for components of signal transduction cascades, such as receptors, kinases and transcription factors.

Example 1 Target gene expression is often exclusively activated through an intra- or extracellular signal and actively repressed in its absence. For this "default repression" miRNAs can act as mediators. For example, during DNA damage, a kinase cascade activates the transcription factor p53, leading to cell cycle arrest, senescence or even apoptosis (Sect. 15.4). In the default state, ubiquitin-mediated degradation inhibits p53. The miRNA miR-125b is essential to complete p53 repression and loss of miR-125b causes p53-dependent apoptosis (Fig. 10.2a). Interestingly, miR-125b belongs to the DNA damage network, as it is downregulated after genotoxic treatments. Thus, **miR-125b establishes a robust DNA damage response (DDR) through a raise in the threshold for p53 activation**.

Example 2 The transcription factors of the SMAD family are the nuclear targets of the TGFβ signal transduction cascade (Sect. 4.3). In addition, *SMAD* mRNAs are also targets of miRNAs (Fig. 10.2b). For example, the miR-23b cluster targets SMAD3, SMAD4 and SMAD5 in developing liver, which results in the inhibition of

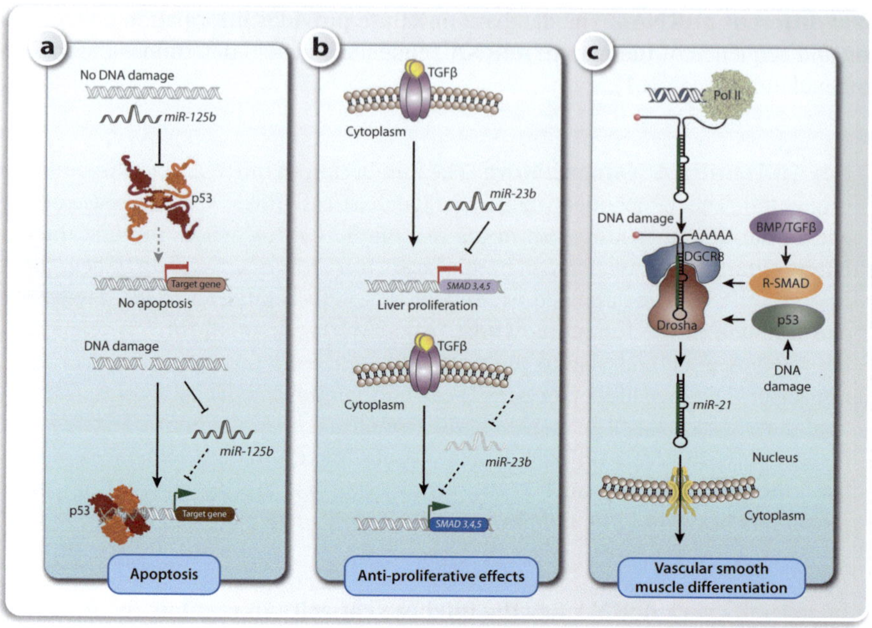

Fig. 10.2 miRNAs in modulating signal transduction cascades. The involvement of miR-125b in the DDR exemplifies how a miRNA can operate as the primary mediator of default repression (**a**). In normal cells (**top panel**), miR-125b targets control remaining p53 activity, in order to avoid apoptosis. Genotoxic effects (**bottom panel**) active p53 and repress miR-125b that results in the induction of apoptosis. The miR-23b cluster targets SMAD3, SMAD4 and SMAD5 and thereby inhibits the antiproliferative response mediated by TGFβ (**b**). When a single miRNA cluster targets several proteins of the same signal transduction cascade, these proteins can amplify their effect. The transcription factors SMAD and p53 bind to the Drosha complex and promote the maturation of many miRNAs to pre-miRNA (**c**). The control of the biogenesis of a limited set of miRNAs by transcription factors may emerge from the recognition of specific Drosha-pri-miRNA complexes

the antiproliferative response via TGFβ and the increase of hepatocyte proliferation. This demonstrates how **a simultaneous attack on a joint set of targets by miRNAs of the same cluster can amplify the biological effect, even if each individual miRNA has only a weak effect**.

Example 3 SMAD proteins stimulate via the association with the Drosha complex a rapid increase of miR-21 expression. Consequently, contractile cell differentiation within vascular smooth muscle is mediated by miR-21 (Fig. 10.2c). In addition, p53 also stimulates the Drosha complex that promotes the conversion of many miRNAs to pre-miRNAs. The control of the biogenesis of a limited set of miRNAs by transcription factors, such as SMAD and p53, emerges then from the recognition of specific Drosha-pri-miRNA complexes.

Signaling pathways are especially relevant in human diseases, in particular in cancer. Important contributions to the understanding of miRNA function in signal transduction arise from the consistent dysregulation of miRNAs in various types of human diseases, in particular in cancer. Since miRNAs are well preserved in body fluids, such as blood serum, liquor or urine and can be quantified more accurately than proteins, they may serve as biomarkers for diverse molecular diagnostic applications. Accordingly, **miRNA profiling became an important method in diverse areas of biology and medicine**.

The regulatory potential of miRNAs resembles on many levels to that of transcription factors. Both families of regulatory molecules have a comparable number (some 1600 transcription factor genes *versus* approximately 2600 miRNA genes) and share a common regulatory logic (Fig. 10.3). Groups of both transcription factors and miRNAs are combinatorial expressed and characterize individual cell types. While transcription factors recognize with their DBDs their specific binding sites within promoter and enhancer regions, the seed sequences of miRNAs bind $3'$-UTR sequences on their target mRNAs. Transcription factors can bind to millions of different locations within the whole human genome, but the very most of them are hidden by chromatin. In contrast, miRNAs have far less different targets within less than 1 kb of the $3'$-UTR of the pool of expressed mRNAs. The accessibility of these miRNA recognition sites is controlled by members of the large family of RNA binding proteins and by secondary structures of the mRNA target. Nevertheless, also miRNAs control hundreds of target genes.

Most, if not all, genes of the human genome are controlled by a combination of several transcription factors (Chap. 4). Thus, **miRNAs provide an additional layer of regulatory complexity and act in most cases as fine-tuners of the action of transcription factors**. Transcription factors can both activate and repress their primary targets, while miRNAs regulate gene expression mostly through repression. Nevertheless, repression is an important mechanism that shapes gene regulation in a cell-specific fashion. Events of transcriptional activation via ubiquitously expressed transcription factors can gain specificity through the action of cell type-specific repressors, such as miRNAs. The repressive mode of miRNAs therefore fits well with the general importance of gene repression.

Because miRNAs control the expression of many transcription factors and in turn the cell type-specific expression profiles of miRNAs are largely under the control of transcription factors, **miRNAs and transcription factors are linked to each other in regulatory networks**. This means that basically every transcription factor-controlled process has also contribution from miRNAs and vice versa. The activity of transcription factors is prominently regulated via posttranslational events, such as phosphorylation, processing or localization. Similarly, miRNAs can be modified by RNA editing. Moreover, proteins being involved in miRNA biogenesis and function, such as Drosha, Dicer and RISC, are subjected to posttranslational modifications.

There are also some significant differences between miRNAs and transcription factors:

- **The knockdown of transcription factor genes has more pronounced phenotypic effects than the deletion of miRNAs**. This may be explained by the redundancy between closely related miRNA family members. Moreover, this indicates

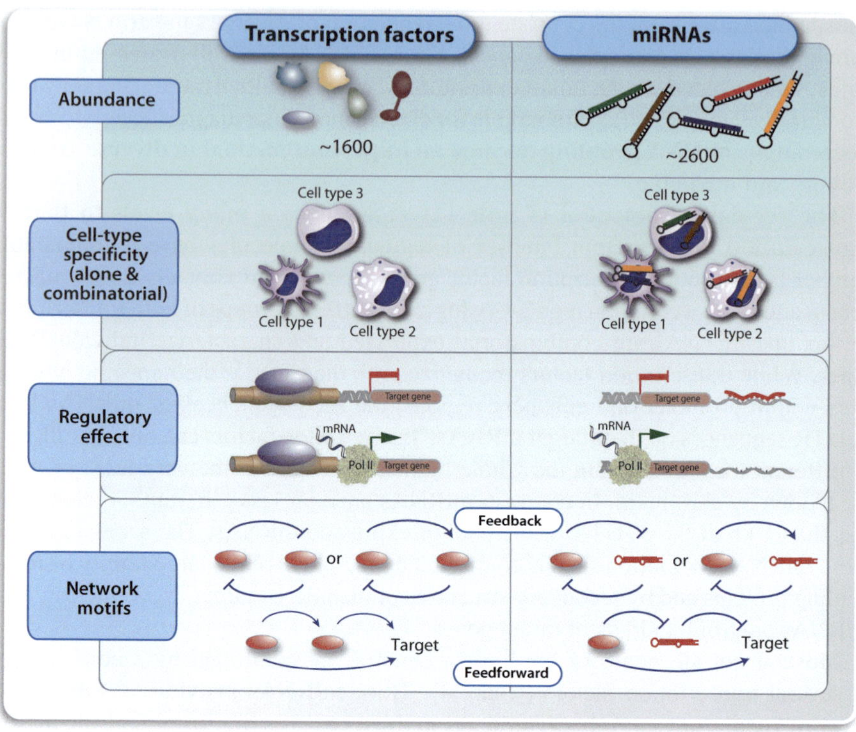

Fig. 10.3 Shared principles of transcription factor and miRNA action. The shared features of transcription factors and miRNAs include abundance (both families of gene regulatory factors contain 1600–2600 members), cell type specificity (both type of regulators act either alone or in combination in a cell type specific fashion), regulatory effects (both can either activate or repress gene expression) and involvement in regulatory networks, i.e., both use of positive and negative feedback loops

that miRNAs control more specific aspects of the terminal differentiation of individual cell types, while transcription factors are more important in earlier steps of development.

- **The action of miRNAs can be compartmentalized within a cell**, in order to rapidly alter local gene expression. For example, in neurons miRNA can control gene expression specifically in synapses. A comparable process is not possible with the action of transcription factors.

The speed of evolutionary changes of miRNAs is faster than that of transcription factors. Only a few new transcription factor families have arisen during vertebrate evolution, while there is continuous emergence of new miRNA families. This suggests that **the increase of complexity in body organization and organs is rather due to miRNA regulation than based on transcription factor action**.

10.3 Long ncRNAs

When ncRNAs are longer than 200 nucleotides, they are called long ncRNA. These ncRNAs are heterogeneous in their biogenesis, abundance and stability as well as they differ in the mechanism of action. Some long ncRNA have a clear function, such as in regulation of gene expression, while others, such as eRNAs, may be primarily side products non-precise of Pol II transcription (Sect. 10.4).

Despite their rather recent discovery, ncRNAs are probably evolutionary older than proteins, i.e., in early cells they mediated most of the regulatory actions, many of which were taken over later by proteins. Long ncRNAs carry out their cellular functions by interacting with proteins to form macromolecular complexes. The complex formation is enabled via elements within ncRNAs, such as short sequence motifs or larger secondary or tertiary structures, that interact specifically with a large set of molecular structures in proteins, RNA and DNA. This allows a large variety of functions, such as the ability to (i) scaffold and recruit multiple regulatory proteins, (ii) localize to specific targets on genomic DNA and (iii) utilize and shape the 3D structure of the nucleus.

A well-known example of an RNA scaffold is the telomerase RNA component (TERC) that assembles the telomerase complex. A number of chromatin modifying and remodeling proteins, such as PRC components, KMT1C, KDM1A, DNMT1 and the SWI/SNF complex, interact with nuclear long ncRNAs. These RNA-protein interactions (i) recruit chromatin regulatory complexes to specific genomic sites, in order to regulate gene expression, (ii) competitively or allosterically modulate the function of nuclear proteins and (iii) combine and coordinate the functions of independent protein complexes (Fig. 10.4).

A key example of how ncRNAs contribute to chromatin organization is the long ncRNA *Xist*, which is the key initiator of XCI in female cells carrying two X chromosomes (Sect. 7.3). *Xist* recruits a series of regulatory complexes at different stages of the XCI process and maintains X chromosome-wide transcriptional silencing (Fig. 10.5). In female ES cells both X chromosomes are actively transcribed and carry markers of active chromatin, such as H3K4me1, H3ac and H4ac. However, in the blastula stage of approximately 100 cells during early embryonic development (Sect. 11.1), XCI is initiated in a random fashion in one of the two X chromosomes by inducing *Xist* expression that gradually spreads over the whole chromosome. XCI leads to cellular mosaicism within and between individuals, so that the two genetically different X chromosomes are expressed in average of all tissues in a similar, so that females are protected from deleterious X-linked mutations. Moreover, some 20% of the genes on the X chromosome escape XCI (defined as being expressed at a level of at least 10% of the non-silent allele) and are biallelically expressed in female cells

Through the interaction with the splicing factor HNRNPU (heterogeneous nuclear ribonucleoprotein U) *Xist* recruits via its A-repeat region SHARP (SMRT/ HDAC1-associated repressor protein), i.e., SHARP is an RNA-binding protein.

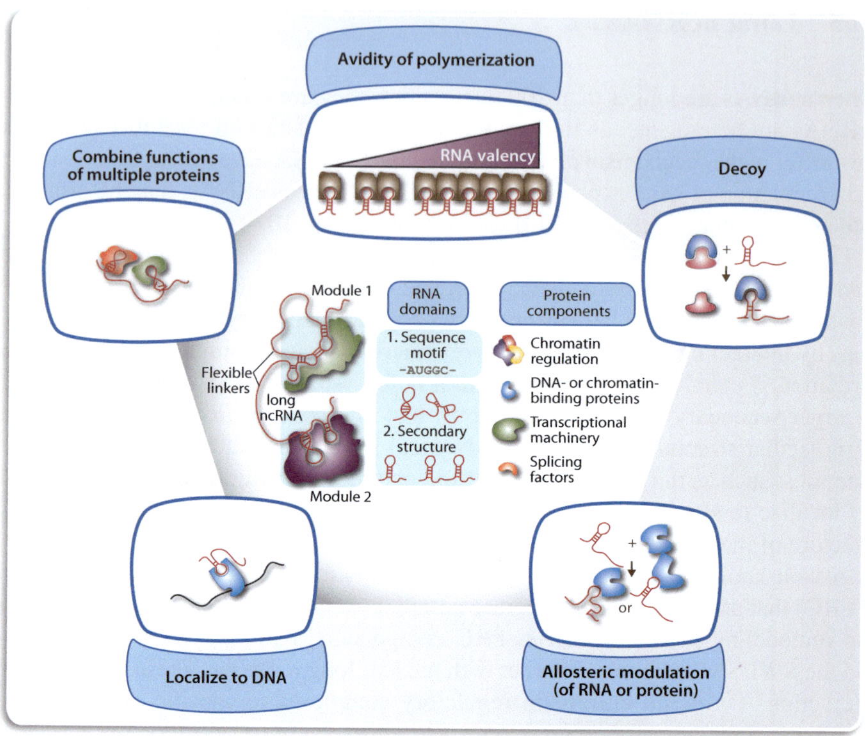

Fig. 10.4 Principles of long ncRNA action. Long ncRNA molecules have various regions for the molecular interaction with distinct protein complexes. These interactions have functions, such as combining the functions of multiple proteins, localizing long ncRNAs to genomic DNA, modifying the structure of long ncRNAs or proteins, inhibiting protein function as decoys and providing a multifunctional platform, in order to increase the avidity of protein interactions or to promote RNA-protein complex polymerization (RNA valency)

HDAC3 is recruited via the corepressor protein NCOR2, which leads to demethylation of H3K4 and ejection of Pol II. In addition, *Xist* recruits the complexes PRC1 and PRC2 that deposit H2AK119ub and H3K27me3 marks, respectively. Moreover, the KMT SETDB1 (SET domain bifurcated histone lysine methyltransferase 1) adds repressive H3K9me2 and H3K9me3 marks. In differentiated cells, XCI is maintained by DNA methylation via DNMTs and the incorporation of the histone variant macroH2A (Box 2.2). In this phase, the repressive marks are sufficient for maintaining silencing of the X chromosome and *Xist* is dispensable.

During XCI the chromatin of the X chromosome undergoes major structural changes (Fig. 10.6). Before the expression of *Xist* both X chromosomes are transcriptionally active, not strongly associated with the nuclear lamina and structurally organized similar to autosomes, i.e., they are subdivided into more than hundred

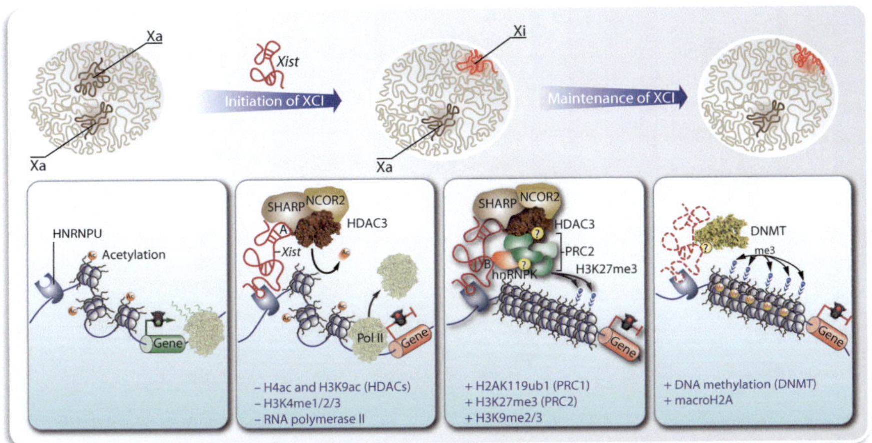

Fig. 10.5 Mechanisms of *Xist*-induced gene silencing. In ES cells both X chromosomes are actively transcribed (Xa) and are marked by H3K4me1, H3ac and H4ac. XCI starts early in embryonic development, when *Xist* expression is initiated on one of the two X chromosomes and gradually spreads across the whole inactive X chromosome (Xi). *Xist* binds to chromatin through interactions with HNRNPU and recruits SHARP, in order to promote histone deacetylation via HDAC3, demethylation of H3K4 and ejection of Pol II. Furthermore, *Xist* recruits PRC1 and PRC2 complexes, which deposit H2AK119ub and H3K27me3 marks, respectively. Moreover, the KMT SETDB1 places repressive H3K9me2 and H3K9me3 marks. XCI is maintained in differentiated cells via DNMT-mediated DNA methylation and the incorporation of the histone variant macroH2A.

TADs. However, once *Xist* is expressed by one allele, it spreads across the X chromosome and interacts with LBR (lamin B receptor), which relocates the chromosome to the nuclear lamina. In this context, active genes are sequestered into the *Xist* compartment and silenced. Moreover, most of the TADs on the X chromosome are lost and two large mega-domains are formed, which have a boundary at the *DANT1* (DXZ4 associated non-coding transcript 1, proximal) locus that associates with the nucleolus.

Other long ncRNAs, such as *HOTAIR* (*HOX* transcript antisense RNA), direct some KDMs, such as KDM1A within the RCOR complex (Sect. 14.3), to their chromatin target sites. RCOR is a large protein complex that also contains HDACs and contributes to transcriptional repression (Sect. 4.2). Long ncRNAs often act as decoys that prevent the access of transcription factors to their genomic DNA binding sites. For example, upon growth factor shortage the long ncRNA *GAS5* (growth arrest specific 5) is induced. A hairpin sequence motif of *GAS5* contains resembles the consensus binding site of the nuclear receptor GR. Thus, upon "shortage" conditions, *GAS5* is induced and acts as a decoy, in order to release GR from its genomic binding sites preventing the expression of its target genes.

The abundance of long ncRNAs in cells is related to their function. Low-abundance long ncRNAs, such as *HOTTIP* (HOXA transcript at the distal tip), that in average has less than 1 copy per cell only regulate genes in their close proximity.

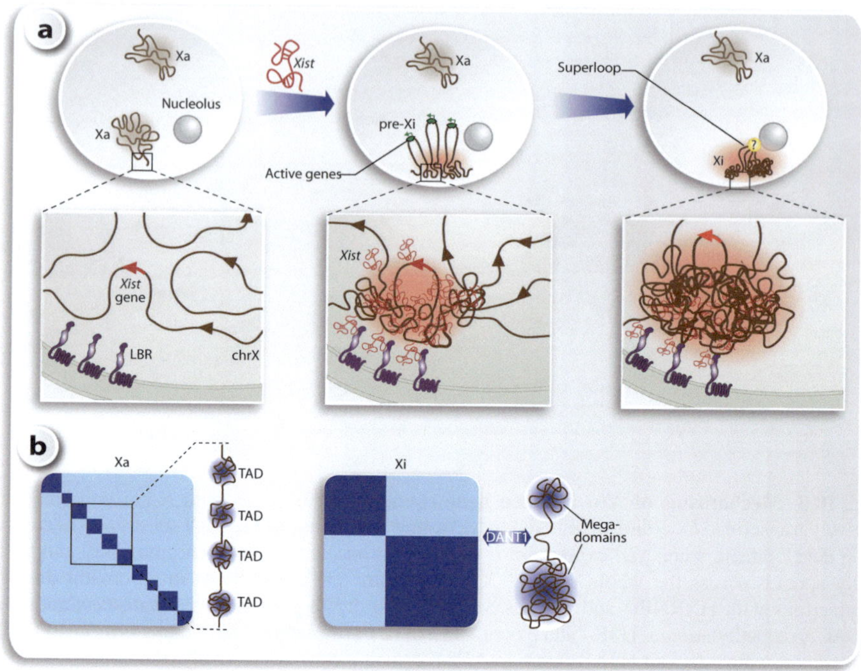

Fig. 10.6 XCI changes the architecture of the X chromosome. During XCI the X chromosome undergoes major structural changes, such as association with nuclear lamina and silencing active genes by sequestering them into the *Xist* compartment (**a**). The heatmap diagram depicts contact frequency between genomic sites on the X chromosome (**b**)

Moderate expression levels, such as 50–100 copies per cell, enable *Xist* to spread across the entire X chromosome but it does not affect other chromosomes. In contrast, the highly abundant (approximately 3000 copies per cell) long ncRNA *MALAT1* (metastasis associated lung adenocarcinoma transcript 1) diffuses throughout the whole nucleus and affects many loci. Genome-wide mapping of *MALAT1* binding loci indicates that it associates with all actively transcribed genes in a dynamic and transcription-dependent manner.

10.4 Enhancer RNAs

Long ncRNAs were found first in connection with repressive chromatin modifying complexes, but they also associate with active chromatin states. Genome-wide patterns of histone modifications and enhancer binding proteins suggest that long ncRNAs are involved also in gene activation, such as the stabilization of local condensates (Sect. 9.3). In addition to its role to interact with TSS regions, Pol II was also found on active enhancer regions. This interaction results in a bidirectional transcription of eRNAs. The FANTOM5 consortium used CAGE technology (Box 3.2) and identified at approximately 44,000 regions within the human genome, which are proven not to contain a TSS, the production of eRNAs (Sect. 10.1). Unlike mRNAs, eRNAs are not polyadenylated, generally short and non-coding and transcribed bidirectionally. Moreover, eRNA levels correlated with mRNA synthesis from nearby genes. Importantly, eRNA transcription requires the presence of a TSS region as primary sites of the genomic attachment (Fig. 10.7). Transcription of eRNAs may contribute to the maintenance of open chromatin at enhancer regions, but can also be a side product of chromatin configuration or looping. Moreover, eRNAs could even be an evolutionary source of new genes.

Since variations in enhancers may be pre-stages in a number of human disorders, modulating their function emerges as novel targeted strategies for preventing and treating these diseases. RNA interference (RNAi) was established as a powerful tool (Box 10.2) for analyzing the function of individual genes. In contrast, the manipulation of enhancer function was considered previously experimentally far more

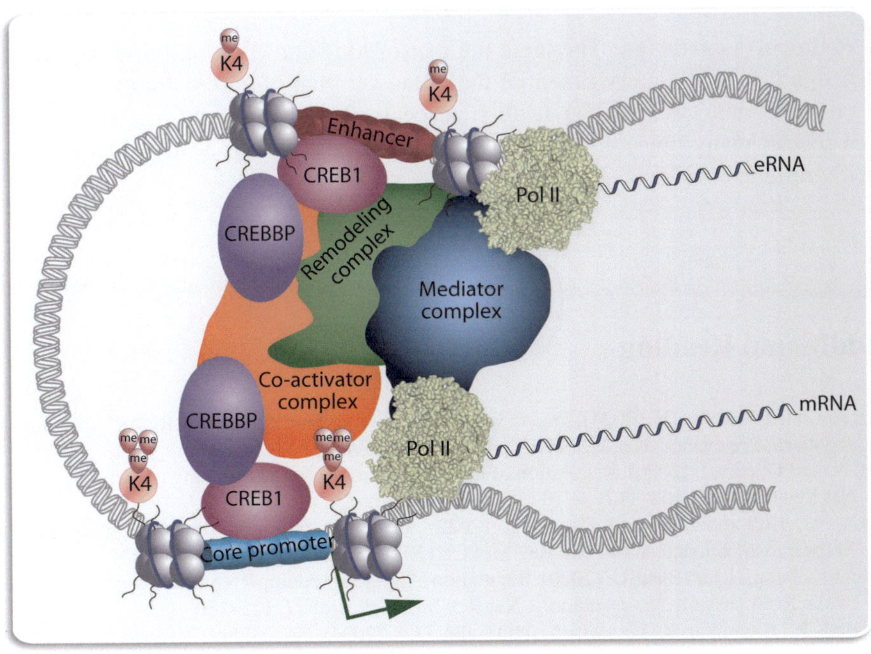

Fig. 10.7 Synthesis of eRNAs as a result of promoter-enhancer interactions. After activation transcription factors and Pol II bind to enhancers and eRNA is synthesized. Simultaneously, Pol II and other components of the basal transcriptional machinery bind to the TSS region and initiate mRNA transcription

demanding. However, for the regulation of target genes, for which eRNAs are necessary, RNAi of eRNAs could be used to inhibit enhancer function. This offers an alternative approach for targeted disruption of gene expression.

Box 10.2: RNAi Small interfering RNAs (siRNAs) are synthetic double-stranded RNA molecules of the size of mature miRNAs (~22 nucleotides). They are transfected into target cells and, like miRNAs, one siRNA strand binds to RISC, thus causing "interference". RNAi are a valuable research tool, both in cell culture and in living organisms, where siRNAs interfere with the action of endogenous mRNAs and selectively and robustly induce suppression of specific genes of interest. RNAi has been used for large-scale screens that systematically knocked down each gene in a cell or organism. This helps to identify the components necessary for a particular cellular process. Thus, **RNAi is a widespread tool in biotechnology and medicine**.

(Clinical) conclusion: The large number of long and short ncRNAs provide a significant regulatory potential that relates to that of chromatin modifying enzymes and transcription factors, respectively. Accordingly, ncRNA play a role in many physiological processes ranging from signal transduction to imprinting.

Additional Reading

Ferrer, J. and Dimitrova, N. (2024). Transcription regulation by long non-coding RNAs: mechanisms and disease relevance. Nat Rev Mol Cell Biol *25*, 396–415.

Gil, N. and Ulitsky, I. (2020). Regulation of gene expression by cis-acting long non-coding RNAs. Nat Rev Genet *21*, 102–117.

Loda, A., Collombet, S. and Heard, E. (2022). Gene regulation in time and space during X-chromosome inactivation. Nat Rev Mol Cell Biol *23*, 231–249.

Nair, L., Chung, H. and Basu, U. (2020). Regulation of long non-coding RNAs and genome dynamics by the RNA surveillance machinery. Nat Rev Mol Cell Biol *21*, 123–136.

Nemeth, K., Bayraktar, R., Ferracin, M. and Calin, G.A. (2024). Non-coding RNAs in disease: from mechanisms to therapeutics. Nat Rev Genet *25*, 211–232.

Shang, R., Lee, S., Senavirathne, G. and Lai, E.C. (2023). microRNAs in action: biogenesis, function and regulation. Nat Rev Genet *24*, 816–833.

Statello, L., Guo, C.J., Chen, L.L. and Huarte, M. (2021). Gene regulation by long non-coding RNAs and its biological functions. Nat Rev Mol Cell Biol *22*, 96–118.

Chapter 11
Epigenetics in Development

Abstract In this chapter, we will present that early embryonic development is very vulnerable to environmental influences. Therefore, from all phases of our life, embryogenesis is the period, where epigenetics has the largest impact. Epigenetics directs the programing of PGCs, induced pluripotency as well as the function of adult stem cells in tissue homeostasis. These key examples demonstrate the impact of epigenetics on the organization of our body in health and disease. Epigenetic regulation is also critical for the normal development and functioning of our brain. Dynamic DNA and histone methylation as well as their demethylation at specific gene loci play a fundamental role in learning, memory formation and behavioral plasticity. Thus, the epigenome of neurons allows a molecular explanation for long-term memories.

Keywords Embryogenesis · ES cells · Cell lineage commitment · PGCs · Cellular reprograming · Induced pluripotency · Pluripotency transcription factors · Gene regulatory networks · Adult stem cells · Hematopoiesis · Neuronal development · DNA methylation · mCH methylation · MECP2

11.1 Epigenetic Changes During Early Human Development

The human body is composed of some 30 trillion ($3 \cdot 10^{13}$) cells forming more than 400 different tissues and cell types. Embryonic development is a tightly regulated process that produces from an identical genome this high diversity of human cell types. In each cell, chromatin serves as a specific filter of genomic information and determines which genes are expressed and which not, i.e., **cellular diversity is based rather on epigenomics than on genomics**. Thus, the differentiation program of embryogenesis is a perfect system for observing the coordination of cell lineage commitment and cell identity specification. Embryogenesis requires the coordination between an increase in cellular mass and the phenotypic diversification of the expanding cell populations. The process is under the control of gene regulatory networks in the context of significant changes in the epigenetic landscape (Sect. 11.2).

C. Carlberg, *Gene Regulation and Epigenetics*,
https://doi.org/10.1007/978-3-031-68730-3_11

At conception, the two types of haploid gametes, oocyte and sperm, fuse and form the diploid zygote (Fig. 11.1). A series of cleavage divisions of the zygote creates the totipotent 16-cell morula stage. A few days after conception cells on the outer part of the morula bind tightly together forming a cavity inside. At the blastocyst stage (50–150 cells) a first differentiation occurs. The outer cells (trophoblasts) are the precursors to extraembryonic cytotrophoblasts that form chorionic villi and syncytiotrophoblasts, which ingress into the uterus, i.e., these cells form the placenta and other extraembryonic tissues. Even before the blastocyst becomes implanted into the uterine wall, its inner cells, the so-called **inner cell mass**, begin to differentiate into two layers, the epiblast and the hypoblast. The epiblast gives rise to some extraembryonic tissues as well as to all the cells of the later stage embryo and fetus. In contrast, the hypoblast is exclusively devoted to making extraembryonic tissues including the placenta and the yolk sac.

Some of the embryonic epiblast cells form PGCs. These cells are the founders of the germ line and provide a link between different generations of an individual's family. During the gastrulation phase the other cells of the embryonic epiblast turn into the **three germ layers ectoderm, mesoderm and endoderm** that are the precursors of all somatic tissues. The cells of these germ layers are only multipotent, i.e., they cannot differentiate to every other tissue. For example, in a series of sequential differentiation steps, ectoderm cells can form epidermis, neural tissue and neural crest, but not kidney (mesoderm-derived) or liver cells (endoderm-derived).

Before conception the CpG methylation rate of the haploid genomes of sperm and oocyte is 90 and 40%, respectively (Fig. 11.2, left). The nucleus of sperm is 10-times more condensed than that of somatic cells and most histones are replaced by protamines, which are small, arginine-rich nuclear proteins that allow denser packaging of DNA. As a consequence, the genome of sperm is transcriptionally silent, while in the oocyte many genes are active. After conception, at the zygote stage,

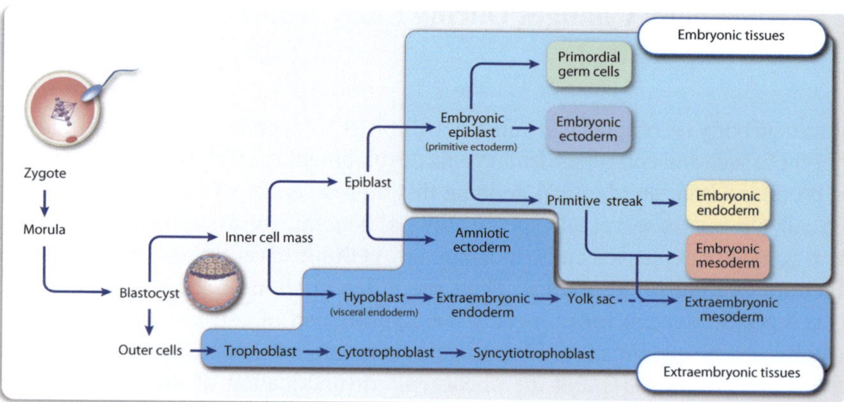

Fig. 11.1 A road map of early human development. Details are provided in the text. The dashed line indicates a possible dual origin of the extraembryonic mesoderm

the two haploid parental genomes initially remain separate in their rather different state of chromatin organization. Then both genomes get extensively demethylated, but the paternal genome much faster than the maternal genome (Fig. 11.2, left). In parallel, the protamines within the paternal chromatin get exchanged back to canonical histones. In the following cell divisions, preceding the blastocyst stage, there is passive demethylation in both parental genomes, until the methylation patterns are reestablished in a lineage-specific manner. In addition, in the paternal epigenome, there is rapid increase in 5hmC and 5fC/5caC, which is a sign of TET-mediated 5mC oxidation (Sect. 7.1). This process accelerates the demethylation of the paternal epigenome. Importantly, there is no complete demethylation of the epigenome and some 5% of the 5mC marks remain active, possibly via protection by methyl-binding proteins (Fig. 11.2, right). **This process serves as the basis for inter- and transgenerational epigenetic inheritance** (Sect. 12.1).

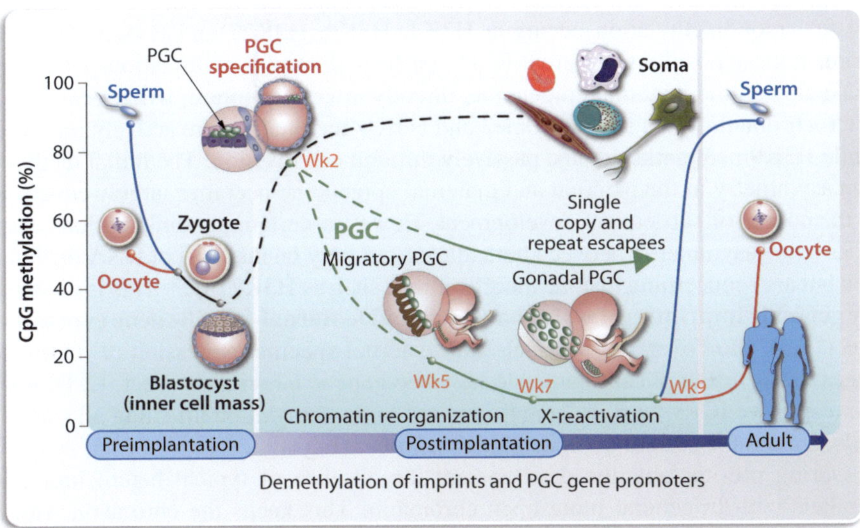

Fig. 11.2   Epigenetic reprograming during embryogenesis. DNA methylation is the most stable epigenetic modification and often mediates permanent gene silencing, during embryogenesis as well as in the adult. Accordingly, there is a hierarchy of events where DNA methylation marks are typically added or removed after changes in histone modifications, i.e., they mostly occur at the end of the differentiation process. Therefore, in this graph the percentage of CpG methylation is indicated as representative marker of epigenome changes. Two waves of global demethylation occur during embryogenesis: the first in the pre-implantation phase (week (wk) 1) affecting all cells of the embryo and the other applies only to PGCs during the phase of their specification reaching a minimum of some 5% CpG methylation at wk 7–9. Some single copy genomic regions and a number of repeat loci remain methylated and are candidates for inter- and transgenerational epigenetic inheritance (Sect. 12.1). Dotted lines indicate the methylation dynamics. Genome-wide de novo DNA methylation takes place after implantation. In PGCs a second wave of de novo DNA methylation happens as early as after wk 9 in males, however, as late as after birth in females. Importantly, **most CpG-rich promoters remain unmethylated at all stages of embryogenesis**, i.e., they are not concerned by these waves of methylation and demethylation

Genome-wide epigenetic reprograming during pre-implantation development resets the zygotic epigenome for naïve pluripotency. This process is far more pronounced in PGCs than in other cells of the embryo in order to erase imprints and most other epigenetic memories (Fig. 11.2). Since DNA methylation is an important epigenetic silencer that modulates gene expression and maintains genome stability (Sect. 7.1), the transient loss of DNA methylation in PGCs bears the risk of causing activation of retrotransposons, proliferation defects and even cell death. Therefore, genome-wide reorganization of repressive histone modifications via pluripotency transcription factors safeguards genome integrity in this phase of embryogenesis.

In early embryogenesis the maternal and paternal epigenomes also differ significantly concerning their histone modification patterns. The global pattern of histone marks within the maternal epigenome resembles that of somatic cells, while the paternal epigenome, due to the protamine-histone exchange process, is hyperacetylated, rapidly incorporates histone variant H3.3 and is devoid of H3K9me3 and H3K27me3 markers of constitutive heterochromatin. On the paternal epigenome the first monomethylation occurs at H3K4, H3K9, H3K27 and H4K20. Furthermore, at these positions, different KMTs perform di- and trimethylation. This takes place also on the maternal epigenome, directly after conception, where features of heterochromatin, such as H4K20me3 and H3K64me3 marks, are actively removed, while H3K9me3 marks are lost passively through cell division. The initial methylation asymmetry in the paternal and maternal epigenome becomes largely equalized in the course of subsequent development. However, certain genomic regions, such ICRs-, stay asymmetric between both alleles, not only on the level of DNA methylation but also concerning histone modifications, such as H3K27me3. This is the **basis for genomic imprinting, i.e., for paternal- and maternal-specific gene expression** (Sect. 7.3). Moreover, the paternal- and maternal-specific expression of imprinted genes is a key example of intergenerational epigenetic inheritance (Sect. 12.1), since the respective ICRs "survive" the first round of demethylation affecting all somatic cells, but not the second germ cell-specific round (Fig. 11.2).

During pre-implantation development the absence of typical heterochromatin parallels with in general more open chromatin. This keeps the chromatin widely accessible and is necessary for epigenetic reprograming, when gamete-specific modifications are removed and new marks are re-established. During the course of further development, such as the blastocyst level, cells of the inner cell mass (forming the embryo) show higher levels of DNA methylation and H3K27 methylation, as well as lower levels of histone H2A and/or H4 phosphorylation than cells of the trophectoderm (forming the placenta). This epigenomic asymmetry is a sign of differentiation of the respective cell types and regulates lineage allocation in the early embryo. Accordingly, **precise and robust gene regulation via the control of access and activity of promoter and enhancer regions is essential**.

11.2 The Epigenetic Landscape

The epigenetic landscape is a very illustrative model for understanding the under-lying molecular mechanisms of cell fate decisions during development. Cellular differentiation happens along lineages and is under natural conditions an irreversible forward-moving process. It results in highly specialized, terminally differentiated cell types. In this model, **cellular differentiation may be compared to a system of valleys of a mountain range**, where a cell (often represented by a ball, Fig. 11.3), e.g., an ES cell, begins at the top and follows existing paths driven by gravitational force. The latter analogy should express that the path of differentiation has a clear path directing the cell into one of several possible fates represented as valleys that get narrower in the trajectory toward terminally differentiated cell types (Fig. 11.3, bottom). Along the downhill path, cell fate decisions need to be taken at bifurcation points. These decisions often depend on the expression of lineage-determining tran-scription factors. Once a cell has taken a decision, it is restricted in its subsequent decisions by the route it has taken.

The developmental potential of stem cells on top of the hill correlates with high entropy (i.e., the potential to take a multitude of cellular stages), which declines during differentiation toward well-defined cell types (Fig. 11.3, left). In contrast, when embryonal pluripotency transcription factors (Sect. 11.2), such as OCT4 or NANOG, are reactivated in terminally differentiated cells, entropy can increase again and a cell may move uphill in the landscape (Fig. 11.3, center). This happens often during tumorigenesis, when (epi)mutations activate transcription factors or other

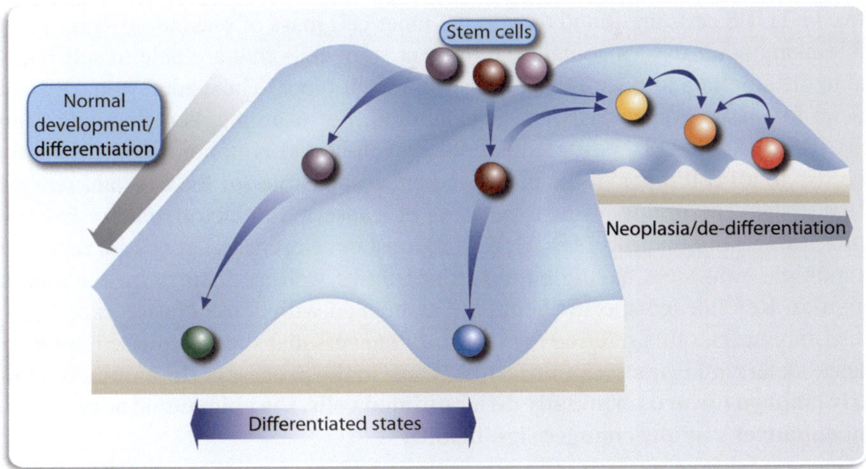

Fig. 11.3 Phenotypic plasticity during cellular reprograming and neoplasia. Waddington's landscape model can be used for the illustration of phenotypic plasticity of cells during normal development (**left**), the creation of iPS cells, i.e., in cellular reprograming (**center**), as well as during the induction of neoplasias, i.e., in tumorigenesis (**right**)

nuclear proteins, such as chromatin modifying enzymes, the activity of which results in gene expression heterogeneity (Chap. 15). The latter discontinues the cell fate choice and the transformed cells reach a state of higher entropy, in which they again proliferate and self-renew, i.e., they are de-differentiated compared with their normal counterparts (Fig. 11.3, right). The epigenetic landscape model is also used to illustrate the phenotypic plasticity of cells during the creation of iPS cells (Box 11.1). Taken together, **the epigenetic landscape is an attractive, intuitively understandable model how the static information provided by the genome is translated dynamically into tissues and cell types.**

> **Box 11.1: Potency of Human Cells** The potency of a cell is its ability to differentiate into other cells. Totipotent cells can form all the cell types in a body including extraembryonic placental cells. Only within the first 6–8 cell divisions after conception embryonic cells are totipotent. Pluripotent cells, such as ES cells, can give rise to all cell types forming our body. Multipotent cells, such as adult stem cells, can develop into more than one cell type but are more limited than pluripotent cells. In contrast, terminally differentiated cells, like more than 99% of those in our body, are unipotent. However, via the overexpression of pluripotency transcription factors unipotent cells can be induced to get pluripotent, i.e., they transform into iPS cells.

Stem cells have the property of both self-renewal, i.e., the ability to run through numerous cell cycles while staying undifferentiated and the capacity to differentiate into specialized cell types. Thus, stem cells need to be either totipotent or pluripotent (Box 11.1). ES cells are found only in the inner cell mass of blastocysts (Fig. 11.2), while many adult tissues contain multipotent stem cells that are able to self-renew and to differentiate into various tissue-specific cell types. Thus, adult stem cells are crucial for tissue homeostasis and regeneration. For example, hematopoietic stem cells (HSCs) (Sect. 13.1) differentiate into myeloid and lymphoid progenitors that give rise to all cell types of the blood and are responsible for the constant renewal of the tissue. The timing of the expression of transcription factors plays a key role in this differentiation processes. Moreover, most developmental genes are regulated by multiple enhancers having both overlapping as well as distinct spatiotemporal activities. Key lineage-specific genes are associated with dense cluster(s) of highly active enhancers, often referred to as super-enhancers that show stepwise binding of lineage-determining transcription factors. Thus, **in the process of development from early embryo toward terminally differentiated cells, the epigenome at promoter and enhancer regions changes significantly.**

Genome-wide analysis demonstrated that histone marks distinguish ES cells from terminally differentiated cells and pluripotency genes from lineage-specific genes. In ES cells (Fig. 11.4, top), enhancer regions of both pluripotency genes are enriched with H3K4me1 and H3K27ac marks. These genes are actively transcribed, because also their TSS regions are marked with H3K4me3 and their gene bodies show H3K36me3 modifications. In contrast, the enhancers of lineage-commitment genes carry H3K4me1 marks and repressive H3K27me3 instead of H3K27ac marks, which keeps the genes in a poised state, even if their TSS regions carry H3K4me3 marks. Thus, **enhancers and promoters of poised genes comprise both activating and repressing histone marks, i.e., they are examples of bivalent chromatin states from which they either get fully activated or repressed**.

After differentiation toward a specific lineage, such as neurons (Fig. 11.4, bottom), only lineage-specific genes are marked by H3K27ac at both enhancer and promoter regions as well as by H3K4me1 at their enhancers. Then Pol II pausing is released and mRNA transcription continues. Genes of other lineages lose marks at their enhancers and obtain repressive H3K27me3 marks at their TSS regions. Moreover, pluripotency genes attain H3K9me3 marks and DNA methylation at their promoter regions in order to keep them stably silenced for the rest of the life. During the differentiation process, heterochromatin regions are marked by H3K9me2 and H3K9me3 modifications, HP1 binding and DNA methylation are expanded so that chromatin becomes more condensed. In repressed genes as well as in intergenic regions H3K27me3 marks also increase. In contrast, during cell lineage commitment, KDMs remove H3K27me3

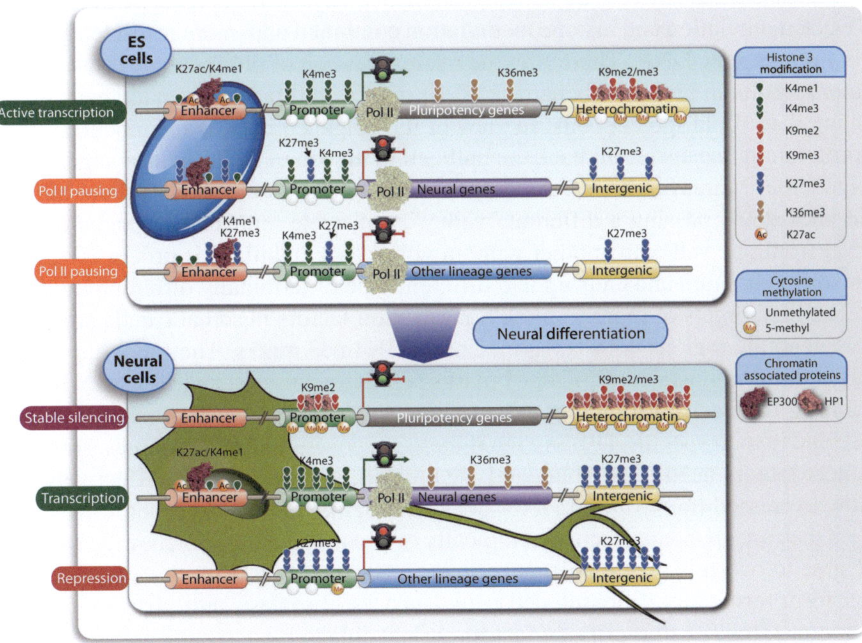

Fig. 11.4 Chromatin states of ES cells in comparison to lineage-specific cells. The chromatin stages at enhancers, promoters, gene bodies and intergenic heterochromatin regions of pluripotent genes, neuronal genes and other lineage genes are compared between ES cells (**top**) and, as an example, neural cells (**bottom**)

marks from specific promoter-associated CpGs, in order to make the respective genes transcriptionally permissive. This also includes depletion of nucleosomes from TSS regions via chromatin remodelers (Sect. 9.2).

Some 20 years ago, the technique of cellular reprograming of terminally differentiated cells into induced pluripotency, i.e., the creation of iPS cells (Sect. 7.2), was invented. The method uses ectopic expression, i.e., an abnormally high expression, of the pluripotency transcription factors OCT4, SOX2, KLF4 and MYC in order to reprogram the somatic epigenome and to induce a stable pluripotent state, similar to that of an ES cell. OCT4, SOX2 and KLF4 cooperatively suppress lineage-specific genes and activate pluripotency genes (Fig. 11.4), while MYC overexpression stimulates cell proliferation, induces a metabolic switch from an oxidative to a glycolytic state and mediates pause release and promoter reloading of Pol II. Nevertheless, the induction of pluripotency is very inefficient and only 0.1–3% of a cell population get fully reprogramed, i.e., there seems to be a number of epigenetic barriers that stabilize the identity of somatic cells and prevent their aberrant transdifferentiation. However, **cellular reprograming has the potential to regenerate diseased organs and also provides further insight into the principles of epigenetic control of cellular differentiation**.

During normal development from stem cells via progenitor cells to terminally differentiated cells, there is a gradual placement of repressive epigenetic marks, such as DNA methylation and histone methylation combined with more restricted accessibility of genomic DNA. Therefore, the overexpression of pluripotency transcription factors results in prominent changes of the epigenetic stage of somatic cells establishing that of pluripotent cells. In view of the epigenetic landscape model, cellular reprograming means to roll back the ball/cell to the top of the hill via changes of the epigenome. Moreover, balls/cells may move only a part of the way up the hill and roll back down passing a different "valley", a discrete number of troughs or even travel from one valley to another without going back uphill. This process is termed transdifferentiation and ends up in a different type of terminally differentiated cell.

The first targets of reprograming transcription factors in somatic cells are accessible genomic regions with H3K4me2 and H3K4me3 marks. The following targets are genomic regions with H3K4me1 marks being in a poised state (Sect. 8.2). **Reprograming transcription factors act as pioneer factors**, i.e., as transcription factors that bind to their specific TFBSs even if these are covered by nucleosomes. However, pioneer factors need to be abundantly expressed, in order to work effectively, i.e., a low expressed transcription factor does not qualify as a pioneer factor. However, when they are abundant, such as ectopically overexpressed pluripotency factors, they are able to override previous epigenetic programming events. In this way, they can transform terminally differentiated cells, such as fibroblast, into pluripotent cells. In contrast, regular transcription factors are not able to bind to nucleosome-covered genomic DNA.

Most pioneer factors have a short DNA binding motif of 4 bp (as a monomer) and contain a structurally flexible DBD. In this way, the can bind to genomic DNA even if it wound around a histone octamer. After pioneer factors have bound to their target regions, they recruit other transcription factors and chromatin modifying enzymes.

For example, pioneer factors can bind to bivalent genes with active H3K4me3 marks and repressive H3K27me3 marks, which represents facultative heterochromatin. The most difficult targets for pluripotency transcription factors are regions with constitutive heterochromatin that carry repressive H3K9me3 marks. These regions need extensive multistep chromatin remodeling in order to get transcriptionally activated. Surprisingly, DNA methylation does not play any essential role in cellular reprograming. In contrast, **DNA demethylation of pluripotency genes, by either active or passive mechanisms, is crucial for effective reprograming**.

11.3 Epigenetic Dynamics During Differentiation

Differentiating cells share accessible chromatin regions with the ES cell they are derived from, but the similarity in the epigenetic landscape (Sect. 11.2) decreases when cells mature. After commitment to a specific lineage, the cellular repertoire expands for accessible regulatory regions that contain motifs for transcription factors being specific to that lineage, whereas it clearly decreases for TFBS of other lineages. Thus, the epigenetic landscape of terminally differentiated cells is constrained by the walls of valleys, the height of which are determined by a gene regulatory network. This network is formed by appropriate levels of DNA methylation and histone modifications as well as by a proper 3D architecture. In this way, cells are prevented from switching states (Fig. 11.5, left). However, **in response to relevant intra- and extracellular signals, the epigenome also allows cell state transitions**. When chromatin homeostasis is disturbed, e.g., by epimutations, cells do not respond appropriately to these signals. Overly restrictive chromatin networks create epigenetic barriers that prevent all types of cell state transitions (Fig. 11.5, center). In contrast, excessively permissive chromatin networks have very low barriers and allow multiple types of cell state transitions (Fig. 11.5, right). For example, deviations from the norm contribute to tumorigenesis (Chap. 15).

Changes in cell identity are reflected by alterations in the usage of the enhancer and promoter regions. Many of the regulatory regions that are active in early embryogenesis lose their activity in later phases of development. This is compensated through the activity of TSS regions and poised enhancers, some of which turn into super-enhancers. Changes in enhancer usage require a chromatin topology that allows a new set of enhancers to interact with their target promoters. In parallel, heterochromatin foci become more condensed and more abundant in differentiated cells than in undifferentiated cells. While in ES cells H3K27me3 marks show only focal distributions, in differentiated cells they largely expand over silent genes and intergenic regions. This results in silencing of pluripotency genes, activating lineage-specific genes and repressing of lineage-inappropriate genes.

On the mechanistic level (Fig. 11.5, top) the scenarios of normal, restrictive and permissive chromatin can be explained, e.g., by the actions of a KMT for repressive H3K27me3 marks, such as EZH2 (enhancer of zeste homolog 2, also called KMT6A) and a KMT for activating H3K4me3 marks, such as KMT2A. EZH2 is the catalytic core of the repressive PRC2 complex and KMT2A belongs to the so-called Trithorax

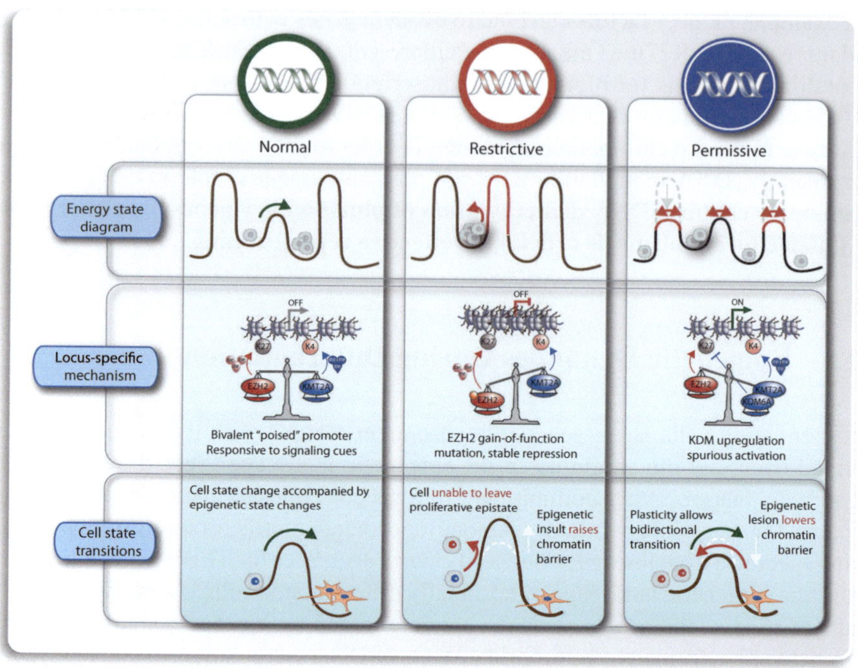

Fig. 11.5 Chromatin structure, cellular identity and cell state transitions. In normal cells (**left**) networks of chromatin proteins stabilize the states of cells but also mediate the response to intra and extracellular stimuli and occasionally allow cell state transitions. However, cells in which the chromatin network is perturbed do not respond appropriately. In restrictive chromatin (**center**) epigenetic barriers prevent cell state transitions, while in overly permissive chromatin (**right**) these barriers are lowered and allow easy transition to other cell states. The scenarios are illustrated via an example of the underlying molecular mechanisms (**top**) or as cell state transitions (**bottom**). Blue nuclei represent normal cells, while red nuclei indicate cancer cells

complex. In normal cells, both KMTs and their histone marks are in balance resulting in bivalent, poised constitutive heterochromatin at TSS regions. This means that their respective target genes are transcribed only in response to appropriate stimuli. In restricted cells, EZH2 may have a gain-of-function epimutation, such as often observed in several forms of lymphoma, resulting in far higher levels of repressive H3K27me3 marks, stable heterochromatin and no gene transcription. In this state, cells may be blocked in differentiation and continue to grow with a high proliferation rate. In contrast, in permissive cells a demethylase, such as KDM6A, inhibits the action of EZH2 and removes H3K27me3 marks. KDMs are often upregulated under stress conditions. In net effect, this leads to the dominance of H3K4me3 marks and to the activation of gene expression, such as of oncogenes, even in the absence of specific stimuli. In the cell state transition diagram (Fig. 11.5, bottom) the barrier between the cell states is either of medium height in normal cells, very high in restricted cells or low in permissive cells.

In the homeostasis of adult tissues, dividing and differentiating resident stem cells replace damaged or dying cells when necessary. For example, human epidermis heavily relies on correct stem cell function, both in homeostasis and after wounding. In the epidermis, stem cell pools reside exclusively in the basal layer and proliferation occurs only there. The basal layer continuously replenishes the whole tissue with fresh cells that migrate and differentiate through the different epidermal layers. During this epidermal differentiation process, the state of chromatin changes dynamically. For example, genome-wide levels of H3K27me3 marks decrease during differentiation, in particular at TSS regions of epidermal differentiation genes. This parallels with decreased binding of PRC2 and increased binding of the H3K27me3 demethylase KDM6B. Moreover, in epidermal differentiation the genome-wide histone acetylation level is decreased and terminally differentiated skin cells show higher expression of HDAC1 and HDAC2. Finally, also genome-wide DNA methylation levels decrease during epidermal differentiation.

11.4   Neuronal Development: The Role of Epigenetics

The frontal cortex region of our brain plays a key role in behavior and cognition. It requires a coordinated interaction of neuronal and non-neuronal cells like supporting glial cells. The highly controlled process of neuronal development and maturation creates the physical structure of our brain. It starts during embryogenesis and continues until the third decade of our life. In parallel, after birth there is first a burst of synaptogenesis and then a pruning of unused synapses during adolescence. This is the cellular basis for experience-dependent plasticity and learning in children and young adults. Its disruption can lead to behavioral alterations and neuropsychiatric disorders. Like other developmental programs of our body (Sect. 11.1), also neuronal development is controlled by precise epigenetic patterns, such as DNA methylation and histone modifications. Thus, **neuroepigenetics is based on the same mechanisms as all other of our tissues and cell types**.

In general, epigenome-wide studies focus on 5mC marks at CpGs (mCG, red in Fig. 11.6a). However, like ES cells, neurons have the special property to carry significant levels of mCH marks (H = A, C or T, Sect. 7.1), some 70% of which are mCA marks. While mCH marks are hardly detectable in fetal brain, they increase during early postnatal development to a maximum of 1.5% of all CH dinucleotides at the end of adolescence (blue in Fig. 11.6a). Interestingly, mCH levels rise most rapidly during the primary phase of synaptogenesis, i.e., within the first two years after birth, and correlate with the increase in synapse density (green in Fig. 11.6a). This widespread methylome reconfiguration occurs in neurons but not in glial cells during the development from fetus to young adult, so that mCH becomes the dominant form of methylation in the neuronal epigenome. Thus, **genome-wide DNA methylation patterns significantly change during brain development and maturation; they are the basis for neuronal plasticity**.

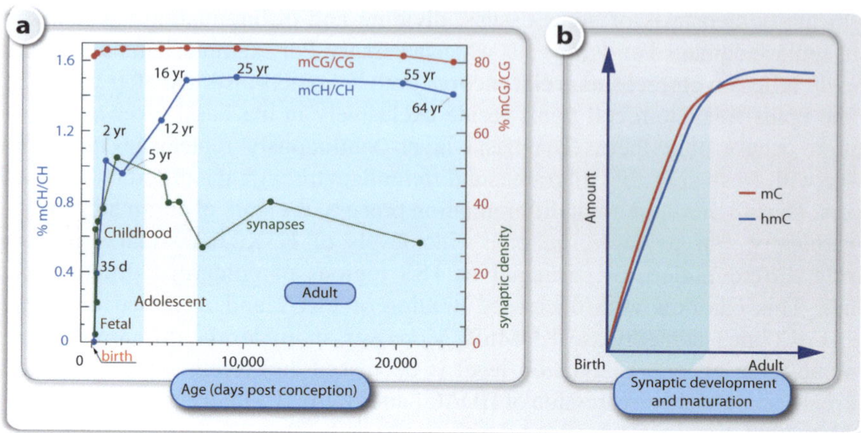

Fig. 11.6 Changes of the neuronal methylome during brain development and maturation. The levels of mCH **a** and 5hmC **b** accumulate in neurons of the human frontal cortex after birth, which coincides with active synapse development (indicated as synaptic density per 100 mm^3) and maturation **a**. Please note that mCH marks cause gene silencing, while 5hmC marks are found at active genes. Data are based on Lister et al. (*Science* 2013, 341:1,237,905)

The frontal cortex develops postnatally in response to various inputs from the environment. Dependent on the sensory input, this leads to neuronal-specific DNA methylation and causes changes in gene expression and synaptic development. The chromatin modifying enzyme DNMT3A sets these mCH marks in particular at CA dinucleotides. Interestingly, although the average methylation percentage at CH is clearly lower than at CG, in the adult brain mCH marks are by number even more abundant than mCG marks, since but CpGs are more rare. Thus, **in the maturing brain mCA is a major epigenetic mark repressing gene expression**.

In general, mCA marks serve as a docking platform for methyl-binding proteins, e.g., at the genomic region of the gene *BDNF* (brain-derived neurotrophic factor). The methyl-CpG-binding transcription factor MECP2 acts as a reader of DNA methylation marks (Fig. 7.2) and also recognizes with high affinity mCA marks. Therefore, genes that acquire mCA enrichment during neurodevelopment recruit MECP2 and are repressed in their transcriptional initiation and elongation. This regulatory process mediates distinct functions in the brains of adult versus that of newborns. Accordingly, the number of mCA marks at gene bodies is more predictive for the level of gene silencing than that of mCG marks at promoters or any measure of chromatin accessibility. Moreover, the impact of mCH marks is further highlighted by the observation that **between individuals mCH marks are more conserved than mCG marks**.

Interestingly, different regions of the brain, such as frontal cortex, hippocampus and cerebellum, show significant age-dependent increase of 5hmC, i.e., of the oxidized forms of 5mC (Fig. 11.6b). This increase is specific to neurons, which in adults have far higher 5hmC levels than any other human tissue or cell type. TET enzymes mediate the 5mC oxidation and active DNA demethylation in neurons. Based on mouse knockout studies, TET1 is most important for neurogenesis and synaptic plasticity. Thus, **5mC oxidation and possibly DNA demethylation are central epigenetic mechanisms of brain development and maturation**.

Like mCA also 5hmC marks recruit MECP2 and other methyl-binding proteins, but in this case the abundance of these transcription factors mostly positively correlates with gene expression. This fits with the observation that the *MECP2* mutation R133C, which occurs in some forms of the autistic spectrum disorder Rett syndrome, specifically disrupts the ability of the protein to bind to 5hmC and results in target gene silencing. Thus, enrichment of 5hmC and depletion of 5mC both increase transcriptional activity and chromatin accessibility (Fig. 11.7). This also suggests that 5mC oxidation may be the key epigenetic mechanism in controlling gene expression in the context of synaptic plasticity. This is important for learning and memory. The latter involves effects on long-term potentiation, excitability and activity-dependent synaptic scaling of neurons. Thus, **via controlling the activity of genes encoding for ion channels, receptors and trafficking mechanisms, the epigenome has the capacity to both sense extracellular signals reaching neurons and to control their output**.

MECP2 interacts with a wide range of proteins, such as heterochromatin protein HP1, the corepressor NCOR1 and the HAT CREBBP. The half-life of MECP2 is only 4 h, as the protein gets rapidly ubiquitinated. The MECP2 protein is ubiquitously expressed in basically all human tissues and cell types, but it shows highest expression in the brain, in particular in neurons. This explains why **MECP2 is critical for brain development**. In neuronal chromatin, MECP2 is a very abundant protein appearing in average at every second nucleosome (Fig. 11.8). MECP2 and the linker histone H1 (Sect. 2.1) compete for binding to the nucleosome. When MECP2 levels rise during neuronal development, the protein replaces histone H1. This causes the reduction of the chromatin repeat length, i.e., the distance from the center of a nucleosome to the center of its neighbor, from 200 base pairs to 165 base pairs (Fig. 11.8). Thus, **MECP2 increases the density of chromatin packaging**. In this configuration, MECP2 is part of tightly folded heterochromatin structures. This suggests that MECP2 would function mainly as a transcriptional repressor. Nevertheless, there are a number of MECP2 interaction partners that stimulate gene expression. This indicates that MECP2 rather is a transcriptional regulator that, depending on its interaction partners, binds to genomic regions of activated or repressed genes. Furthermore, MECP2 has opposite roles when binding to promoter regions or gene bodies. DNA methylation at TSS regions recruits MECP2 in complex with corepressor proteins and HDACs and results in transcriptional repression. In contrast, the bodies of transcribed genes is methylated and recruits MECP2, which in turn prevents the binding of the repressive histone variant H2A.Z (Box 2.2). Thus, **the genomic location of MECP2 is of critical importance of its function**.

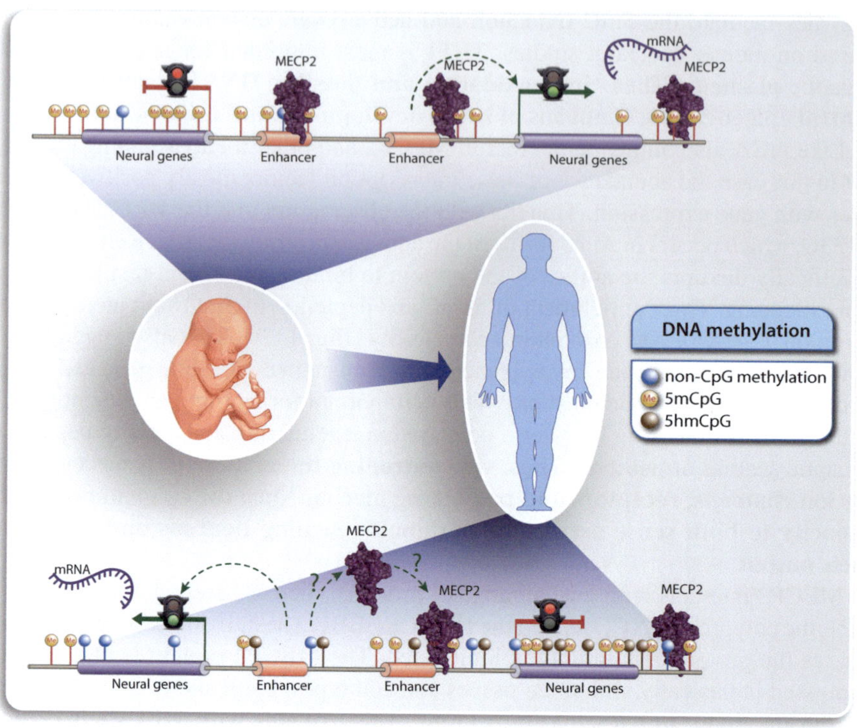

Fig. 11.7 The neuronal methylome during brain development. 5hmC marks label enhancers in the fetal brain (**top**) that will become demethylated and active in the adult brain (**bottom**). The DNA-binding pattern of methyl-DNA-binding proteins, which have different affinity for 5hmC and 5mC, can change. This affects the neuronal transcriptome

11.5 Epigenetic Basis of Memory

The acquisition of new information is defined as **learning**, while the ability to retain information for later reconstruction is **memory**. Memories provide the basis for our behavior, since they allow us to reliably navigate throughout our life and to interact with our (social) environment. For the consolidation of new information into memory, distinct gene expression profiles are activated in neurons, which are based on changes of the epigenome. In contrast, impairments in learning and memory can have devastating consequences for an individual's ability to function independently in society. In addition, the persistence of undesirable memories, such as those of trauma (e.g., war veterans) or violence (e.g., child abuse), can contribute to mental illness and may lead to social inhibition. Thus, **(social) intelligence is based on our neuronal epigenome**.

The epigenomic status of a tissue or a cell type depends on an effective interplay between environment and chromatin. In principle, **any perturbation of cellular**

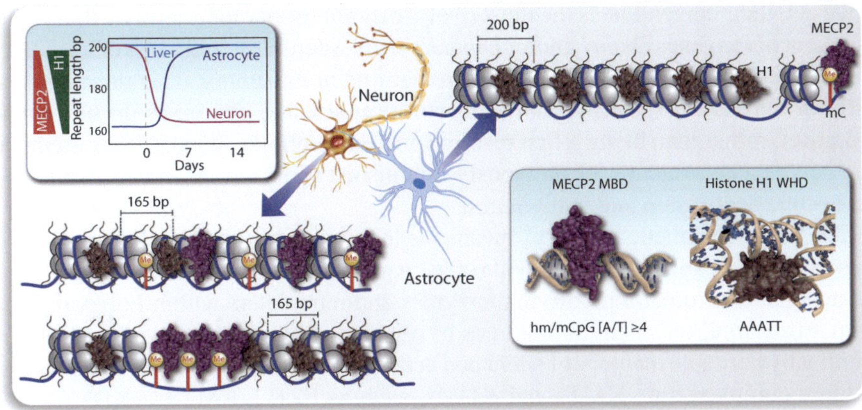

Fig. 11.8 Binding of MECP2 to chromatin of neurons and astrocytes. Astrocytes have lower MECP2 levels than neurons resulting in a regular chromatin repeat length of 200 base pairs compared to 165 base pairs for neurons. In neurons, MECP2 is evenly distributed throughout the chromatin, binds to sites of methylated DNA and replaces the linker histone H1, which decreases the repeat length. Changes of chromatin repeat length during development (in days, observed in a mouse model) before and after birth are indicated for astrocytes, neurons and as a reference liver tissue. A higher proportion of MECP2 in relation to histone H1 results in a shorter repeat length (**insert top left**)

homeostasis can result via epigenetic changes in long-lasting effects of the phenotype, in particular when the perturbed cells are either self-renewing stem cells or long-lived, terminally differentiated cells, such as neurons or memory T cells. These perturbations result in the activation of signal transduction pathways that often reach the nucleus. Within the nucleus activated transcription factors communicate with chromatin modifying enzymes and remodelers and in this way create changes in epigenetic signatures, such as histone modifications, DNA methylation and 3D chromatin architecture (Chap. 2). Some of these epigenetic changes are very transient (lasting from minutes to hours), while others can stay far longer (days, months or even years). Accordingly, the **epigenome serves as a storage facility of cellular perturbations of the past**. Speaking in numbers: each diploid cell contains approximately 30 million nucleosomes, each of which carries more than 130 different possibilities for posttranslational modifications (Sect. 8.1). With the approximately four billion (130 × 30 million) "letters" within the histone code of a single cell, it can store via histone modifications a tremendous amount of data.

Memories of our childhood start from an age of approximately three years, i.e., they last for many decades. DNA is the only molecule in our body that has a comparable half-life, i.e., only the human genome can be the molecular basis for a long-term memory. Since memories do not change the sequences of the genome, the neuronal epigenome remains as the exclusive device for information storage. Accordingly, **DNA methylation not only silences gene expression but also serves as a mediator of memory acquisition and storage**. Since neurons do not divide and get as old as we get, their epigenome is the most likely location for our long-term memory. The knockout of the genes *Dnmt1* and/or *Dnmt3a* in mice demonstrated a loss of long-term potentiation and consecutively deficits in learning and memory. This indicates that **without active DNA methylation, information storage processes do not work**. Moreover, rather than being a first responder to extracellular signals, DNA methylation acts as a consolidator of previously established short-term memory acquisition via histone acetylation and methylation.

The term "epigenetic memory" means the long-term stability of cell identities via preserved gene expression states and robust gene regulatory networks. For example, the actions of chromatin modifying enzymes and remodelers within Polycomb and Trithorax complexes result in hundreds of regions within the genome in long-term, mitotically heritable memory of silent and active gene expression states. Key proteins of these complexes are KMTs and KDMs, such as EZH2, KMT2A, KDM5C and KDM6B, while HATs and HDACs are missing. This reiterates the principle that **short-term "day-to-day" responses of the epigenome are primarily mediated by non-inherited changes in the histone acetylation level, while long-term decisions, e.g., concerning cellular differentiation, are stored in form of histone methylation marks**.

The epigenome is able to memorize lifestyle events in basically every tissue or cell type. Thus, not only neurons store the memory of an individual, but also the immune system (Chap. 13) memorizes encounters, e.g., with microbes and metabolic organs, such as skeletal muscle, fat and liver, remember lifestyle choices on diet and physical activity (Chap. 16). In contrast, **a disruption of the epigenetic memory can lead to the onset of cancer** (Chap. 15).

In summary, each of the approximately 100 billion (10^{11}) neurons within our brain have a tremendous data storage capability, since each of it contains some 700 million cytosine that may be methylated and some 30 million nucleosomes comprised of 240 million histone proteins that can be posttranslationally modified in more than 100 individual ways.

(Clinical) conclusion: It is undisputable that epigenetics has most impact during early embryogenesis. However, epigenetic landscapes that are created within daily differentiation processes in the bone marrow, skin and intestine are essential for stabilizing the functional identity of terminally differentiated cells. Furthermore, epigenetic changes are also the molecular basis for memorizing life-long experiences, such as in neurons but also in most other cell types.

Additional Reading

Arzate-Mejia, R.G. and Mansuy, I.M. (2023). Remembering through the genome: the role of chromatin states in brain functions and diseases. Transl Psychiatry *13*, 122.

Dieckmann, L. and Czamara, D. (2024). Epigenetics of prenatal stress in humans: the current research landscape. Clin Epigenetics *16*, 20.

Hekselman, I. and Yeger-Lotem, E. (2020). Mechanisms of tissue and cell-type specificity in heritable traits and diseases. Nat Rev Genet *21*, 137–150.

Lim, B., Domsch, K., Mall, M. and Lohmann, I. (2024). Canalizing cell fate by transcriptional repression. Mol Syst Biol *20*, 144–161.

Yadav, T., Quivy, J.P. and Almouzni, G. (2018). Chromatin plasticity: a versatile landscape that underlies cell fate and identity. Science *361*, 1332–1336.

Chapter 12
Epigenetics and Aging

Abstract In this chapter, we will discuss that the epigenome has a memory function in both somatic and germ cells. The latter is the basis for inter- and transgenerational inheritance. The key example of the intergenerational epigenetic inheritance concept is the agouti mouse model. However, also different types of human cohorts as well as "natural experiments" allow studying the impact of epigenetics on a population level. The progressive decline in the function of cells, tissues and organs associated with aging is affected by both genetic and epigenetic factors, i.e., there are characteristic epigenome-wide changes during aging acting as **epigenetic clocks** over years and decades. In contrast, **epigenetic circadian clocks** in the brain as well as in peripheral tissues coordinate a large set of physiological functions over a day.

Keywords Epigenetic memory · Epigenetic drift · Transgenerational epigenetic inheritance · Human cohorts · Natural experiments · Aging · Epigenetic clock

12.1 Transgenerational Epigenetic Inheritance

The epigenome is able to preserve the results of cellular perturbations by environmental factors in form of changes in DNA methylation, histone modifications and 3D organization of chromatin. Thus, **the epigenome has memory functions** (Sect. 11.5). Changes in epigenomic patterns, such as DNA methylation maps, are called **epigenetic drifts**. They primarily describe the lifelong information recording ("experience") of somatic cell types and tissues, but may also be inherited to daughter cells, when the cells are proliferating. In case of germ cells, epigenetic drifts may be, at least in part, even transferred to the next generation, i.e., the drifts may become heritable epigenetic marks (Sect. 11.1). Inheritance to the immediate offspring (F1 generation) is termed "intergenerational", while persistence of the epigenetic mark over many generations (F2, F3, …) is called "transgenerational". This leads to the concept of **inter- and transgenerational epigenetic inheritance**, which suggests that the lifestyle of the parent and grandparent generation, such as daily habits in food intake or physical activity, can affect their offspring.

C. Carlberg, *Gene Regulation and Epigenetics*,
https://doi.org/10.1007/978-3-031-68730-3_12

The key example of the transgenerational epigenetic inheritance concept is the agouti mouse model. This transgenic mouse carries the retrotransposon IAP (intracisternal A particle) within the regulatory region of the gene *Asip* (agouti signaling protein) (Fig. 12.1). This creates a dominant allele of the *Asip* gene (termed Agouti viable yellow (A^{vy})), the expression of which is depending on the methylation level of IAP, i.e., on its epigenetic status. Since the methylation of the IAP retrotransposon happens stochastically, A^{vy} behaves as a so-called metastable epiallele. In this way, the CpG methylation levels can vary considerably between individuals but stay relatively constant between tissues of a single person. The *Asip* gene encodes for a paracrine signaling peptide hormone that stimulates hair follicle melanocytes to synthesize yellow pheomelanin pigments instead of black/brown eumelanin pigments. Interestingly, the peptide hormone is involved also in the neuronal coordination of appetite. Thus, yellow coat A^{vy} mice become obese and hyperinsulinemic. Heterozygous A^{vy}/a mice vary in their coat colors from yellow via mottled to wild-type dark coat color. When the IAP retrotransposon is methylated, the synthesis of the yellow pigment is downregulated and a dark coat color appears. In contrast, unmethylated IAP allows ubiquitous *Asip* gene expression leading to both yellow coat color and obesity. **When A^{vy}/a mice inherit the A^{vy} allele maternally, *Asip* gene expression and coat color correlate with the maternal phenotype**. Interestingly, the mottled phenotype indicates that IAP methylation is mosaic, i.e., the *Asip* gene is not expressed in all cells. This suggests that the methylation pattern of the IAP retrotransposon is established early in development and the coat color provides an easy phenotypic readout of the epigenetic status of IAP throughout life. This makes the A^{vy} *Asip* mouse an ideal in vivo model for the investigation of a mechanistic link between environmental stimuli, such as nutrition and epigenetic states of the genome. A similar mouse model is that of Axin fused ($Axin^{Fu}$) alleles that result in the phenotype of a kinked tail.

The agouti mouse model was used for the following experiment: two weeks before mating with male A^{vy}/a mice, female wild-type a/a mice were either supplemented or not with methyl donors, such as folate, vitamin B_{12} and betaine (Fig. 12.1). The supplementation was continued during pregnancy and lactation. While the F1 generation of non-supplemented mothers displayed the expected number of yellow color phenotypes, the offspring of supplemented mothers shifted toward a brown coat color phenotype. This suggests that maternal methyl donor supplementation leads to increased A^{vy} methylation in the offspring. Furthermore, this means that an environmentally induced epigenetic drift in the mothers was inherited to their children. The inheritance of an epigenetic programing to the next generation indicates that at least metastable epialleles, such as IAP, are able to resist the global demethylation of the genome before pre-implantation (Sect. 11.1).

Interestingly, when A^{vy} mice were fed with a soy polyphenol diet causing changes in their DNA methylation patterns, their offspring were protected against obesity and T2D across multiple generations. In another mouse model, maternal undernutrition leads to low birth weight and glucose intolerance in male and female F1 offspring.

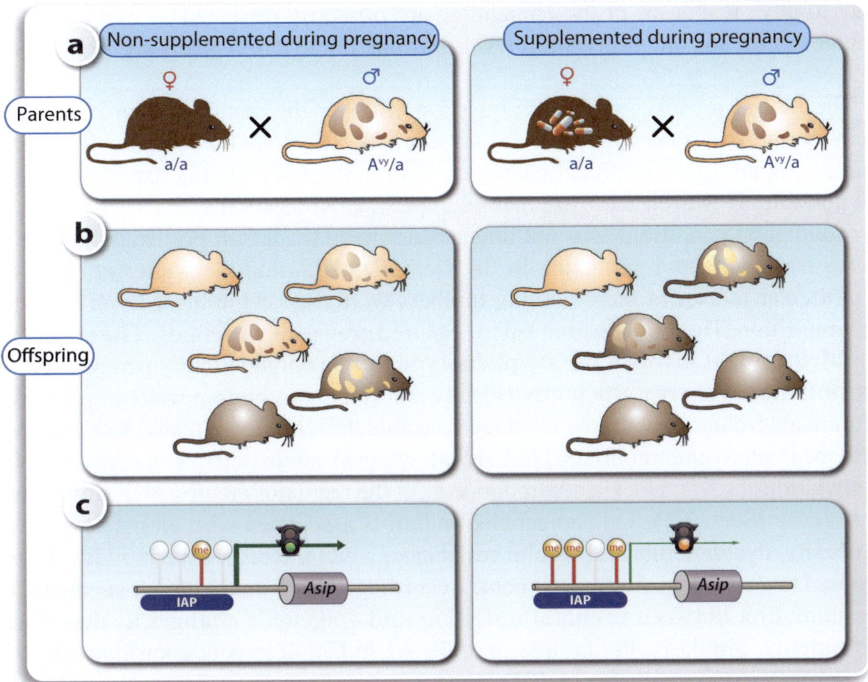

Fig. 12.1 Maternal dietary supplementation affects the phenotype and epigenome of A^{vy}/a offspring. The diets of female wild-type a/a mice are either not supplemented (**left**) or supplemented with methyl-donating compounds (**right**), such as folate, choline, vitamin B$_{12}$ and betaine, for 2 weeks before mating with male A^{vy}/a mice and ongoing during pregnancy and lactation **a**. The coat color of offspring that are born to non-supplemented mothers is predominantly yellow, whereas it is mainly brown in the offspring from mothers that were supplemented with methyl-donating substances **b**. About half of the offspring does not contain an A^{vy} allele and is therefore black (a/a, not shown here). Molecular explanation of DNA methylation and *Asip* gene expression: maternal hypermethylation after dietary supplementation shifts the average coat-color distribution of the offspring to brown by causing the IAP retrotransposon upstream of the *Asip* gene to be more methylated on average than in offspring that are born to mothers fed a non-supplemented diet **c**. White circles indicate unmethylated CpGs and yellow circles are methylated CpGs

Exposure to suboptimal nutrition during fetal development in utero leads to changes in the germ cell DNA methylome of male offspring, even when these males were nourished normally after weaning. These phenotypic differences are transmitted through the paternal line to the F2 offspring. In the F1 sperm genome of maternally undernourished male offspring more than 100 regions were found to be hypomethylated compared to controls. This indicates that PGCs from nutritionally restricted fetuses did not completely remethylate their DNA. Accordingly, **epigenetic memory can be passed from one generation to another by inheriting the same indexing of chromatin marks**. From the different types of chromatin marks, DNA methylation seems to be designed in particular for a long-term cellular memory, while short-term "day-to-day" responses of the epigenome are primarily mediated by non-inherited changes in the histone acetylation level. Histone methylation levels are in between both extremes.

The mouse models impose the question whether the concept of an epigenetic memory and inheritance is also valid for humans. There are no comparable natural human mutants and for ethical reasons human embryonal feeding experiments are not possible. However, there are natural experiments, where the exposure to severe environmental conditions was not under experimental control. For example, in the *Dutch Hunger Winter* occurring in the Netherlands during the winter of 1944/45 in particular fetuses of their starving mothers were exposed in utero to an extreme undernutrition. This malnutrition led to impaired growth of the fetuses. **The resulting low birth weight favors a thrifty phenotype that is epigenetically programed to use nutritional energy efficiently** (Fig. 12.2). Thus, the children were prepared for a future environment with low resources during adult life. Even many decades after birth the in utero undernourished individuals showed subtle (<10%) changes in DNA methylation at several loci in adulthood, e.g., at the regulatory region of the imprinted gene *IGF2* (Sect. 7.3). This epigenetic pattern is associated with an increased risk of obesity, dyslipidemia and insulin resistance, when the respective individuals are exposed to an obesogenic environment. Accordingly, **for humans there seems to be the same link between prenatal nutrition and epigenetic changes as described for rodents**. Similarly, the famine of 1959–61 in China largely contributed to the over-proportionally high rise of T2D in the country (Sect. 16.3). These examples led to the Developmental Origins of Health and Disease (DOHaD) concept indicating that early developmental events, such as perturbations of the nutritional state in utero, have significant effects on disease risk as adult. Thus, **environmental exposures of humans, in particular during early life, can be stored as epigenetic memory**.

Epigenetic information can also be transmitted via the germ line to the next generation. Although some 95% of DNA methylation marks are erased throughout the two rounds of demethylation during PGC generation (Sect. 11.1), some single-copy genomic regions escape both demethylation waves and remain methylated in gametes. In case the DNA methylation pattern at these escape regions is susceptible to environmental influences, such as dietary molecules, **lifestyle choices of an individual may be transmitted to subsequent generations and may lead to phenotypic consequences**.

Interestingly, in humans the amount of brown adipose tissue, which is used for thermogenesis, correlates with the month when we were conceived. Individuals conceived in cold months have significant differences in brown fat characteristics and metabolic phenotypes than those, who were conceived in warm months or environment. An analogous mouse experiment confirmed the observation in rodents and demonstrated that only the cold exposure of the fathers before conception affected the amount of brown adipose tissue in the offspring.

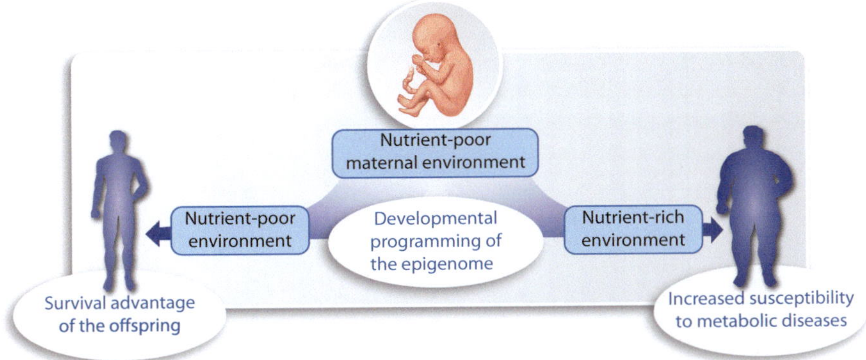

Fig. 12.2 The DOHaD concept. Intrauterine stressors, including maternal undernutrition or placental dysfunction (leading to impaired blood flow with consecutively hypoxia or reduced nutrient transport) can initiate abnormal patterns of development, histone modifications and DNA methylation. Additional post-natal environmental factors, including accelerated post-natal growth, obesity, inactivity and aging further contribute to the risk for T2D, potentially via changes in histone modifications and DNA methylation patterns of metabolic tissues. Obviously, epigenetic changes during embryogenesis have a much greater impact on the overall epigenetic status of an individual than that of adult stem cells or somatic cells, since they affect far more following cell divisions

12.2 Population Epigenetics

The field of epigenetic epidemiology, i.e., the study of epigenetics in populations, combines epigenome-wide methods (Sect. 10.2) with population-based epidemiological approaches. Epigenetic changes can occur at any time during life, although increased sensitivity exist during early embryogenesis (Sect. 11.1, Fig. 12.3a). **Our epigenome primarily changes due to environmental exposures but also based on stochastic epigenetic drifts associated with aging** (Sect. 12.3). Epigenetic epidemiology studies different types of human cohorts with the goal to identify both the causes as well as the phenotypic consequences of epigenomic variations. These are cohorts of natural experiments, such as the *Dutch Hunger Winter* (Sect. 12.1), longitudinal birth cohorts, longitudinal studies on monozygotic twin cohorts, prenatal cohorts and in vitro fertilization (IVF) conception cohorts (Fig. 12.3b).

Studies of families are well suited to investigate epigenetic changes in the offspring that may be based on environmental exposures of parents during gametogenesis. Birth cohorts and IVF cohorts track life from as early as peri-conception (i.e., around the time of conception) and allow the study of epigenetic changes based on the prenatal environment and their association with disease phenotypes early in life. Cohorts based on natural experiments, i.e., when the exposure to severe environmental conditions was not under experimental control, enable the investigation of the link of environmental exposures in early life with the onset of disease phenotypes decades later. Prospective cohorts, in particular those involving monozygotic twins having identical genomes, parents, birth date and gender, study in a longitudinally

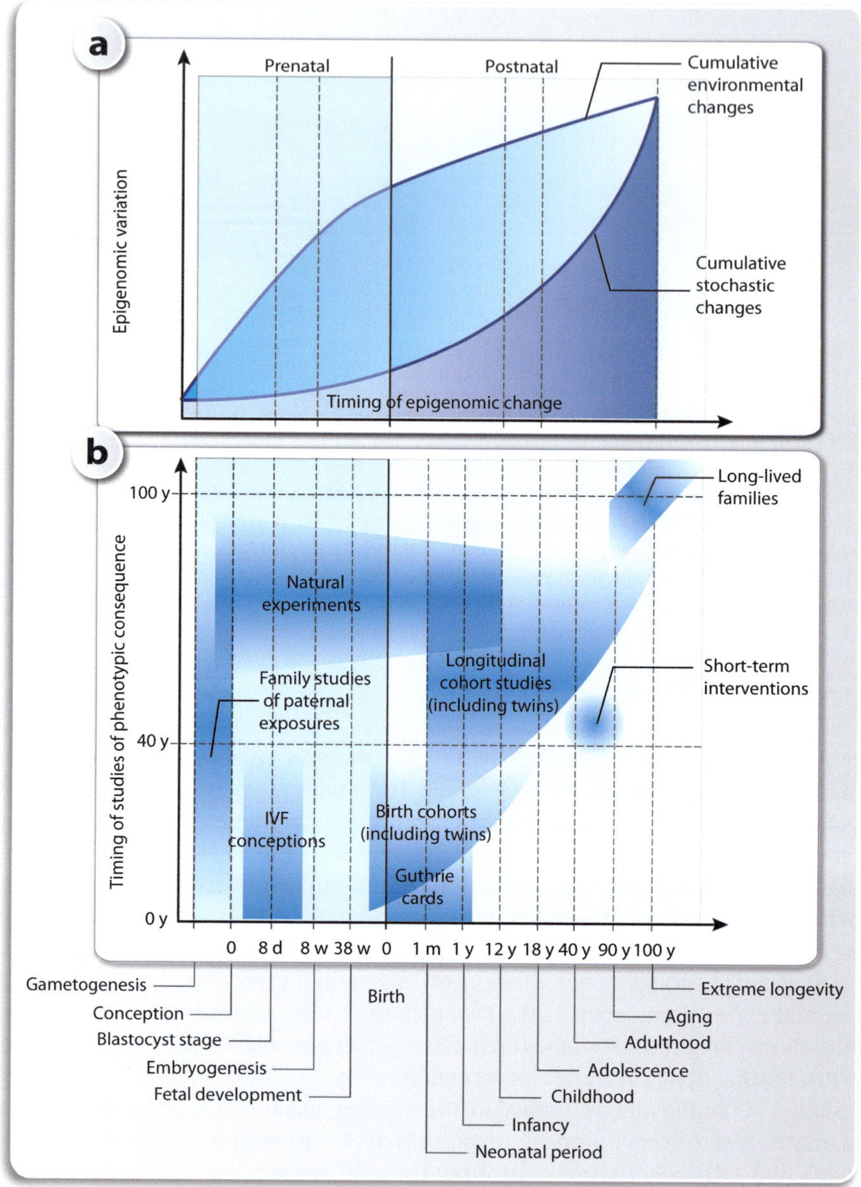

Fig. 12.3 Epigenetic variation in populations. Epigenetic changes can occur at any time during life, but there is significantly increased sensitivity during early prenatal development **a**. Prenatal epigenetic changes may be investigated using IVF cohorts, archived newborn metabolic screening tests (formerly called Guthrie cards) and birth cohorts tracking life from as early as periconception **b**. Historical famines represent the few opportunities to link the prenatal environment to health outcomes later in life. Longitudinal cohort studies (especially involving twins) sample peripheral tissues and take biopsies from disease-relevant tissues

way the contribution of age-related epigenetic modifications in common diseases, such as T2D, autoimmune diseases, cancer, Alzheimer's and autistic spectrum disorders. Interestingly, twin studies demonstrated that epigenetic variations significantly increase across lifespan. Short-term interventions, such as dietary studies, can identify specific environmental exposures that lead to tissue-specific epigenetic changes. Finally, the follow-up of long-lived families helps to identify the impact of epigenetics for healthy aging (Sect. 12.3).

Epigenetic drifts, such as hypermethylation of CpG islands close to the regulatory regions of tumor suppressor genes, contribute to the risk for cancer and other diseases (Fig. 12.4). In particular the risk for diseases that are related to the exposure with environmental factors, such as microbes causing inflammation (Sect. 13.4) or overeating leading to obesity and T2D, have a large epigenetic contribution. Of special interest are diseases that have their onset long time before the phenotype emerges, i.e., where accumulations of epigenetic changes stepwise increase disease susceptibility.

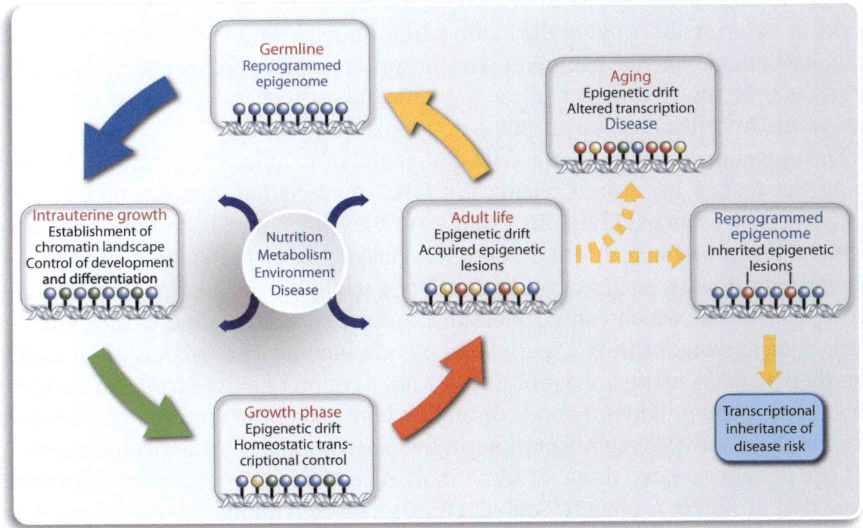

Fig. 12.4 Epigenetic drift and transgenerational epigenetic inheritance. During embryogenesis epigenetic marks, such as DNA methylation and histone modifications, are established in order to maintain cell lineage commitment. After birth, this epigenetic landscape stays dynamic throughout lifespan and responds to nutritional, metabolic, environmental and noxious signals. Epigenetic drifts are part of homeostatic adaptations and should keep the individual in good health. However, when an adverse epigenetic drift compromises the capacity of metabolic organs to adequately respond to challenges as provided by nutrition and chronic inflammation, the susceptibility to diseases, such as T2D or cancer, increases. Some of these acquired epigenetic marks can be inherited to subsequent generations when they escape epigenetic reprograming during gametogenesis

12.3 Epigenetics of Aging

Individuals have a personal rate of aging that depends on gender (women tend to live longer than men), lifestyle choices, such as smoking or physical inactivity and many environmental factors. There is a genetic basis for longevity, but the non-genetic contribution to aging is estimated to be more than 70%. For example, some molecular markers of age, such as telomere length or the expression of genes in metabolic and DNA repair pathways, are sensitive to environmental stress. Moreover, various animal models of aging suggest that calorie restriction, low basal metabolic rate, increased stress response and reduced fertility play a major role in determining the lifespan of individuals. The molecular basis of all non-genetic factors of aging is cellular perturbations that result in the modulation of signal transduction pathways affecting the epigenome, i.e., **epigenetic changes are a major contributor to the aging process**. Thus, not only diseases but also aging result in epigenetic drifts. Changes in the epigenome also can occur spontaneously (i.e., stochastically) without the contribution of a cellular perturbation. Nevertheless, the likelihood for stochastic changes of the epigenome is increased by chemicals disrupting DNA methylation marks or by errors in copying the methylation status during DNA replication. Such chemicals are metals like cadmium, arsenic and mercury, peroxisome proliferators, air pollutants, such as black carbon and benzene, as well as endocrine disruptors, such as diethylstilbestrol, bisphenol A and dioxin.

The epigenome is able to preserve the results of cellular perturbations by environmental factors in form of changes in DNA methylation, histone modifications and 3D organization of chromatin. Changes of the epigenome, in particular the DNA methylome, are associated with the chronological age of an individual as well as with age-related diseases, such as cancer. The DNA methylation status at a few hundred key CpG islands, which can be measured from easily accessible tissues and cell types, such as skin or PBMCs (peripheral blood mononuclear cells), is often used as a biomarker. DNA methylome profiling of a large cohort of individuals spanning over a wide age range provides a good correlation between chronological and biological age, but there are also significant interindividual variations. At a given chronological age the investigated tissue of some individuals has a far "younger" epigenome, while that of others is already "older" (Fig. 12.5). Accordingly, it can be expected that the latter individuals may have an earlier onset of age-related diseases, such as cancer, from the respective tissue and eventually may die at younger chronological age than the former individuals. This has been observed with individuals suffering from premature aging syndromes. In contrast, the blood of the offspring of super-centenarians, i.e., individuals who reached an age of at least 105 years, has a lower epigenetic age than that of age-matched controls. Thus, **epigenetic signatures can serve as biomarkers of aging**.

The biological age of a specific tissue, e.g., that of the liver of an obese person, may be significantly higher than that other tissues from the same person. The DNA methylation-based age of a reference tissue, such as PBMCs, can be considered as an epigenetic clock. This epigenetic clock is a more accurate predictor for mortality

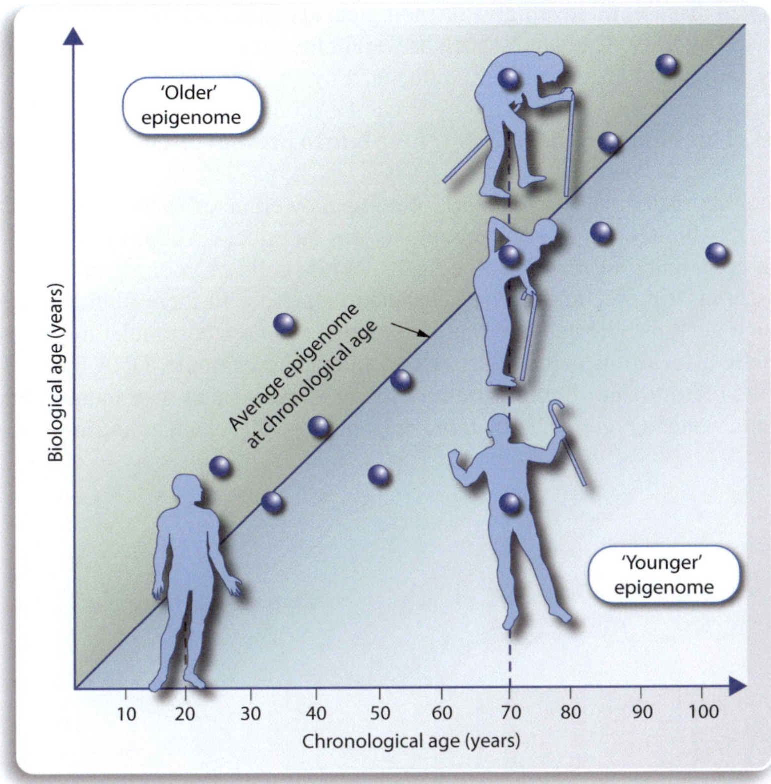

Fig. 12.5 Epigenetic biomarkers of age. Epigenome-wide patterns not only monitor cellular identities but also cellular health and age. For example, changes in the DNA methylation status of CpG islands, as measured in white blood cells, taken from individuals of different age (balls), can serve as a sensor for chronological age. However, there is significant deviation from the postulated linear fit (diagonal line) suggesting that methylation patterns also represent biological age

by all causes in later life than other biomarkers of aging, such as changes in telomere length. For comparison, ES and iPS cells show in these assays as "ageless". Moreover, sperm cells might be classified "younger" than somatic cells from the same individual. Epigenetic clocks of mice (average lifespan some 2 years) tick faster than those of humans (average lifespan some 80 years). Moreover, a quantitative model of the aging methylome is able to distinguish relevant factors in aging, including gender and genetic variants.

Aging clocks of the first generation used primarily the parameter chronological age to train machine learning models, whereas second generation aging clocks improved their predictive power through the implementation of additional individual-specific information, such as calculated mortality risk factors, clinical protein markers and cigarette pack-years. The difference between biological and chronological age is the "age gap". Persons with a positive age gap have a higher probability to develop an age-related disease. Interestingly, the biological age is a dynamic parameter that can be reduced by lifestyle changes, such as weight loss, increased physical activity and healthy eating, but also negatively influenced by an unhealthy lifestyle (Fig. 12.6). The dataset indicates that the ageotype can be modulated by individual's lifestyle changes. **This indicates that it is never too late to prevent diseases and slow down aging**.

Genomic instability is a hallmark of aging as well as of cancer. The accumulation of DNA mutations or even aneuploidy has an effect on both the transcriptome and the epigenome of the concerned cells. While cells of young individuals show a robust transcriptome and normal chromatin states, with increasing age the transcriptome gets instable and aberrant chromatin states are accumulating. Therefore, **epigenetic alterations are a hallmark of aging**. For example, DNA damage stimulates the recruitment of chromatin modifying enzymes that may induce abnormal chromatin states (Fig. 12.7). In turn, epigenome-wide changes during aging can

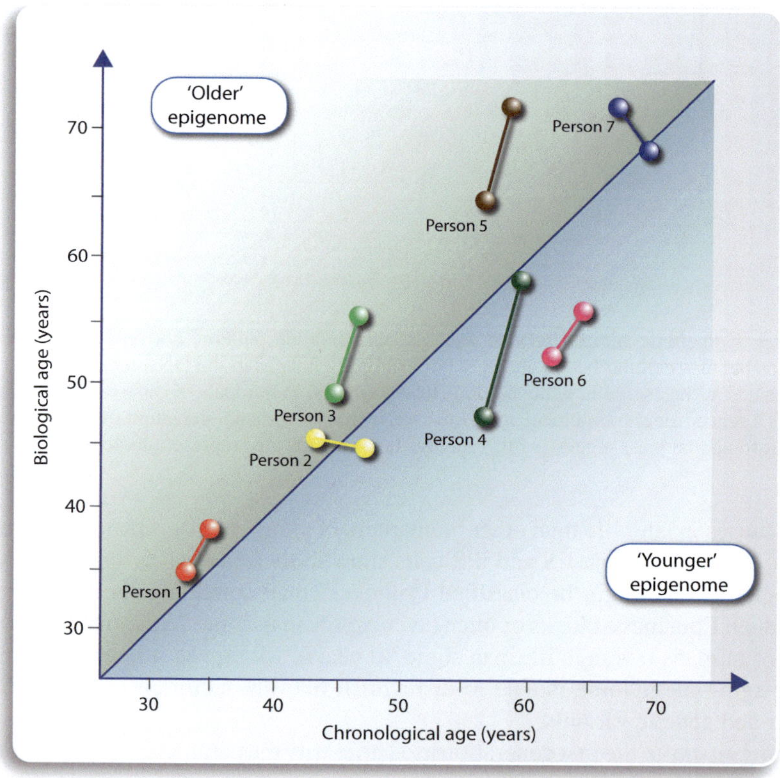

Fig. 12.6 Phenotypic age regression on the chronological age of individuals. The gray trend line is the overall regression line for the cohort. The colored lines are the fitted regression lines for 7 representative individuals out of 43 persons investigated by Ahadi et al., Nature Medicine **26**, 83–90 (2020)

increase the susceptibility of the genome to mutations and in parallel reduce the precision of transcription. Moreover, errors in DNA repair and failure to correctly replicate the genome and epigenome not only increase the number of DNA mutations but also of epimutations (Sect. 15.1). Thus, since genome surveillance and epigenetic remodeling influence each other, **environment-induced epigenome instability throughout life is an important driver of the aging process**.

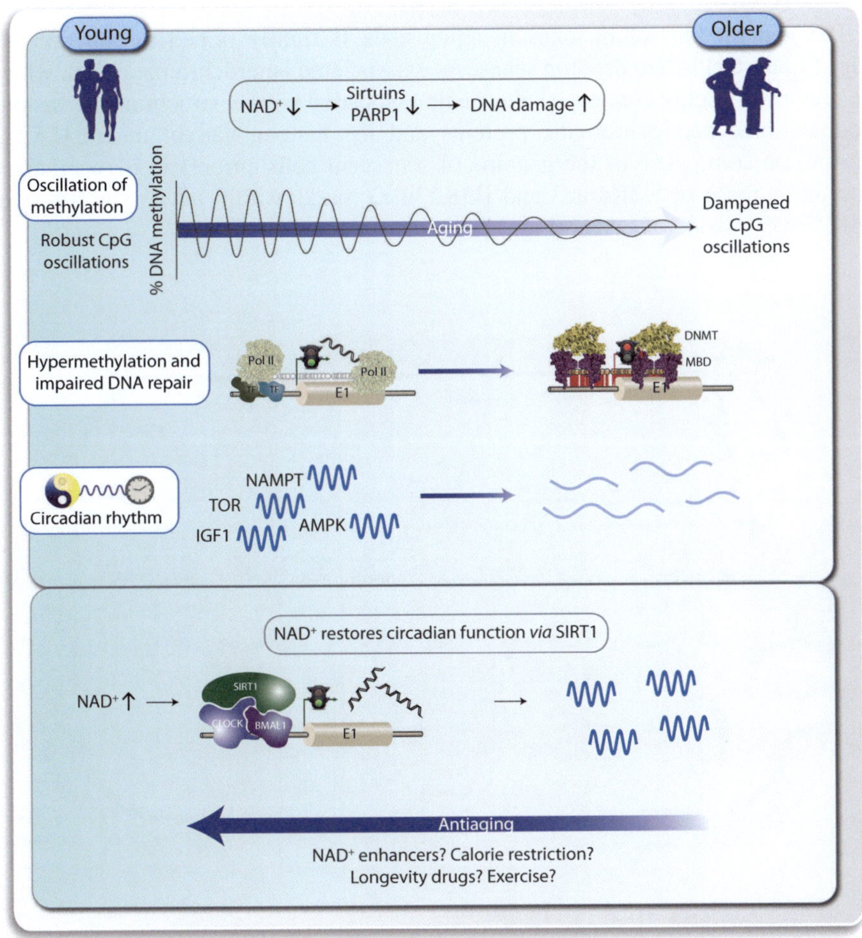

Fig. 12.7 Mechanisms of epigenetic aging. The epigenetic clock starts to tick when tissue stem cells undergo asymmetric division and differentiate into non-stem cells. CpGs exhibit robust circadian oscillations in cells of young individuals (**top left**) but are dampened with age due to reduced levels of NAD+, sirtuins and poly(ADP-ribose) polymerase 1 (PARP1) and increased levels of DNA damage. altered activity of TETs and DNMTs (**top right**). NAD+ enhancers, dietary restriction, longevity compounds or exercise may restores CpG oscillations (**bottom**)

In addition to changes in DNA methylation the epigenetic hallmarks of aging (Fig. 12.8) include:

- a general loss of histones due to local and global chromatin remodeling
- an imbalance of activating and repressive histone modifications
- site-specific loss and gain in heterochromatin
- significant nuclear reorganization
- transcriptional changes.

The general loss of histones in aging cells is tightly linked to cell division (Fig. 12.8a). Cells then develop senescence-associated heterochromatin foci, which are regions of highly condensed chromatin associated with heterochromatic histone modifications, heterochromatic proteins and the histone variant macroH2A. In general, on some 30% of the genome of senescent cells chromatin is reorganized with an increase of H3K4me3 and H3K27me3 marks within LADs and a loss of H3K27me3 outside of LADs (Fig. 12.8b).

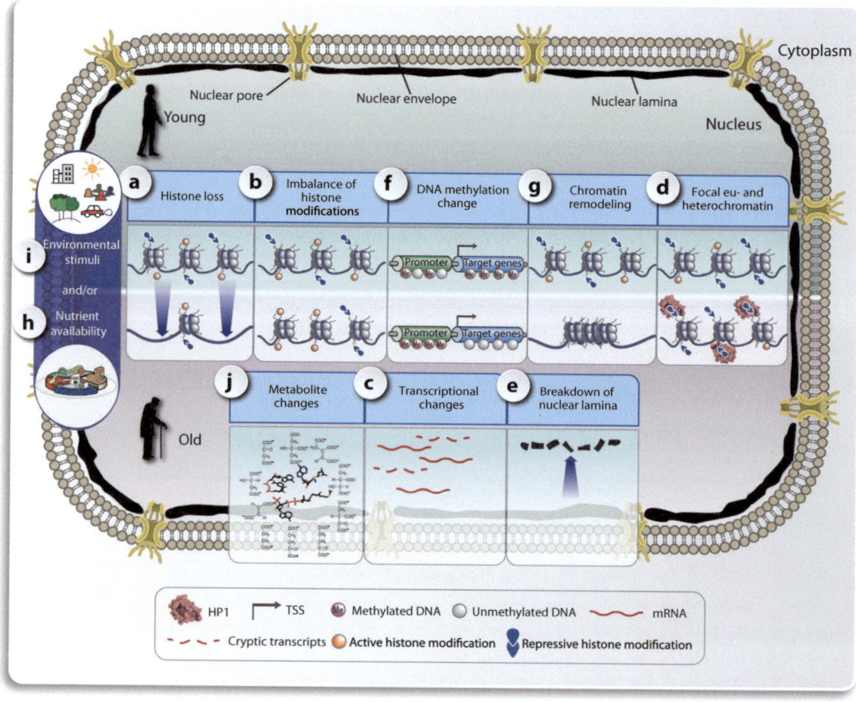

Fig. 12.8 Epigenetics of senescence and aging. The epigenetic hallmarks of senescence and aging are a loss of histones **a**, imbalance of activating and repressive modifications **b**, changes in gene expression **c**, losses and gains in heterochromatin **d**, breakdown of nuclear lamina **e**, global hypomethylation and focal hypermethylation **f** as well as chromatin remodeling **g**. These changes are heavily dictated by environmental stimuli **h** and nutrient availability **i** that in turn alter intracellular metabolite concentrations **j**

The most significant molecular consequence of the loss of repressive histone marks and the gain of activating marks during aging is a change in gene expression, such as the upregulation of genes related to cell adhesion and ribosomal proteins, while genes related to cell cycle regulation, DNA repair and DNA replication are downregulated (Fig. 12.8c). The lack of some of the repressed genes negatively affects longevity, while some of the activated genes are detrimental to lifespan. Moreover, the transcriptional reprogramming happening during the onset of senescence also occurs through altered activity of chromatin modifying enzymes and remodelers. Constitutive heterochromatin at telomeres, centromeres and peri-centromeres is established during embryogenesis and is thought to be maintained throughout lifespan. However, senescent cells loose some of these regions of constitutive, chromatin resulting in an increase of euchromatic regions (Fig. 12.8d). Moreover, a loss of the nuclear lamina (Fig. 12.8e) stimulates the breakdown of heterochromatin organization and the relocalization of heterochromatic proteins to regions in the genome where they contribute to the formation of region-specific foci. The global hypomethylation and local hypermethylation of genomic DNA during aging fits with the observation of global heterochromatin deregulation in combination with focal increase at some genomic regions (Fig. 12.8f). DNA methylation is primarily lost at repetitive genomic regions that are in constitutive heterochromatin, while hypermethylation mostly occurs at CpG islands close to promoter regions. SWI/SNF chromatin remodelers are associated with gene activation and they seem to promote aging, while repressive chromatin remodelers support longevity (Fig. 12.8g).

Highly conserved proteins, such as insulin and IGF receptors, the amino acid-sensing kinase TOR (target of rapamycin) and the NAD^+-sensing HDAC SIRT1, belong to the nutrient signaling pathways that integrate metabolic signals into chromatin responses (Sect. 16.2). They inform the epigenome on nutrient availability and their misregulation is one of the hallmarks or aging (Fig. 12.8h). Similar principles apply to the sensors of other environmental inputs (Fig. 12.8i) or intracellular metabolites (Fig. 12.8j), such as steroid hormones like estrogen and testosterone via their nuclear receptors ESR1 and AR or α-ketoglutarate via TET enzymes, that affect longevity via changes in the chromatin landscape.

12.4 Epigenetics of the Epigenetic Clock

Light-sensitive organisms, such as humans, synchronize their daily behavioral and physiological rhythms with the rotation of the Earth around its axis, i.e., they display circadian (i.e., approximately one day) activity cycles, such as sleep/wake and fasting/eating. These circadian rhythms are under control of a molecular clock, which is a hierarchical network of transcription factors and associated nuclear proteins that adapts to environmental changes. These rhythms are generated by the suprachiasmatic nucleus (SCN) of the hypothalamus (Fig. 12.9a). The SCN is composed of only 15–20,000 neurons, which autonomously oscillate in a 25 h rhythm. Via a direct connection with the retina this central clock is adjusted to the daily light–dark cycle.

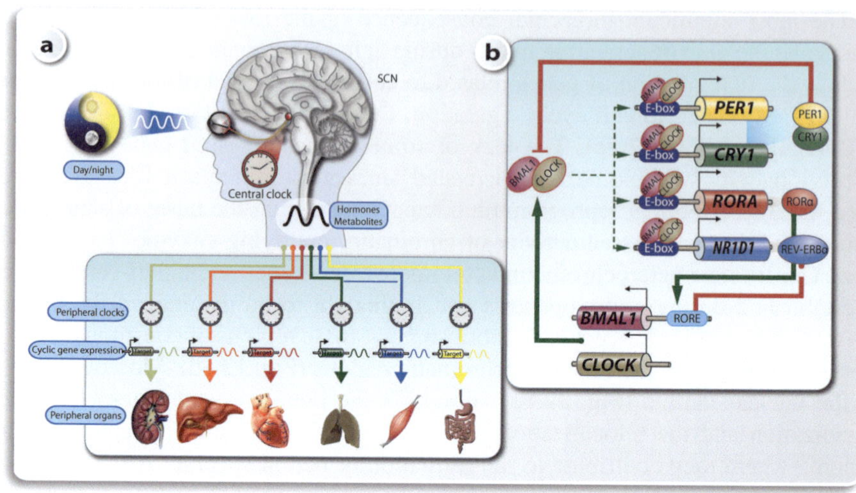

Fig. 12.9 The circadian clock. Electrical and humoral signals from the SCN synchronize phases of circadian clocks in peripheral organs, which then generate time-dependent rhythms in gene expression, metabolism and other physiological activities **a**. In the feedback loop of the molecular circadian oscillator positive elements, such as the transcription factors BMAL1, CLOCK and ROR, are shown in green and negative elements, such as PER1, CRY1 and REV-ERBα, in red **b**. The combined actions of hundreds of BMAL1-CLOCK target genes provide a circadian output in physiology

The SCN is the main driver of circadian fluctuations, e.g., in blood glucose levels via scheduling food ingestion to the activity phases. Moreover, the SCN directs the peripheral clocks in all other tissues and cells of our body via synchronizing rhythmic food intake. The output of the oscillating system is coordinated physiology via the control of various processes in metabolism and behavior. Thus, **the circadian clock is a critical interface between nutrition and homeostasis directing and maintaining proper rhythms in metabolic pathways**.

The molecular core of the circadian clock is a series of transcription-translation feedback loops of the transcription factors BMAL1 (basic helix-loop-helix ARNT like 1) and CLOCK (clock circadian regulator) and their corepressor proteins. The BMAL1-CLOCK complex activates in a circadian fashion the expression of hundreds of genes both in the brain and in peripheral metabolic tissues, including also the genes *PER1* (period circadian clock 1) and *CRY1* (cryptochrome circadian clock 1). The PER1-CRY1 corepressor complex inactivates BMAL1-CLOCK, but phosphorylation and ubiquitylation of CRY1 during the night initiates the proteosomal degradation of the repressors and reactivates BMAL1-CLOCK. The genes encoding for the nuclear receptors REV-ERBα and RORα (Sect. 5.1) are further targets of BMAL1-CLOCK. REV-ERBα negatively and RORα positively regulates the expression of the *BMAL1* gene, i.e., these nuclear receptors form additional feedback loops in the control of the circadian clock (Fig. 12.9b). **In total, the expression of some 10–15% of genes in all organs and tissues displays a circadian rhythm**.

The circadian clock can be modulated by metabolites, in particular by those representing energetic flux (Box 12.1). For example, the AMP sensor AMPK (adenosine monophosphate-activated protein kinase connects the internal clock function to the nutrient state via phosphorylation and subsequent proteasomal degradation of the BMAL1-CLOCK repressor CRY1. In parallel, the cyclical activity of BMAL1-CLOCK is modulated by the chromatin modifying enzyme KDM5A (Sect. 8.3), which in turn is linked via its cofactors iron and α-ketoglutarate to cellular redox and mitochondrial energetics. The bidirectional interaction between circadian and metabolic signaling is the inhibition of BMAL1-CLOCK by the NAD-dependent HDAC SIRT1. This represents another feedback control of the circadian clock, since the gene encoding for the critical enzyme for NAD synthesis, *NAMPT* (nicotinamide mononucleotide phosphoribosyltransferase, also known as visfatin), is a direct BMAL1-CLOCK target. Since NAD$^+$-dependent SIRTs are important regulators of metabolic pathways in response to calorie restriction, the **link between the circadian clock and SIRT activity has implications for aging**.

Box 12.1: Modulating the circadian clock by metabolic systems The proteins of the circadian clock are not only expressed in the SCN but also at peripheral sites, such as the liver. Entrainment of the liver clock to feeding involves glucocorticoid signaling, temperature via HSF1 (heat shock transcription factor 1) and ADP-ribosylation. In this way, the circadian clock coordinates daily behavioral cycles of sleep–wake and fasting-feeding with anabolic and catabolic processes in the periphery. The central and peripheral clocks are also synchronized via posttranslational modifications of transcription factors and histones that tune gene expression rhythms into changes of the metabolic state. Therefore, in addition to the transcription-translation-based feedback systems, mammals and other species use NAD$^+$ oscillation, redox flux, ATP availability and mitochondrial function, in order to influence acetylation and methylation reactions. For example, the redox-based clock represents oscillations in the redox state of the family of peroxiredoxin antioxidant enzymes that rhythmically anticipate the generation of ROS. Moreover, NAD$^+$ is an electron shuttle in oxidoreductase reactions and also acts as a cofactor in HDAC and ADP-ribosylation modifications.

In the absence of external stimulation, the release of the stress hormone cortisol as well as some peptide hormones, such as thyrotrophin and growth hormone 1 (GH1), follow a circadian rhythm, i.e., the respective endocrine glands are under the influence of circadian clocks. Food intake is organized into distinct meals during the daily cycle, i.e., in the active part of the day energy stores are replenished, while the sleep phase represents a daily period of fasting and mobilization of energy stores. Epigenetics of the circadian clock coordinate daily behavioral cycles of sleep–wake and fasting-feeding with anabolic and catabolic processes in the periphery. NAD^+ oscillation, redox flux, ATP availability and mitochondrial function are used to influence acetylation and methylation via chromatin modifying enzymes (Sect. 8.3). Dietary restriction is known to extend the lifespan in many model organisms, such as yeast, worms, flies and even primates. Gene expression changes related to dietary restriction promote global preservation of genome integrity and chromatin structure, such as maintenance of heterochromatin (Fig. 12.10). Nutrition signaling pathways via insulin, TOR and SIRTs integrate metabolic signals into chromatin responses. They inform the epigenome on nutrient availability and thus have a key role in determining lifespan. **Bad lifestyle decisions are able to reprogram the circadian clock in a negative way** (Box 12.2). For example, eating at night, artificial light, shift work, travel across time zones and temporal disorganization have disrupted for many humans the alignment between the external light–dark cycle and their internal clock. This is of disadvantage for metabolic health. Longitudinal population studies and clinical investigations both indicated an association between shift work and diseases, such as T2D, gastrointestinal disorders and cancer, that can be modulated by changes in the circadian rhythm. Furthermore, the habit of altering bedtime on weekends, the so-called "social jet lag", has been associated with increased body weight.

> **Box 12.2: Circadian disruptions and complex diseases** The disruption of circadian rhythms and sleep homeostasis increase the incidence and severity of various diseases, such as T2D, obesity, cancer, depression and neurodegenerative diseases. Thus, circadian disruptions, such as behavioral changes in sleep–wake cycles like occurring in the context of shift work, affect human health. Other desynchronosis like jet lag, social jet lag (mostly at weekends) or irregular food uptake have also a negative contribution to health. For example, misalignments between circadian timing like daily sleep–wake behavior and food intake contributes towards the development of insulin resistance. Moreover, stress often causes circadian disruptions and can alter the physiological stress responses and may cause metabolic impairments. This involves long lasting epigenetic changes.

Physical activity promotes healthy aging, prevents cognitive decline and is associated with a 30% reduction in all-cause mortality. Moreover, it induces changes in the chromatin of skeletal muscles, such as increased H3K36ac levels and the cellular localization of HDAC4 and HDAC5. Thus, **physical activity has direct effects on the epigenome**. Interestingly, the signaling of pheromones, which are compounds triggering a response in other members of a given species, works via chromatin modifying enzymes, i.e., pheromones have an effect on the epigenome. The

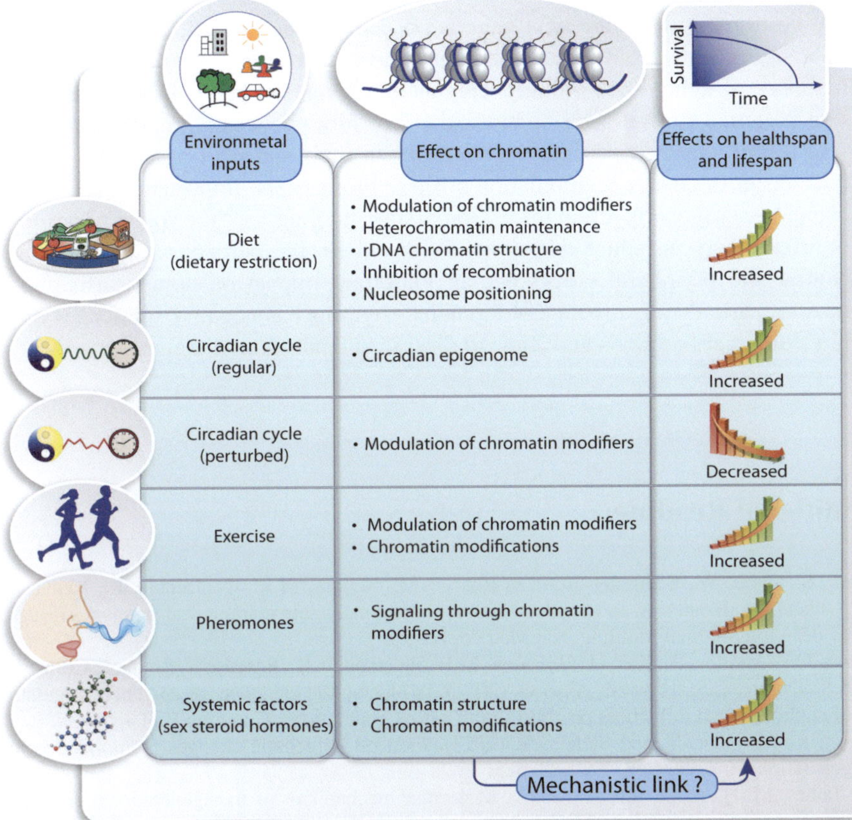

Fig. 12.10 Effects of environmental inputs on longevity and chromatin. Many environmental signals that modulate lifespan also affect chromatin. These are dietary restriction, the circadian cycle, physical activity and sex steroid hormones. More details are provided in the text

systemic levels of sex steroid hormones (Fig. 12.10), such as estrogens in females and androgens in males, decline with age. For example, estrogens reduce the risk for age-related diseases, such as osteoporosis, sarcopenia (muscle weakness), CVD, reduced immune function and neurodegeneration, while lower levels of steroid hormones increase the prevalence for these diseases.

(Clinical) conclusion: Epigenetics contributes to phenotypes inherited across generations, since both somatic as well as germ cells show a life-long accumulation of epigenetic changes. Thus, the cells are able to memorize any type of perturbation that they were exposed to and a part of this epigenetic memory can be transferred to the next generation. Epigenetic alterations during lifetime significantly contribute to the aging process and represent an important hallmark of aging. Accordingly, epigenetic changes, such as DNA methylation patterns, can be used as biomarkers of biological age, i.e., they represent a biological clock. Individuals differ in their speed of aging, i.e., they have faster or slower accumulation of these biomarkers. Importantly, lifestyle changes significantly affect these biomarkers of aging, i.e., the **speed of aging is affected by personal decisions**. Furthermore, the perception of time via central and peripheral clocks has an important impact on homeostasis and health, while the disturbance of circadian rhythms has negative influences on physiological functions and leads to disease and accelerated aging.

Additional Reading

Ahadi, S., Zhou, W., Schussler-Fiorenza Rose, S.M., Sailani, M.R., Contrepois, K., Avina, M., Ashland, M., Brunet, A. and Snyder, M. (2020). Personal aging markers and ageotypes revealed by deep longitudinal profiling. Nat Med *26*, 83–90.

Breton, C.V., Landon, R., Kahn, L.G., Enlow, M.B., Peterson, A.K., Bastain, T., Braun, J., Comstock, S.S., Duarte, C.S., Hipwell, A., *et al.* (2021). Exploring the evidence for epigenetic regulation of environmental influences on child health across generations. Commun Biol *4*, 769.

Cheng, S., Mayshar, Y. and Stelzer, Y. (2023). Induced epigenetic changes memorized across generations in mice. Cell *186*, 683–685.

Fitz-James, M. H., & Cavalli, G. (2022). Molecular mechanisms of transgenerational epigenetic inheritance. Nat Rev Genet *23*, 325–341.

Kabacik, S., Lowe, D., Fransen, L., Leonard, M., Ang, S.-L., Whiteman, C., Corsi, S., Cohen, H., Felton, S., Bali, R., *et al.* (2022). The relationship between epigenetic age and the hallmarks of aging in human cells. Nature Aging *2*, 484–493.

Kempfer, R. and Pombo, A. (2020). Methods for mapping 3D chromosome architecture. Nat Rev Genet *21*, 207–226.

McMahon, M., Forester, C. and Buffenstein, R. (2021). Aging through an epitranscriptomic lens. Nature Aging *1*, 335–346.

Oh, E.S. and Petronis, A. (2021). Origins of human disease: the chrono-epigenetic perspective. Nat Rev Genet *22*, 533–546.

Seale, K., Horvath, S., Teschendorff, A., Eynon, N. and Voisin, S. (2022). Making sense of the ageing methylome. Nat Rev Genet *23*, 585–605.

Zimmet, P., Shi, Z., El-Osta, A. and Ji, L. (2018). Epidemic T2DM, early development and epigenetics: implications of the Chinese Famine. Nat Rev Endocrinol *14*, 738–746.

Chapter 13
Epigenetics and Immunity

Abstract In this chapter, we will discuss the prominent impact of epigenetics on the proper function of the immune system. This involves the epigenetic programming of HSCs and progenitor cells during hematopoiesis, in order to produce in response to the respective immunological challenge the appropriate amounts of innate and adaptive immune cells. Furthermore, in the context of the process of inflammation epigenetic programming of innate immune cells leads to appropriate response to microbe exposure creating epigenetic memory referred to as trained immunity. Nuclear receptors provide an important link of nutrition and endocrinology with the immune system. They modulate the differentiation process of immune cells and counteract inflammation, which is primarily mediated by NFκB and its primary and secondary targets. In this way, metabolism and inflammation are linked via gene regulatory actions of transcription factors and epigenetic programming affecting basically all common diseases.

Keywords Hematopoiesis · HSCs · Innate immunity · Inflammation · Epigenetic programming · Trained immunity · Nuclear receptors · VDR · NFκB

13.1 Epigenetics of Blood Cell Differentiation

Cells of the immune system have a rapid turnover and are therefore able to show a maximal adaptive response to environmental changes. In an adult, all circulating blood cells, including immune cells, are produced in the bone marrow by a process termed hematopoiesis. Exceptions are tissue-resident macrophages, which origin from fetal liver and colonized the organs during embryogenesis. Examples for such tissue-resident macrophages are microglial cells in the CNS (central nervous system), Kupffer cells in the liver or alveolar macrophages in the lungs.

Every day HSCs give rise to some 10^{11} cells, i.e., over our lifespan the bone marrow produces far more cells than any other tissues of our body. This highly dynamic developmental process includes the self-renewal of multipotent HSCs as well as their differentiation in a hierarchical cascade. Hematopoiesis leads to

C. Carlberg, *Gene Regulation and Epigenetics*,
https://doi.org/10.1007/978-3-031-68730-3_13

11 major lineages of mature blood cells including some 100 phenotypically distinct cell subtypes (Fig. 13.1a). Most of these cells belong to the immune system. HSCs differentiate into immature progenitor cells, such as multipotent progenitors (MPPs), which then give rise to the progenitors of the myeloid or lymphoid lineages, called common myeloid progenitors (CMPs) and common lymphoid progenitors (CLPs), respectively. CMPs further differentiate into megakaryocyte-erythrocyte progenitors (MEPs) and granulocyte-monocyte progenitors (GMPs). In further differentiation steps of the myeloid line there is progressive commitment to erythrocytes (red blood cells essential for oxygen transport), megakaryocytes creating platelets (needed for blood coagulation after injuries), mast cells, basophils, eosinophils, neutrophils and monocytes. The latter cells can further differentiate into macrophages or dendritic cells (DCs) (Sect. 13.2). With the exception of erythrocytes and megakaryocytes these myeloid cells belong to the innate immune system. Differentiation of the lymphoid line results in B and T cells of the adaptive immune system as well as NK (natural killer) cells and ILCs (innate lymphoid cells) of the innate immune system.

A number of extrinsic and intrinsic factors, such as growth factor-stimulated signal transduction pathways, transcription factors and chromatin modifying enzymes, regulate the equilibrium between self-renewal and differentiation of HSCs.

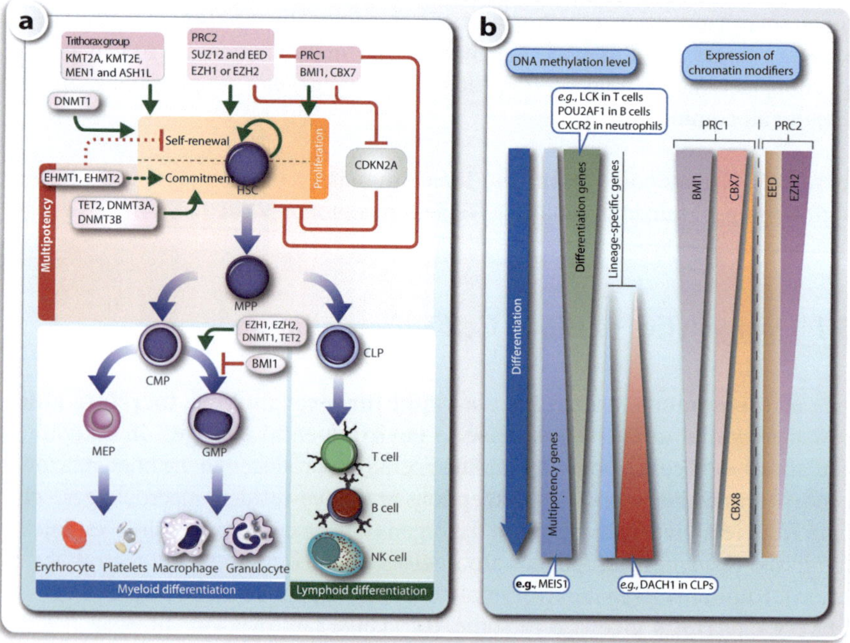

Fig. 13.1 Epigenetics of hematopoiesis. Key chromatin modifying enzymes of HSC self-renewal and their differentiation during hematopoiesis are indicated (**a**). During hematopoietic lineage commitment DNA methylation levels change dynamically (**b, left**). The expression of key epigenetic regulators alters during hematopoiesis (**b, right**). More details are provided in the text

In contrast, a disruption or misregulation of this process can lead to hematological disorders, such as leukemia, lymphoma and myeloma. In these diseases the excessive production of white blood cells in the bone marrow significantly raises their levels in blood circulation. Moreover, **a disruption in any stage of hematopoiesis affects the production and function of blood cells and may have severe consequences, such as the inability to fight against infections or the risk of uncontrolled bleeding**.

During hematopoiesis the genome-wide DNA methylation pattern of differentiating blood cells changes dynamically and is very locus specific, i.e., at some genomic regions there is a rise in methylation, while it decreases at other regions (Fig. 13.1b, left). The latter correlates with the upregulation of cell-specific genes and their encoded proteins, such as the tyrosine kinase LCK (LCK proto-oncogene, SRC family tyrosine kinase) in T cells, the cofactor POU2AF1 in B cells and the chemokine receptor CXCR2 (C-X-C motif chemokine receptor 2) in neutrophils. In general, the commitment to a specific cell lineage increases the level of DNA methylation, since a larger set of genes is not anymore needed in these terminally differentiated cells, such as the transcription factor MEIS1 (meis homeobox 1), which maintains the undifferentiated state, or the myeloid-specific transcription factor DACH1 (dachshund family transcription factor 1) in lymphoid cells. Interestingly, the myeloid lineage is the default outcome of hematopoiesis, since its differentiation requires less correction by increased DNA methylation than that of the lymphoid lineage. Thus, **hematopoietic cells can be easily segregated based on their DNA methylation profile**.

Parallel to alterations in the DNA methylation pattern, also key chromatin modifying enzymes are changing their expression during hematopoiesis. For example, the expression of most members of the Polycomb family changes during HSC differentiation (Fig. 13.1b, right). The PRC1 components CBX7 and BMI1 (BMI1 proto-oncogene, Polycomb ring finger), which are responsible for the recognition and monoubiquitination of H2AK119 (Sect. 8.3), are highly expressed in HSCs but are downregulated during lineage commitment. In contrast, the CBX7 competitor CBX8 is upregulated. The PRC2 component EED (embryonic ectoderm development) does not change during hematopoiesis, while the H3K27-specific KMT EZH2 is downregulated. Similarly, members of the Trithorax group family, such as KMT2A, KMT2E, ASH1L (ASH1 like histone lysine methyltransferase) and MEN1 (menin 1), contribute to hematopoiesis. Furthermore, the KMTs EHMT (euchromatic histone-lysine N-methyltransferase) 1 and 2 deposit repressive H3K9me2 marks to the epigenome of HSCs (Fig. 13.1a). Accordingly, **a misregulation of the genes encoding for these chromatin modifying enzymes can result in hematopoietic failure**, such as HSC cell cycle arrest, premature differentiation, apoptosis and defective self-renewal and finally leads to hematological malignancies.

In addition to chromatin modifying enzymes, some transcription factors, such as CEBPα, SPI1 and GATA2, have a key role in hematopoiesis. They act as pioneer factors that directly bind nucleosomal DNA to prime enhancers for activation. The pioneer factors then recruit chromatin remodeling and modifier complexes, which in turn facilitate removal and posttranslational modification of nucleosomes at these genomic regions. For example, in myeloid progenitors the sustained expression of CEBPα generates macrophages, whereas under sustained expression of GATA2 mast cells are created. However, when initially CEBPα and afterward GATA2 are expressed, eosinophils are produced, while the reversed order leads to basophils. Furthermore, CEBPα interacts with the DNA demethylating enzyme TET2, so that its target genes get demethylated during hematopoiesis. The activity of TET2 may be the key mechanism why myeloid cells are closer to HSCs than lymphoid cells. This fits with the observation that TET2 is mutated in several myeloid malignancies. Moreover, TET2 could link environmental conditions, such as nutrient availability (Sect. 16.1), to myeloid differentiation, since the metabolite 2-hydroxyglutarate inhibits TET2 activity and leads to DNA hypermethylation.

Taken together, hematopoiesis is an important process ensuring key function of our body by constantly providing new cells needed for oxygen transport (erythrocytes), blood coagulation after injuries (platelets) and immunity (leukocytes). **Transcription factors and chromatin modifying enzymes work together in creating appropriate epigenetic profiles on the level of DNA methylation and histone modifications, which determine the respective functions of the more than 100 different cell types of the hematopoietic system.**

13.2 Epigenetics in Immune Responses

Most immune cells are very mobile and experience many different microenvironments throughout our body. This leads to a wide range of different signals and respective **adaptive epigenetic programing of the enhancer repertoire of these cells**. There is an equilibrium between the persistence of an epigenome instructed by previous stimuli and the reprograming in response to a changing environment. For example, when the body is perturbed by low oxygen levels (e.g., at high altitudes), an infection or the onset of an inflammation-related common disease, the bone marrow responds by increased production of erythrocytes, monocytes or neutrophils, respectively, in order to achieve an adapted state of homeostasis. In turn, deviations from the average blood counts for different types of leukocytes are regularly used in clinical practice as **disease biomarkers**. Due to persistent hematopoiesis (Sect. 13.1) most cells of the immune system are replaced every few days to weeks. This also implies that their epigenetic training via encounters with microbes and other antigens is an ongoing learning event leading to **transient epigenetic memory in short-lived cells as well as to persistent memory in long-lived memory B and T cells.**

Monocytes are produced from GMPs in the bone marrow (Fig. 13.2), released into the blood stream and migrate within 1–3 days through the endothelium of blood vessels into tissues. In tissues, monocytes differentiate into macrophages or DCs. This differentiation process is based on **epigenome changes in response to contacts with antigens, such as infectious microbes. These epigenetic changes create a memory of the microbe encounter**. In response to their activation by pathogens or metabolites, macrophages secrete a number of signaling proteins, such as cytokines, chemokines and growth factors, which affect the migration and activity of other immune cells. This response is called **acute inflammation**, no matter whether it is caused by infection or injury. Acute inflammation is often associated with erythema, hyperthermia, swelling and pain, but it resolves within a few days to weeks. Inflammation can also derive from changes in the concentration of nutrients and metabolites. In this case, the immune system cannot cope the primary stimulus, so that **chronic inflammation** develops.

Most inflammatory lesions are initially dominated by monocyte-derived macrophages. The altered gene expression profile of these macrophages is based on changes of their epigenome in response to extracellular signals. The stimuli are classified into PAMPs (pathogen-associated molecular patterns), such as surface proteins of pathogenic bacteria and DAMPs (damage-associated molecular patterns), such as excess of saturated fatty acids and other molecules representing metabolic stress (often called alarmins). The fact that DAMPs can induce the same cascade of inflammatory reactions like PAMPs explains why a wide variety of molecules and events cause inflammation and respective **specific changes in the epigenome of macrophages**.

The direct or indirect contact of immune cells with microbes and other molecules with antigenic pattern results in effects on gene expression that are often stronger than in any other tissue or cell type of our body. The strong reaction is necessary, since bacteria proliferate far faster than human cells and may represent immediate danger to our body. The strength and specificity of the response of the immune cells, such as different populations of macrophages, depends on their epigenomic profile before encountering microbes. This implies that the **proper epigenomic programing of immune cells before contact with antigen is essential for an optimal response**.

The more than 100 genes of the inflammatory cascade differ in their kinetics, i.e., there are fast responding primary target genes, delayed responding secondary targets and late responding tertiary targets. This is reflected by the underlying epigenetic changes in enhancer and promoter regions of the respective genes of macrophages. The promoter regions of primary target genes typically contain H3K4me3 and H3K27ac marks of active chromatin (Fig. 13.3, center). Moreover, these promoters often carry non-methylated CpGs. In contrast, the promoter regions of secondary target genes are first labeled by repressive H3K27me3 marks. The removal of H3K27me3 marks and CpG demethylation from these promoter regions after PAMP exposure and the introduction of H3K4me3 and H3K27ac marks take time. This explains the delayed response in the expression of the respective genes. The stimulation of macrophages with PAMPs also changes the chromatin status at enhancer regions, such as the de novo deposition of H3K4me1 and H3K27ac marks (Fig. 13.3, left). The activation of some of these enhancers is only transient, while others are marked more persistently and retain in this way a memory of the PAMP exposure, i.e., of a contact with microbes.

Antigen binding to receptors of T and B cells, i.e., of cells of the adaptive immune system, activates signal transduction pathways that potently trigger the expression of cytokine genes. Specific epigenetic changes to enhancer and promoter regions are involved in this antigen recognition process (Fig. 13.3, right). For example, the

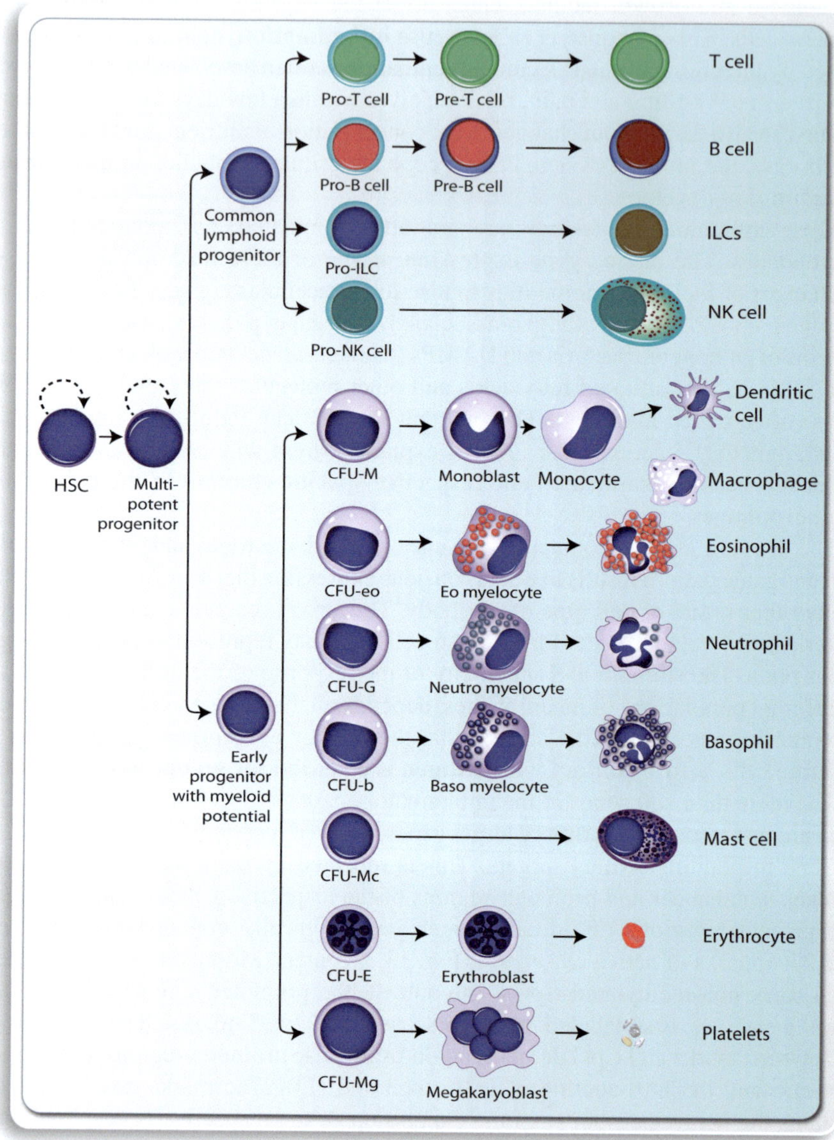

Fig. 13.2 The hierarchy of hematopoiesis. The scheme illustrates the hematopoietic tree of the development of the major lineages of blood and immune cells. HSCs differentiate into immature progenitor cells, such as multipotent progenitors, which then give rise to the progenitors of the myeloid or lymphoid lineages. CFU = colony forming unit, ILC = innate lymphoid cell

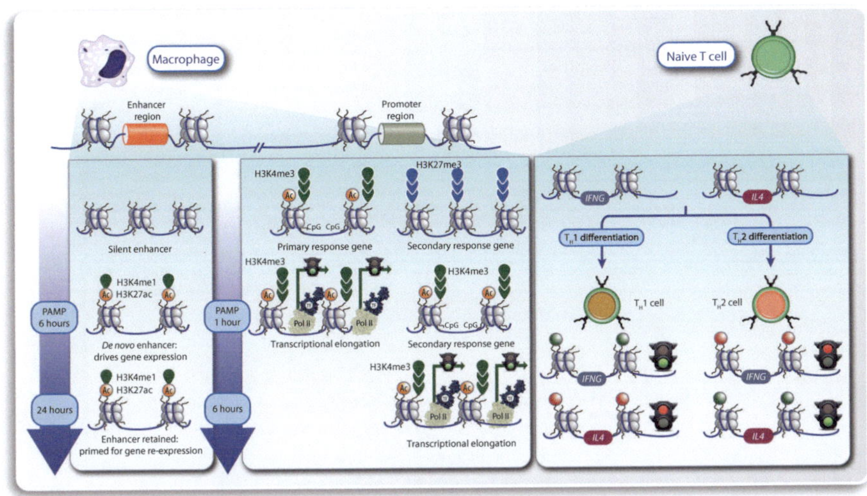

Fig. 13.3 Epigenetic modifications in immune cells. Enhancer (**left**) and promoter regions (**center**) of macrophages show a differential response after the exposure with PAMPs, such as LPS. Primary responses (after 1 h) and secondary responses (after 6 h) are distinguished. Even after removal of PAMP stimulation some enhancer regions can keep their activation status for 24 h or longer. This memory effect is part of trained immunity. Differential epigenetic programing of the regulatory regions of key cytokine genes, such as *IFNG* and *IL4*, is the key event in the polarization of T_H cell subtypes (**right**). In T_H1 cells the *IFNG* gene carries marks of active chromatin (green) and the gene is induced after antigen exposure. In contrast, in the same cells the *IL4* gene carries repressive histone markers (red) and stays repressed. In T_H2 cells the reverse process applies, i.e., the *IL4* gene is induced and the *IFNG* gene stays repressed

differentiation of T helper (T_H) cells into T_H1 and T_H2 subtypes involves epigenetic programing that primes these cells to increase after antigen exposure either the expression of the genes encoding for the cytokines interferon (INF) γ or IL4, respectively. While the T_H1 response results in antigen clearance, the T_H2 response is typical for allergic reactions, i.e., the differential epigenetic programing of these cells causes a clearly different physiological response. The epigenetic changes involve histone modifications but also the appearance of 5hmC marks at promoter regions and demethylation via TET2. The latter modifications are very stable and can last over 20 and more replication cycles of long-lived memory T cells.

The impact of epigenetic changes in the context of regulatory T cell proliferation is demonstrated at the example of the *FOXP3* gene (Fig. 13.4). In this case, the transcription factors REL (Sect. 13.4), CBFB (corebinding factor subunit β) and RUNX1 (runt-related transcription factor 1) both bind enhancers downstream of the *FOXP3* TSS region and stimulate *FOXP3* mRNA expression. The binding of the transcription factors to the enhancers leads to their rapid demethylation allowing FOXP3 protein binding for stable auto-regulation. Thus, **transcription factor-mediated local demethylation of an enhancer region creates epigenomic memory that stabilizes T cell lineage progression over multiple cell divisions.**

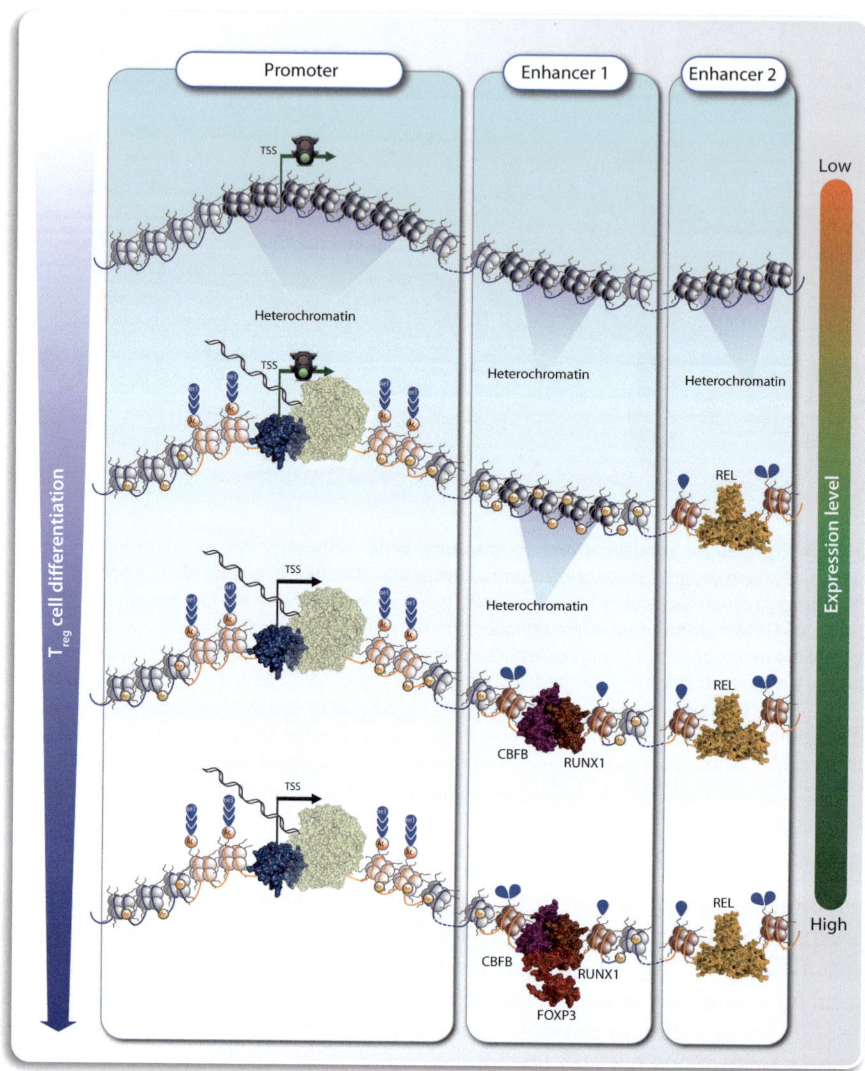

Fig. 13.4 Epigenetic memory of transcriptional activity. During differentiation of regulatory T cells, the *FOXP3* gene must show a stable and strong expression. A homodimer of the transcription factor REL binds to a downstream enhancer and stimulates *FOXP3* mRNA expression and demethylation of the promoter region. *FOXP3* gene expression is stabilized through the binding of the transcription factors CBFB and RUNX1 to a more proximal enhancer. The latter induces local demethylation and permits binding of the transcription factor FOXP3. Thus, in an auto-regulatory fashion FOXP3 ensures constitutive activity of the promoter of its own gene. The demethylation of enhancer and promoter regions represents epigenetic memory

The fact that alternative cell-specific enhancers and cell-specific transcription factors can regulate a given gene explains the differential expression of genes in cells of the immune system. For example, the *FANTOM5* project (Sect. 6.1) identified some 44,000 enhancers from their large collection of primary human tissues and cells types, i.e., there are approximately two enhancers per protein-coding gene. Thus, a given gene can be regulated in B cells via B cell-specific transcription factors to a B cell-specific enhancer, while in macrophages macrophage-specific transcription factors controls same gene via a macrophage-specific enhancer. Furthermore, the *BLUEPRINT* (www.blueprint-epigenome.eu) consortium demonstrated that cells of the innate immune system, such as monocytes/macrophages and NK cells, also have a memory function, referred to as trained immunity (Fig. 13.5). This rather short-term epigenetic memory monitors the close relationship between immune challenges and effects on chromatin. **Trained immunity is based on epigenetic changes, such as DNA methylation and histone modifications, as well as on the actions of miRNAs and long ncRNAs**. The rather long half-life of the latter molecules makes them well suited for a persistent programing of the epigenome.

Trained immunity enables innate immune cells to react with a quantitatively different response, i.e., a higher magnitude of gene expression when they are rechallenged with a pathogen (Fig. 13.6, bottom left). This response can in part also be qualitatively different, such as via the expression of an alternative PRR (pattern-recognition receptors) (Fig. 13.6, bottom left). A key mechanism in trained immunity is enhancer poising (Sect. 8.2), i.e., the addition of persistent histone marks, such as H3K4me1, enabling a strong response after restimulation (Fig. 13.6, bottom right).

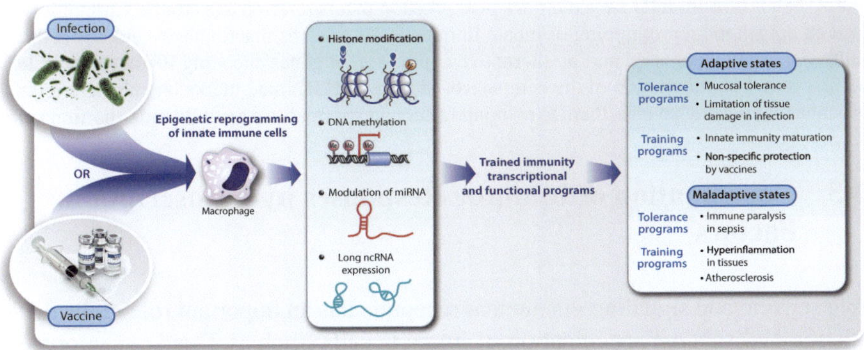

Fig. 13.5 Epigenetics of innate immune cells. The activation of the innate immune cells, such as monocytes, macrophages, or NK cells, leads to their epigenetic reprograming, known as trained immunity. This innate immune memory leads to adaptive states that protect the host during and after infections. In certain situations, however, trained immunity can result in mal-adaptive states, such as immune paralysis after sepsis or hyperinflammation

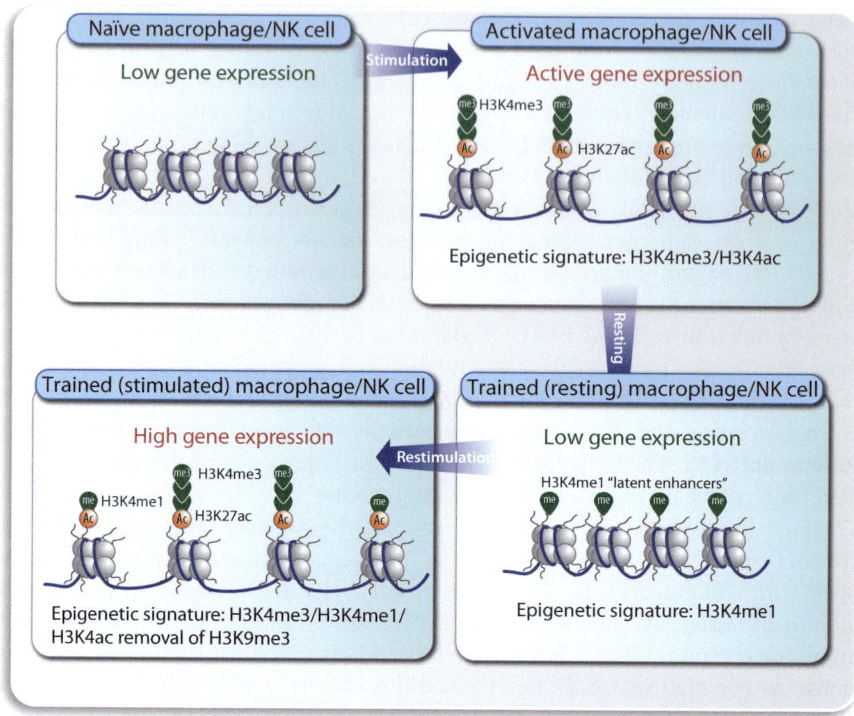

Fig. 13.6 Trained immunity. Enhanced inflammatory and antimicrobial properties of innate immune cells (**bottom left**) are a memory phenomenon that is referred to as trained immunity. It is based on epigenetic reprograming of innate immune cells, such as macrophages and NK cells and results, e.g., in the increased and/or alternative expression of genes encoding PRRs (**bottom left**). The first round of stimulation of the cells leaves persistent H3K4me1 marks on enhancer regions. This enhancer poising enables them to respond faster and stronger to a restimulation (bottom **right**)

13.3 Coordination of Immune Responses by Transcription Factors

Lipid sensing and signaling via nuclear receptors has an important role in the differentiation and subtype specification of immune cells, such as T cells, macrophages and DCs. Importantly, these cells are very mobile and are found in a wide range of subtypes nearly everywhere in our body. This also applies to metabolic tissues and disease scenarios, such as obesity. Thus, **macrophages and DCs as well as their precursors, monocytes, coordinate metabolic, inflammatory and general stress-response pathways via changes of their transcriptome profile and respective subtype specification**. Nuclear receptors, such as VDR, RAR, LXR and PPAR, have central functions in sensing these endogenous and exogenous stimuli as well as in adapting the respective gene expression profiles of immune cells. As a representative of all micro- and macronutrient-sensing nuclear receptors, in the following we will focus on VDR and its ligand $1,25(OH)_2D_3$.

Vitamin D_3 and its most abundant metabolite, 25-hydroxyvitamin D_3 ($25(OH)D_3$), either derive from diet, such as fatty fish, or from endogenous production of vitamin D_3 in response to UV-B exposure of the skin. Since vitamin D_3 and its metabolites can be stored in adipose tissue, there are rather seasonal than daily variations in the vitamin D status of our body. **Worldwide more than one billion people are vitamin D deficient**, i.e., their $25(OH)D_3$ serum levels are below 50 nM. Bone malformations, such as rickets and osteomalacia, are extreme examples of the effects of vitamin D deficiency. Since vitamin D is involved in a broad range of physiological processes, it can increase the risk for various diseases and the susceptibility for infectious diseases, such as tuberculosis and COVID-19 (coronavirus disease 2019). Moreover, living at higher latitudes, i.e., in geographic regions of significant seasonal variations of UV-B exposure, increases the risk of the autoimmune diseases T1D, multiple sclerosis (MS) and Crohn's disease.

VDR is expressed in all important cell types of the immune system, i.e., these cells are sensitive to changes in $25(OH)D_3$ serum levels. Importantly, macrophages and DCs express the enzyme CYP27B1 that converts $25(OH)D_3$ into the VDR ligand $1,25(OH)_2D_3$, i.e., in these cells vitamin D can act autocrine or paracrine (Fig. 13.7). Interestingly, while *CYP27B1* expression in the kidneys is negatively regulated by a number of signals, such as Ca^{2+}, parathyroid hormone, phosphate and $1,25(OH)_2D_3$, antigen-presenting cells do not respond to these inhibitory signals but rather further upregulate *CYP27B1* expression after stimulation with cytokines and TLR (Toll-like receptor) ligands.

TLRs and other PRRs detect pathogens on the surface of macrophages and initiate an immune response, e.g., against the intracellular bacterium *M. tuberculosis*. In macrophages, the TLR-triggered increased expression of VDR target genes encoding for antimicrobial peptides, such as *CAMP* (cathelicidin) and *DEFB4* (defensin, β4A), which efficiently kill intracellular *M. tuberculosis* (Fig. 13.7). This vitamin D-dependent antimicrobial mechanism explains, why sun or artificial UV-B exposure are efficient in the supportive treatment of tuberculosis. Vitamin D deficiency is associated with more aggressive tuberculosis and people with dark skin living distant from the equator show an increased susceptibility to tuberculosis infection.

Vitamin D-induced cytokine production of T cells and monocytes modulate *CAMP* expression. Thus, **the availability of the micronutrient vitamin D is essential for an appropriate response to infections**. Like many other nuclear receptors, VDR can antagonize, via transrepression of transcription factors, such as AP1 (Sect. 4.2), NFAT (Sect. 4.3) and NFκB (Sect. 13.4), the inflammatory response of immune cells. The resulting decreased expression of cytokines, such as IL2 and IL12, demonstrates the antiinflammatory potential of vitamin D metabolites.

Interestingly, VDR is a key transcription factor controlling the differentiation of myeloid progenitors into monocytes and macrophages. In contrast, in DCs vitamin D inhibits differentiation, maturation and immuno-stimulatory capacity via the repression of the genes encoding for the different variants of major histocompatibility

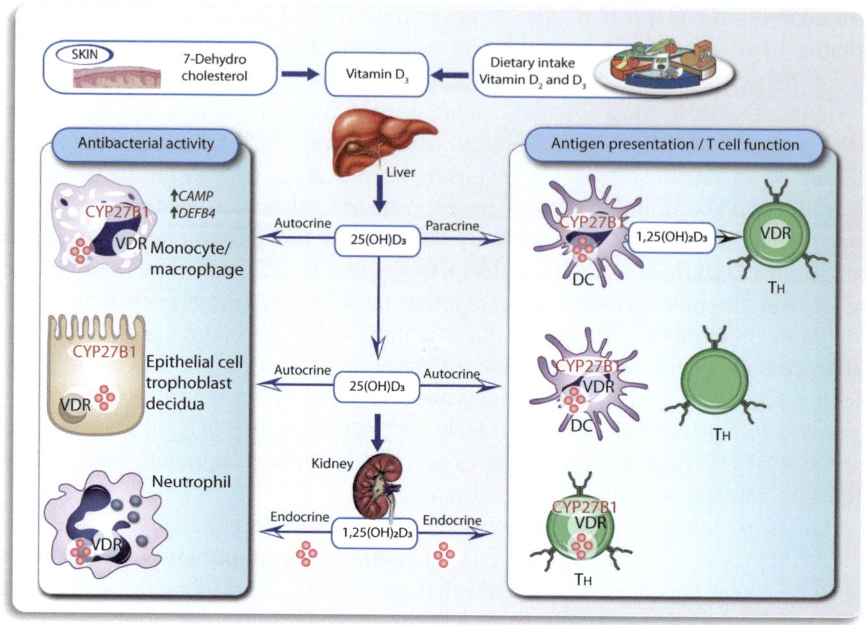

Fig. 13.7 **Innate and adaptive immune responses to vitamin D**. Macrophages and DCs express the vitamin D-activating enzyme CYP27B1 and VDR can then utilize 25(OH)D$_3$ for autocrine and paracrine responses via localized conversion to active 1,25(OH)$_2$D$_3$. In monocytes and macrophages, this promotes the response to infection via the antibacterial peptides CAMP and DEFB4. 1,25(OH)$_2$D$_3$ inhibits DC maturation and modulates T$_H$ cell function. Intracrine immune effects of 25(OH)D$_3$ may also occur in *CYP27B1/VDR*-expressing epithelial cells. In contrast, most other cells, such as T$_H$ cells and neutrophils, depend on the circulating levels of 1,25(OH)$_2$D$_3$ that are synthesized by the kidneys, i.e., they are endocrine targets of 1,25(OH)$_2$D$_3$

complex (MHC) and its costimulatory molecules CD (cluster of differentiation) 40, CD80, CD86 as well as the upregulation of inhibitory molecules, such as CCL22 (chemokine C–C motif ligand 22) and IL10. This tolerogenic (i.e., immune tolerance inducing) phenotype of DCs is associated with the induction of regulatory T (T$_{REG}$) cells (Fig. 13.7).

13.4   Inflammation, NFκB and Epigenetics

The five members of the NFκB family, REL (REL proto-oncogene, NFκB subunit) A (also called p65), RELB, REL, NFκB1 (also called p50) and NFκB2 (also called p52) are defined by the amino-terminal REL-homology domain that mediates DNA binding and homo- and heterodimerization (Fig. 13.8). The proteins p50 and p52 are obtained from their respective precursors p105 and p100. RELA, RELB and REL contain a carboxy-terminal transactivation domain. The dimeric NFκB complexes are retained in the cytoplasma by proteins called inhibitors of NFκB (IκBs). The three principal IκBs, IκBα, IκBβ and IκBγ, mask the conserved nuclear localization sequence (NLS) of the NFκB family members. For the activation and translocation of NFκB different types of IκB kinases (IKKs) phosphorylate IκBs that leads to their degradation. In contrast, p50 and p52 homodimers often evade regulation by IκBs. They are found constitutively in the nucleus and interact there with the IκB family member BCL3 (BCL3 transcription coactivator) that acts as a coactivator. NFκB target genes control numerous cellular processes, ranging from apoptosis, adhesion, proliferation, innate immune responses including inflammation, stress responses to tissue remodeling. However, in most cases the respective genes not only responsive to NFκB, but are also targets to a number of other transcription factors and signal transduction pathways. Thus, **the outcome of NFκB activation depends very much on the cellular context**.

The most frequently observed way of NFκB activation is the classical pathway that is induced in response to inflammatory stimuli, such as the cytokines TNF

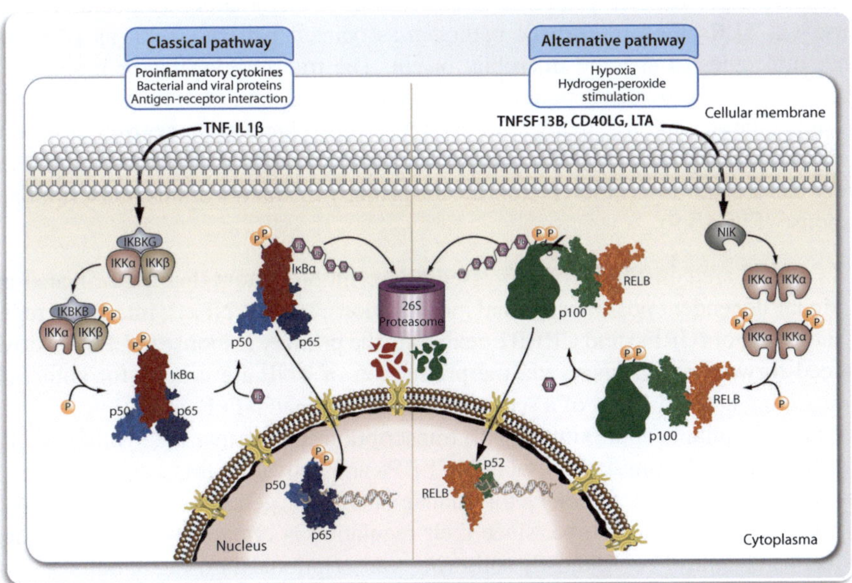

Fig. 13.8 **Pathways leading to the activation of NFκB**. The induction of the classical NFκB activating pathway by TNF, IL1β and many other immune-related stimuli is mostly mediated by IKK activation. This results in the phosphorylation (P) of IκBα, its ubiquitynation (Ub) and subsequent proteosomal degradation. Release of the NFκB complex allows the p50-p65 heterodimer to translocate to the nucleus. Genotoxic stress can cause IKK-dependent activation of NFκB. The alternative pathway represents the activation of IκBα by NIK, followed by phosphorylation of the p100 NFκB subunit by IKK1. This causes processing of p100 to p52 in the proteosome and leads to the activation of p52-RelB heterodimers targeting distinct genomic NFκB-binding sites. TNFSF13B = TNF superfamily member 13B

and IL1β, or exposure to bacteria-specific molecules, such as lipopolysaccharide (LPS) (Fig. 13.8, left). In this pathway, IκBα is rapidly phosphorylated, ubiquitinated and degraded at the proteasome. IκB phosphorylation is mediated by the activated IKK-complex that consists of the catalytic subunits IKK1 and 2 as well as IKBKG (inhibitor of NFκB kinase regulatory subunit gamma, also called NEMO). The key step in NFκB signaling is the activation of IKBKG. Interestingly, IKBKG often locates in the nucleus, where it "senses" via sumoylation and phosphorylation genotoxic stress and translocates then to the cytoplasm, where it activates NFκB. In contrast, some stimuli for NFκB activation, such as CD40LG (CD40 ligand) and LT (lymphotoxin) A, activate the alternative pathway (Fig. 13.8, right). This pathway is characterized by the activation of IKK1 by the NFκB inducing kinase (NIK) leading to the formation of p52 from p100. p52-RELB heterodimers have a higher affinity for some NFκB-binding sites and regulate a distinct subset of NFκB target genes. Once a dimeric NFκB complex is bound to its target sequences in the nucleus, the post-translational modification of its subunits, such as phosphorylation of RelA, defines its interaction with either coactivators or corepressors. This then leads to either target gene activation or repression.

Macrophages are the central mediators of the inflammatory response as they sense via TLRs the presence of pathogen-associated patterns, such as LPS and other molecules of specific microbial origin. The transcription factor program in response to LPS provides insight into the transcriptional control of inflammation. As a result of the activation of many different transcription factors, **the transcriptome of macrophages significantly changes within the first hours after LPS stimulation**. Three classes of transcription factors are the primary mediators of this transcriptional response (Fig. 13.9):

- In class I are constitutively expressed transcription factors that are activated by signal-dependent posttranslational modifications, such as NFκB, interferon regulatory factors (IRFs) and CREB1 mediating the primary response to LPS. Positive feed-forward mechanisms via the production of TNF are crucial for autocrine signaling and induction of a second wave of sustained NFκB activation.
- Class II contains approximately 50 transcription factors that are synthesized de novo after LPS stimulation, such as CEBPδ and activating transcription factor 3 (ATF3). These transcription factors induce subsequent gene expression waves over a prolonged period of time, since their regulation is often subjected to positive feedback control being mostly following transcriptional auto-regulation.
- The expression of class III transcription factors is induced during macrophage differentiation, such as SPI1, CEBPβ, RUNX1 and IRF8. Their combinatorial expression determines the detailed phenotype of the macrophages. The class III transcription factors activate constitutively expressed genes, remodel chromatin structure at genomic loci of inducible genes and silence genes that are critical for alternative cell stages.

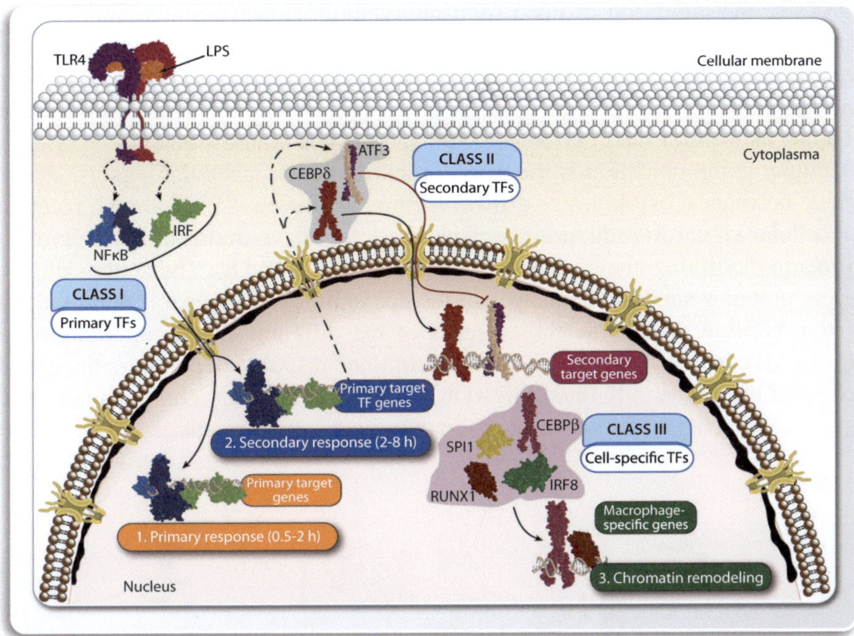

Fig. 13.9   Primary and secondary LPS-responding genes are regulated by three classes of transcription factors. Class I contains transcription factors that are activated directly by TLR signaling, such as NFκB and IRF proteins. Class II transcription factors, such as CEBPδ and ATF3, have class III transcription factors, such as SPI1, CEBPβ, RUNX1 and IRF8, as their targets. The latter category is not a direct target of LPS but induced during macrophage differentiation

The transcription factors of these three classes do not function independently, but act coordinately in the control of LPS-induced transcriptional response of macrophages. A transcriptional network that consists of transcription factors NFκB, the repressor ATF3 and the pioneer factor CEBPδ mediates the sustained expression of several inflammatory genes. This also illustrates how NFκB is able to play a major role in the specific regulation of inflammatory gene expression. Furthermore, the latter critically depends on cofactor proteins. For example, corepressor complexes contain HDACs and other proteins with activities for inhibiting gene expression. The stimulus-dependent dissociation of these proteins from regulatory genomic regions of inflammatory genes is known as de-repression. Furthermore, many nuclear receptors, such as GR, LXR, VDR and PPARs, have an antiinflammatory profile that is mediated largely via the inhibition of NFκB and AP1 activation. Most mechanisms of the repression of NFκB involve direct interactions between NFκB and nuclear receptor proteins. This leads to blocking of NFκB proteins, so that they do not activate their target genes (Fig. 13.10). The interaction of nuclear receptors with NFκB target genes can have also a number of other effects, such as the recruitment of

HDACs or the inhibition of Pol II phosphorylation. However, there are also indirect mechanisms, such as induction of *NFKBIA* gene expression and competition for coactivators, such as EP300 and CREBBP (Sect. 8.2). Interestingly, a number of these mechanisms are not specific for the interaction of NFκB with nuclear receptors, but apply as well for the interference with p53 or JUN kinase signaling.

Cellular communication is mediated primarily via extracellular signals, such as peptide hormones, cytokines or growth factors, activating via membrane receptors intracellular signal transduction cascades that often have transcription factors and chromatin modifying enzymes as end points (Chaps. 4 and 8). The actions of these nuclear proteins cause local changes of the epigenome, which enable and modulate the transcription of specific target genes of the different signals affecting a cell (Fig. 13.11, center). This epigenome/transcriptome response depends on the genetic predisposition of the individual, lifestyle decisions like dietary choices and amount of physical activity and epigenetic programming of immune cells.

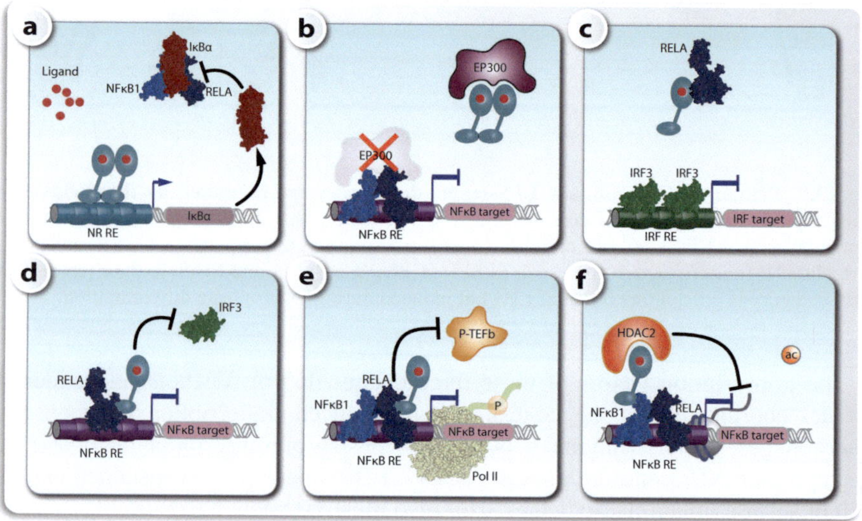

Fig. 13.10 Crosstalk between the NFκB and nuclear receptor signaling. Nuclear receptors repress the NFκB pathway via multiple mechanisms. Some pathways of nuclear receptor-mediated repression are indirect and involve either induction of *NFKBIA* expression (**a**) or competition for coactivator proteins, such as CREBBP and EP300 (**b**). However, most mechanisms involve the direct interaction of the nuclear receptor with NFκB and are referred to as trans-repression. Direct interaction with nuclear receptors can result in blocking of NFκB. This inhibits the IRF3-dependent regulatory region that uses the NFκB subunit RELA as a coactivator (**c**). Conversely, interaction of nuclear receptors with RELA prevents IRF3 from acting as a coactivator at some NFκB-regulated genomic regions (**d**). RELA-dependent recruitment of nuclear receptors to regulatory regions can lead to transcriptional repression by other mechanisms, such as inhibition of Pol II phosphorylation (P) by the positive transcription elongation factor (P-TEFb) (**e**) or recruitment of HDACs (**f**). NR = nuclear receptor

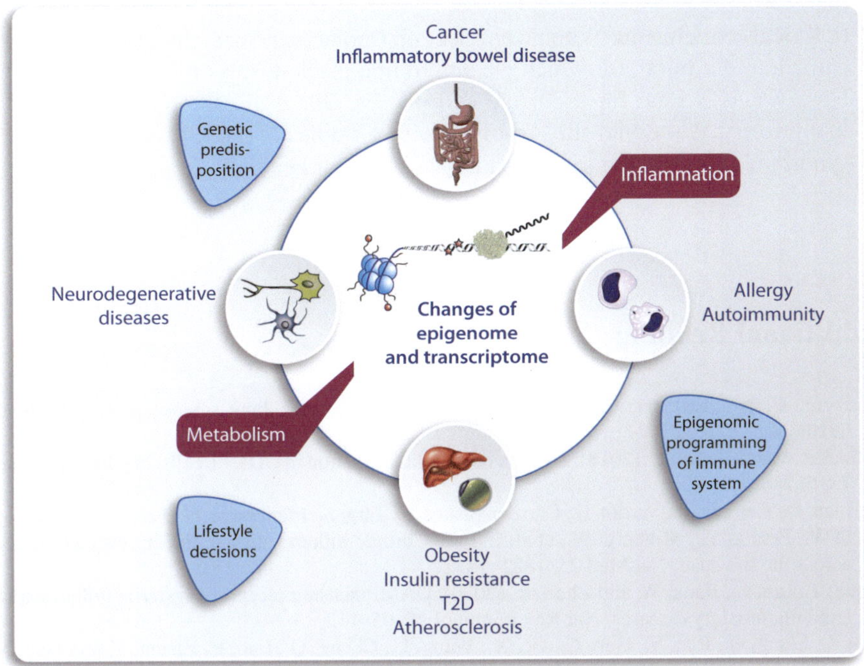

Fig. 13.11 Immune-mediated pathologies as key driver processes of diseases in various target organs. Inflammation and cellular metabolism are closely linked via coordinated changes in the epigenome and transcriptome of target tissues and cell types. More details are provided in the text

Immune reactions in general and inflammation in particular are related to cellular metabolism. The proliferation of immune cells as well as their action in defense and tissue repair both require high levels of energy metabolism. In turn, metabolic stress, which is often caused by lipid overload in the blood and in adipose tissue, stimulates **chronic inflammation**. This is the central causal mechanism behind many lifestyle-related diseases, such as obesity, insulin resistance, T2D and atherosclerosis, all of which are contributing to the metabolic syndrome (Sect. 16.3). Moreover, also neurodegenerative diseases, such as Alzheimer's disease, most types of cancer, allergy, autoimmune diseases and inflammatory bowel diseases, such as Crohn's disease and ulcerative colitis, are closely linked to inflammation (Fig. 13.11).

The immune system uses the same mechanisms in its fight against viruses as in the battle against thousands of transformed cancer cells arising every day in each human body. This is another indication of the key role of immunity in basically all aspects of health. Moreover, the surveillance of the immune system for cancer cells is only one of multiple environmental influences on the process of tumorigenesis. One can control many of these pro- and anticancer effects of the environment through **lifestyle decisions**, such as retaining from smoking, selecting healthy food and being physically active. Thus, **cancer preventive interventions are the most effective way to fight against cancer** (Chap. 15).

(Clinical) conclusion: When we reflect on (molecular) medicine, first disease come into our mind, for which we aim to find mechanisms of their treatment. However, it is time for a change of mindset, where disease prevention has first priority. We should shift from molecular studies of defects, disorders and syndrome to investigations on maintenance of our health.

Additional Reading

Carlberg, C. and Velleuer, E. (2022). Molecular immunology: how science works. Springer Textbook.

Ellmeier, W. and Seiser, C. (2018). Histone deacetylase function in CD4+ T cells. Nat Rev Immunol *18*, 617–634.

Furman, D., Campisi, J., Verdin, E., Carrera-Bastos, P., Targ, S., Franceschi, C., Ferrucci, L., Gilroy, D.W., Fasano, A., Miller, G.W., et al. (2019). Chronic inflammation in the etiology of disease across the life span. Nat Med *25*, 1822–1832.

Gong, T., Liu, L., Jiang, W. and Zhou, R. (2020). DAMP-sensing receptors in sterile inflammation and inflammatory diseases. Nat Rev Immunol *20*, 95–112.

Liu, Z., Liang, Q., Ren, Y., Guo, C., Ge, X., Wang, L., Cheng, Q., Luo, P., Zhang, Y. and Han, X. (2023). Immunosenescence: molecular mechanisms and diseases. Signal Transduct Target Ther *8*, 200.

Netea, M.G., Joosten, L.A., Latz, E., Mills, K.H., Natoli, G., Stunnenberg, H.G., O'Neill, L.A. and Xavier, R.J. (2016). Trained immunity: a program of innate immune memory in health and disease. Science *352*, aaf1098.

Park, M.D., Silvin, A., Ginhoux, F. and Merad, M. (2022). Macrophages in health and disease. Cell *185*, 4259–4279.

Wilfahrt, D. and Delgoffe, G.M. (2024). Metabolic waypoints during T cell differentiation. Nat Immunol *25*, 206–217.

Zhu, X., Chen, Z., Shen, W., Huang, G., Sedivy, J.M., Wang, H. and Ju, Z. (2021). Inflammation, epigenetics and metabolism converge to cell senescence and ageing: the regulation and intervention. Signal Transduct Target Ther *6*, 245.

Chapter 14
Epigenetics and Disease

Abstract In this chapter, we will emphasize the main message of this book: **diseases are largely preventable by an appropriate lifestyle**. We will elaborate the relation of phenotypes, genetic predisposition, epigenetic programming and lifestyle decisions of dietary choices and physical activity. A central mechanism linking most common diseases is chronic inflammation that is often caused by metabolic stress.

Keywords Genomics · Epigenomics · Prenatal epigenetic programming · T2D · Obesity · Metabolic syndrome · Healthy aging · Inflammation · Metabolism · Lifestyle · Cancer

14.1 Genetics, Epigenetics and Environment

Specific expression of genes is shaping the phenotype of cells and tissues (Chap. 2). The regulation of gene expression, i.e., their up- and downregulation, is the fundamental aspect of nearly all processes in physiology, both in health and in disease. These processes are in part very dynamic and respond to divergent daily challenges, such as incoming diet, altered cells or infections. The tremendous technological push by NGS methods, which are all based on the knowledge of the reference human genome (Sect. 6.2), allow us to have an unbiased view on changes in gene expression in the context of every biological process that we are interested in. Thus, **epigenomics can be considered as the "second dimension" of genomics**, as it is the global, comprehensive view of processes that modulate gene expression patterns in a cell independent from the genome sequence. These patterns are primarily DNA methylation states (Sect. 7.2) and covalent modification of histone proteins (Sect. 8.2) that organize the architecture of the nucleus, restrict or facilitate transcription factor access to genomic DNA and preserve an epigenetic memory of past gene regulatory activities. Epigenetics plays an extremely important role in the first weeks of life (Sect. 11.1) and **intrauterine exposures of the fetus, such as under- and over-nutrition, have huge impact on the risk for obesity, T2D and the metabolic syndrome** (Fig. 14.1).

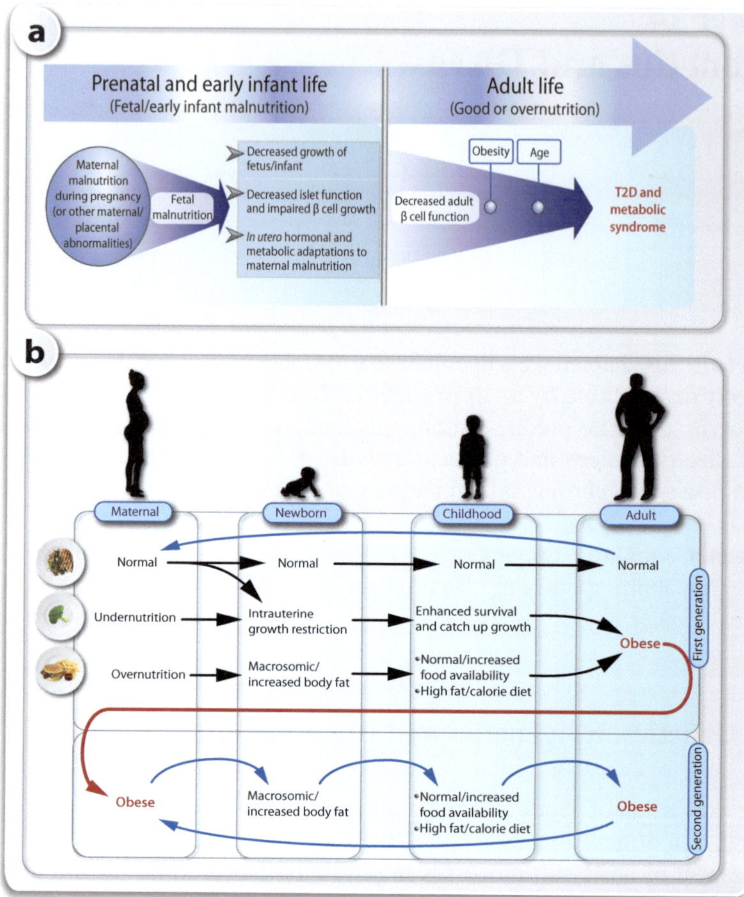

Fig. 14.1 Transgenerational view on T2D and obesity. T2D and the metabolic syndrome can be the outcome of maternal–fetal malnutrition, such as poor maternal diet, low maternal fat stores or reduced transfer of nutrients because of placental abnormalities (**a**). The fetus adapts to this environment by being nutritionally thrifty, resulting in decreased fetal growth, islet function and β cell mass and other hormonal and metabolic adaptations. A transition to overnutrition later in adult life exposes the impaired islet function to increased metabolic stress, which is further enhanced by obesity and age, so that T2D results. Non-obese mothers usually give birth to non-obese children, which develop into adults with a normal metabolic profile and a normal body fat content (**b**). However, undernutrition combined with improved neonatal survival, formula feeding and exposure to a Western postnatal diet increases the incidence of prematurity and intrauterine growth restriction. This results in increased obesity of the offspring and higher risk for developing the metabolic syndrome, when prenatally exposed to Western diet. Some obese mothers may give birth to newborns with increased body fat, as a result of consumption of a high-fat diet. All these processes contribute to a shift of the population toward an obese phenotype. This also includes that second-generation obese women have an increased risk to give birth to infants with increased body fat content and a further increased risk to develop obesity and the metabolic syndrome

Epigenetic programming happens not only during the prenatal phase, but occurs also in the postnatal and adult phases of life. For example, T2D patients maintaining intense glucose control remain at increased risk of macrovascular complications and organ damage even years after the initial diagnosis. This "glycemic memory" is an epigenetic effect, i.e., histone methylation changes within human aortic endothelial cells in response to increased glucose exposure. Individuals that carry an epigenome, which during their anthropologic development was programed by suboptimal nutrition in utero, despite normal postnatal nutrition, transgenerationally transmit a predisposition for obesity (Fig. 14.1).

It is not clear how many generations epigenetic marks can be inherited, but it is obvious that epigenetic inheritance is not comparably persistent as genetic inheritance (Sect. 12.1). Within more or less one generation (between 1981 and 2014) the worldwide prevalence for both obesity and T2D doubled. Thus, **populations that made a transition from famine to food surplus just within 1–2 generations are under significantly higher risk for obesity, T2D and the metabolic syndrome**, than those that were improving their nutritional conditions over many generations. For example, Polynesians show an overproportionally high prevalence for T2D. One possible explanation is that their ancestors experienced a number of challenges, such as cold stress and starvation. This happened a few hundred years ago during long open-ocean travels over the Pacific, which may have let only the initially most obese members of the group survive. These ancestors may have been evolutionary selected for energetic efficiency, referred to as thrifty metabolism. Accordingly, present-day Polynesians have inherited an increased T2D susceptibility, because their ancestors went through an evolutionary bottleneck. In contrast, populations, which did not experience long periods of starvation in the past centuries, such as the Europeans, have a lower prevalence for T2D. Similar explanation may apply to other indigenous populations, such as the Pima tribe of Native Americans in Arizona who had adapted to life of deprivation in the desert. When exposed to Western diet both Polynesians and Pima Native Americans get far more likely obese than Europeans. Thus, **populations who are born and live in countries that had particularly rapid changes in urbanization and economic development have an increased risk of the different features of the metabolic syndrome**.

Probably the most dominant external signal that we are exposed to is our diet (Sect. 16.1). Many dietary compounds, such as unsaturated fatty acids, directly activate transcription factors or cause changes in the levels of metabolites, such as α-ketoglutarate, which modulate the activity of chromatin modifying enzymes. The actions of these nuclear proteins cause local changes of the epigenome, which enable and modulate the transcription of specific target genes of the different signals affecting a cell. Most of these changes are transient but some may leave permanent marks on the epigenome. In this way, **our epigenome can memorize environmental events**, such what and how much we have eaten, whether we had been in contact with microbes or whether we had been stressed in any other way. The dynamic component of epigenome implies that **some epigenetic programing events are reversible**. For example, if an unhealthy lifestyle paired with food excess and physical inactivity over many years causes epigenetic programing of metabolic tissues that results in insulin resistance, this process may be reversed by significant lifestyle changes leading to a **reprograming of the dynamic component of the epigenome**. This reprograming of our epigenome can have significant consequences for our health, i.e., as long no irreversible tissue damage has happened, we may have it in our own hands to reverse a disease condition. Thus, not only our mental memory, such as memories of our childhood, is a learning process that is based on epigenetic programing of neurons (Sect. 11.5), but each cell of our body has an epigenetic memory recording the perturbations that the cell had been exposed to.

Figure 14.2 provides a summary of genetic and epigenetic effects throughout the life from "womp to tomp". The genetic composition of all cells of our body is determined at the moment of conception and will stay like this, in case we do not get cancer. Thus, we cannot change the genetic contribution to our life, but fortunately, in very most cases **the contribution of the human genome to the risk for diseases is far lower than expected**. Although GWAS and whole genome sequencing data have indicated for basically all common diseases a clear genetic predisposition, **all identified risk SNPs explain in total less than 20% of the genetic risk for common diseases** and other traits describing the functions and properties of our body. Some of the missing heritability may be explained by rare variants with high ORs (Sect. 1.2). However, most of our individual disease risk is modulated by environmental triggers, which are majorly influenced by lifestyle decisions. Thus, environmental exposures, including those that we are experiencing as fetus, affect the epigenome and may explain large parts of the missing heritability (Sect. 12.2). This insight can be summarized in the formula: phenotype = genetic + epigenetic + environment (Fig. 14.2). This conclusion implies that **we have it to a large extent in our own hands to stay healthy and to avoid getting major common diseases**.

14.2   DNA Methylation and Disease

The approximately 100 imprinted genes in human have important roles during development. Accordingly, changes in their expression and function can lead to imprinting disorders. For example, the Silver-Russell syndrome, which is a disease leading to undergrowth and asymmetry, and the Beckwith-Wiedemann syndrome, which is a disorder leading to overgrowth, are based on epigenetic errors in the 11p15 locus (Fig. 7.7). Thus, both imprinting disorders have opposite growth phenotypes, where affected children may be exposed to abnormally high (Beckwith-Wiedemann syndrome) or abnormally low (Silver-Russell syndrome) levels of placental hormones. Importantly, individuals with the Beckwith-Wiedemann syndrome have a 1000-times increased chance of getting kidney tumors (mostly Wilms' tumors) and embryonal tumors that arise from fetal cells and persist after birth, i.e., **epigenetic changes precede and increase the risk of cancer rather than arise after tumor formation**. Most of the patients with Beckwith-Wiedemann syndrome lost the methylation at ICR2, resulting in the expression of the *KCNQ1OT1*

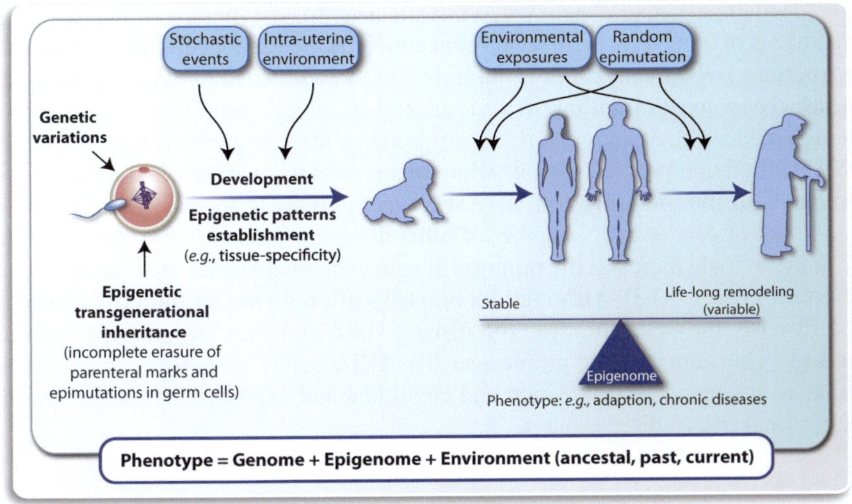

Fig. 14.2 The molecular story of life: genome, epigenome and environment. Details are provided in the text

ncRNA on both alleles (biallelic) and aberrant repression of *CDKN1C*, i.e., in reduced cell cycle repression. Other Beckwith-Wiedemann syndrome patients show overexpression of *IGF2* caused by deletions in ICR1 on the maternal allele and disrupted CTCF binding leading to biallelic *IGF2* expression and loss of *H19* expression. In contrast, many individuals with Silver-Russell syndrome have an opposite epigenetic phenotype, where ICR1 is unmethylated, resulting in biallelic *H19* expression and loss of *IGF2* expression. Other imprinting disorder are based on the control of the imprinted genes MAGEL2 (MAGE family member L2) and UBE3A (ubiquitin protein ligase E3A), such as the neurodevelopmental disorders Prader-Willi and Angelman syndromes. The disorders are caused by loss of paternal gene expression (Prader-Willi syndrome) and maternal gene expression (Angelman syndrome). Thus, **imprinting disorders are clearly epigenetic diseases**.

While imprinting disorders are very rare, perturbations in the DNA methylome are observed in the most cases of cancer (Sect. 15.3). Well-known examples are the hypermethylation of the CpG-rich promoters of tumor suppressor genes, such as *TP53* and *RB1* (RB transcriptional corepressor 1) leading to their transcriptional silencing. Thus, DNA methylation leads to the silencing of genes that are essential for the inhibition of tumorigenesis progression. Moreover, in about 25% of adult cases of acute myeloid leukemia (AML) the *DNMT3A* gene is mutated, which disturbs the methylome and makes the regulatory landscape of pre-leukemic blood stem cells more vulnerable for additional mutations. Similarly, mutations in the genes of other chromatin modifying enzymes can enhance the phenotype of cancer-driven mutations in transcription factor genes or their genomic binding sites. Therefore, **epigenetic alterations belong to the hallmarks of cancer**. Interestingly, the first approved epigenetic drug, decitabine (5-aza,2′-deoxy-cytidine), is used for the therapy of leukemia and other forms of blood cancer, in which hematopoietic progenitor cells do not maturate.

Aberrant DNA methylation is not only a well-established marker of cancer and disturbed genomic imprinting but also can lead to general instabilities of the genome through reduced heterochromatin formation on repetitive sequences (Fig. 7.4a, right). DNA methylation profiles, e.g., of white blood cells that can be obtained from test persons with minimal invasion, may serve as biomarkers for evaluating the individual risk of cancer and a number of other diseases, such as T2D. Moreover, the DNA methylome indicates the progress of aging showing significant interindividual differences (Sect. 12.3). **Although biomarkers often do not explain the causality of a disease, they can monitor the disease state and may suggest appropriate therapy**. Thus, epigenomic profiles, such as DNA methylation patterns, in combination with genetic predisposition and environmental exposure may be prognostic for personal risk of disease onset.

14.3   Epigenetic Basis of Neurological Diseases

Chromatin modifying enzymes play central roles in cognitive disorders (Table 14.1). At least seven proteins are known to be mutated in X chromosome-linked intellectual disabilities. These proteins, such as MECP2, are either methyl-binding proteins or methyl modifying enzymes. During brain development neurons acquire epigenetic marks, such as 5mC (at CG and CH) and 5hmC, that stabilize the neuronal phenotype over our lifespan. **Although neurodevelopmental disorders are generally thought to be irreversible, the plastic nature of the dynamic part of the neural epigenome via 5hmC and demethylation has major implications for a possible epigenetic reprograming in the context of these diseases.** In fact, some neuroleptics are pan-HDAC inhibitors like valproic acid.

The disruption of the *MECP2* gene leads to a special form of autism, referred to as Rett syndrome. More than 95% of all cases of the Rett syndrome can be explained by mutations within the *MECP2* gene, i.e., this special form of autism is largely a monogenic disease. This means that **the mechanistic understanding of the Rett syndrome is based on dysfunctional MECP2 proteins in neurons**. In addition, also functional alternations of MECP2 in astrocytes and in microglia affect the disease phenotype, although MECP2 is much lower expressed in these cell types. Individuals with Rett syndrome are heterozygous for a mutation in the *MECP2* allele. Since the *MECP2* gene is located on the X chromosome, the disease is mostly embryonal lethal for males and almost exclusively females are affected with an incidence of approximately 1 in 10,000 persons. The syndrome mostly occurs through de novo mutations in the paternal germline. Affected females have an apparent normal early postnatal development, which may be due to the fact that during brain maturation mCH levels gradually increase (Sect. 11.4). However, between the age of 6–18 months the syndrome develops, such as that the individuals lose the ability to retain communication function and motor skills. Moreover, physical growth becomes retarded and

Table 14.1 Epigenetic mechanisms of neurological disorders

Function or disorder	Mechanism(s) implicated
Rett syndrome	*MECP2* mutations
Autism spectrum disorders	MECP2 overexpression/increased dosage, aberrant DNA methylation
Alzheimer's disease	MECP2 decrease, histone modifications, aberrant DNA methylation
Parkinson's disease	Loss of MECP2
Huntington's disease	MECP2 dysregulation
Fragile X syndrome	Aberrant DNA methylation
Rubinstein-Taybi syndrome	HAT deficiency
Friedreich's ataxia	Reduced histone acetylation
Angelman syndrome	Genomic imprinting (DNA methylation)
Addiction and reward behavior	MECP2 decrease, histone modifications, DNA methylation, miRNAs
Posttraumatic stress disorder	Histone modifications, DNA methylation
Depression and/or suicide	DNA methylation
Schizophrenia	Increased MECP2 binding, histone methylation, aberrant DNA methylation
Epilepsy	MECP2 upregulation, histone modifications, aberrant DNA methylation

Neuronal disorders and functions that are affected by epigenetics are listed, such as mutations or overexpression of MECP2, aberrant DNA methylation and/or histone modifications

microcephaly establishes. Thus, Rett syndrome is the **first human neuronal disease, for which a significant impact of epigenetics could be demonstrated** and it serves as a key example for the impact of neuroepigenetics on disease.

The example of the Rett syndrome leads to the question, whether epigenetics plays also a role in complex multigenic disorders of the nervous system, such as neurodegenerative diseases. MECP2 is involved in controlling the secretion of the neurotransmitters GABA (gamma-aminobutyric acid), dopamine and serotonin from respective neurons and affects the number of synapses of glutamatergic neurons. This is, at least in part, mediated by the interaction of MECP2 with the growth factor BDNF, which acts as a modulator on glutamatergic and GABAergic synapses. Thus, **the tightly regulated expression of MECP2 is critical for neuronal homeostasis**.

Since both too high as well as too low MECP2 protein levels trigger opposite effects in synaptic transmission, dysregulation of MECP2 protein expression contributes to many neuro-pathological disorders, such as Alzheimer's disease, Huntington's diseases, schizophrenia and epilepsy (Table 14.1). In parallel, aberrant DNA methylation has been observed in some autistic spectrum disorders, Alzheimer's disease, epilepsy and schizophrenia. In general, **epigenetic mechanisms may be particularly relevant to complex diseases with low genetic penetrance that use epigenetic mechanisms of development and learned behavior**, such as drug addiction, posttraumatic stress disorder, epilepsy and schizophrenia.

Histone acetylation is the best-understood epigenetic modification. Several neurodegenerative diseases involve disruptions in the HAT/HDAC balance, i.e., patients with these disorders have abnormal histone acetylation levels (Fig. 14.3). A key example is the Rubinstein-Taybi syndrome, which is characterized by short stature, mental retardation, moderate to severe learning difficulties, distinctive facial features as well as broad thumbs and big toes. The Rubinstein-Taybi syndrome is a monogenic disease that is based on mutations of the *CREBBP* gene, which encodes for a HAT. The resulting low histone acetylation levels may be counterbalanced by the inhibition of HDACs, i.e., HDAC inhibitors are a therapeutic option for patients with the syndrome (Sect. 14.4).

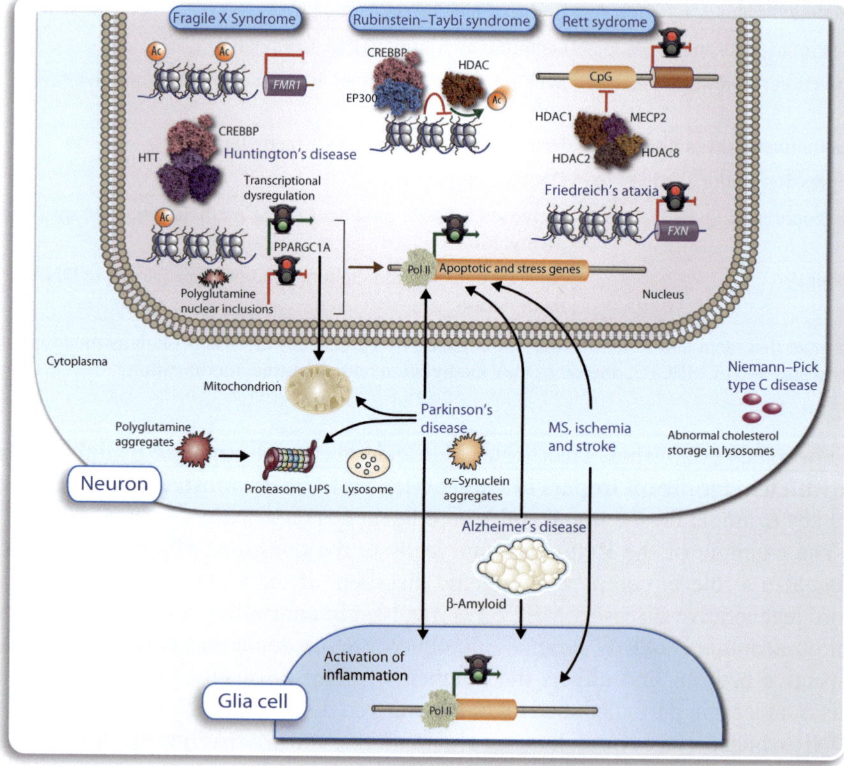

Fig. 14.3 Role of histone acetylation in neurodegenerative diseases. The level of histone acetylation depends on the balance of HATs and HDACs and is related to several neurodegenerative diseases. For example, decreased acetylation activity of the HAT CREBBP is associated with the Rubinstein-Taybi syndrome and polyglutamine diseases, such as Huntington's disease. The transcriptional silencing of the *FXN* gene in Friedreich's ataxia or of the *FMR1* gene in fragile X syndrome can be released by HDAC inhibition. Other examples of neuronal disorders are schematically depicted and their possible treatment by HDAC inhibitors is indicated

Another example of a monogenic neurodegenerative disease is Friedreich's ataxia, which results from the degeneration of nervous tissue in the spinal cord, in particular in sensory neurons that are essential for directing muscle movement of arms and legs. In this disorder the expansion of a triplet repeat region within an intron of the gene *FXN* (frataxin) leads to the loss of H3ac and H4ac marks and gain of H3K9me3 marks, heterochromatin formation and finally transcriptional silencing of the gene (Fig. 14.3). In a mouse model of this disease, HDAC inhibitors increased H3 and H4 acetylation and corrected the *FXN* expression deficiency.

Similarly, mental retardation associated with the fragile X syndrome is caused by a CGG-triplet repeat expansion in the 5′-UTR of the gene *FMR1* (fragile X mental retardation 1), which leads to extensive DNA methylation at CpGs close to the gene's TSS and gene silencing (Fig. 14.3). Also in this case, a treatment with HDAC inhibitors resulted in reactivation of *FMR1* expression. In particular SIRT1 inhibitors were able to increase acetylation and decrease methylation of histones at the *FMR1* gene locus.

Finally, Huntington's disease is based on polyglutamine repeats in the 5′-coding region of the gene *HTT* (huntingtin) causing progressive motor and cognitive decline. The disease involves perturbations in many aspects of neuronal homeostasis, such as abnormal histone acetylation and chromatin remodeling as well as aberrant interactions of the HTT protein, e.g., affecting the function of the gene *PPARGC1A* (Fig. 14.3). In different animal models of Huntington's disease HDAC inhibitors showed a neuroprotective effect.

Neurodegenerative diseases have different underlying causes and pathophysiology, but they all involve impaired cognition, neuronal death and the dysregulation of the transcription factor REST (RE1-silencing transcription factor). REST is a repressing transcription factor that has an effect on both acetylation and methylation levels of histones. It is widely expressed throughout embryogenesis, but at the end of neuronal differentiation it gets downregulated in order to acquire the neuronal phenotype. The transcription factor serves as the DNA-binding platform of a large protein complex containing the corepressors RCOR1, HDAC1 and 2, MECP2, the KMT EHMT2 and the KDM LSD1 (lysine specific demethylase 1, also KDM1A). In addition, the REST complex can also recruit the DNA methylation machinery. This leads to deacetylation and methylation of local nucleosomes at position H3K9 and demethylation at H3K4me2 marks. Thus, the local chromatin at TSS regions of REST target genes gets very effectively silenced. In total there may be up to 2000 primary REST target genes within the human genome. However, the set of active REST targets is cell type- and context-dependent and varies with developmental and disease stage. Probably, different epigenetic landscapes in various brain regions and disease states largely influence the selection of REST for a subset of its target genes.

Since most cases of Alzheimer's disease are sporadic and develop over time, there is a significant contribution of environmental factors to the onset of this neurodegenerative disorder. While in neurons of the prefrontal cortex and the hippocampus of the healthy aging brain REST silences genes involved in apoptosis and oxidative stress, this is lost in patients with mild or severe cognitive impairment and Alzheimer's disease. With the occurrence of misfolded proteins that are characteristic for the onset of Alzheimer's disease, such as Aβ and tau, the rate of autophagy increases. Under these conditions autophagosomes engulf not only Aβ and tau complexes but also REST, which results in depletion of the transcription factor from the nucleus. The loss of REST results in an increase in expression of previously repressed genes being involved in oxidative stress and neuronal death. Thus, **REST deprivation increases the loss of neurons in the respective brain regions and promotes the progression of Alzheimer's disease**.

Taken together, neuroepigenetics provides alternative and/or additional mechanistic explanations for the onset of neurodevelopmental and neurodegenerative diseases. Thus, **neuroepigenetics has the potential to guide the design of novel therapeutic strategies for improving the harmful consequence of cognitive deficits and neurodegeneration**.

14.4   Epigenetic Therapy

Most epigenomic modifications are reversible, which implies a significant therapeutic potential. Accordingly, **epigenomics is one of the most innovative research areas in modern biology and biomedicine, where the molecular hallmarks of epigenetic control can be used as targets for medical interventions and treatments**. The promises of epidrugs increased the number of compounds that are in preclinical or clinical trials. Basically, all molecules designed for epigenetic cancer therapy are inhibitors of enzymes affected by gain-of-function mutations. In addition to writers and erasers, now also chromatin modifying enzymes of the reader class, such as chromatin remodeling proteins with bromodomain or methyl-binding proteins, are addressed. Oncology is currently the main focus of clinical epigenetics, but the large number of clinical trials or preclinical studies in other medical areas indicates that **clinical epigenetics expands far beyond cancer** (Fig. 14.4).

During aging, histone acetylation and methylation of many genomic regions changes, most likely because SIRTs can promote gene silencing and longevity. Similarly, the epigenome of cancer cells is also reprogramed during the transformation process from normal cells. The mapping of active and repressed chromatin regions in cancer cells allows more accurate prognosis and even may facilitate therapy. For example, HDAC inhibitors reactivate the transcription of tumor suppressor genes, such as *CDKN1A*, by increasing histone acetylation, but they also mediate the acetylation of non-histone proteins, such as p53 and stabilize their activity. In this way, they have a wide impact on cancer cells and can induce apoptosis, cell cycle arrest and many other anticancer actions. Currently, the HDAC inhibitors vorinostat, romidepsin, belinostat, panobinosta and valproic acid are approved for treatment of different types cancer, such as leukemia, T cell lymphomas and multiple myeloma. Valproic acid, which primarily used for the treatment of epilepsy, is thereby redefined for the treatment of childhood cancers. As many other HDAC inhibitors are currently under testing in clinical trials, it is expected that more HDAC inhibitors will be developed for the treatment of solid tumors, in particular for brain cancers.

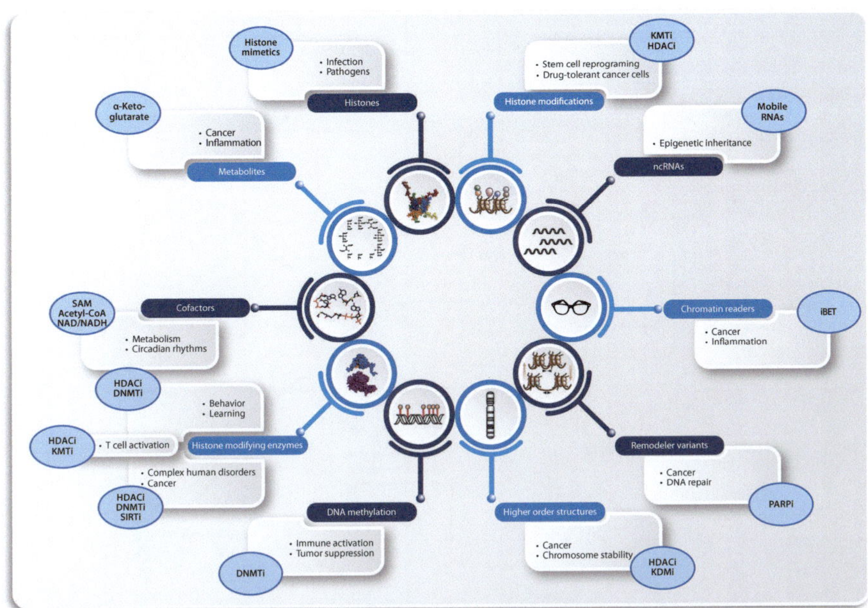

Fig. 14.4 Prognostic and therapeutic potential of epigenomics. Examples for the impact of epigenomics in normal development and in disease are indicated. The circles each represent key epigenomic mechanisms and the mainly associated nuclear proteins. Dysregulated epigenomics may be reversed by pharmacological intervention with small molecule inhibitors, such as HDAC inhibitors (HDACi), DNMT inhibitors (DNMTi), SIRT inhibitors (SIRTi), metabolic cofactors, such as SAM and α-ketoglutarate, KMT inhibitors (KMTi), BET inhibitors (iBET), poly(ADP-ribose) polymerase inhibitors (PARPi) and KDM inhibitors (KDMi)

In addition, numerous psychiatric disorders, such as anxiety and depression, can be treated with HDAC inhibitors. Interestingly, these small molecule inhibitors also enhance the efficacy of immunotherapeutic agents, such as a blockage of the interaction between the surface inhibitory receptor PDCD1 (programmed cell death 1, also called PD1) on cytotoxic T cells and CD274 (CD274 molecule, also called PDL1) on cancer cells (Fig. 14.5). At present, immunotherapy is the most promising therapy of cancer, since it takes advantage of the general immune surveillance function of cytotoxic T cells. In these cells a PDCD1-induced signal transduction cascade would inhibit their activation, which is prevented via blocking of PDCD1. This immune checkpoint blockage can boost the antitumor immune response of the host, when the cytotoxic T cells start again their growth and effector function.

DNMT inhibitors are either nucleoside analogs that after incorporation into the DNA covalently trap DNMTs or non-nucleoside analogs that directly bind to the catalytic region of DNMTs. These molecules prevent DNA methylation leading to reduced promoter hypermethylation and re-expression of silenced tumor suppressor genes. The two nucleoside analogs azacitidine and decitabine have already been approved more than 30 years ago. Both compounds induce the expression of genes

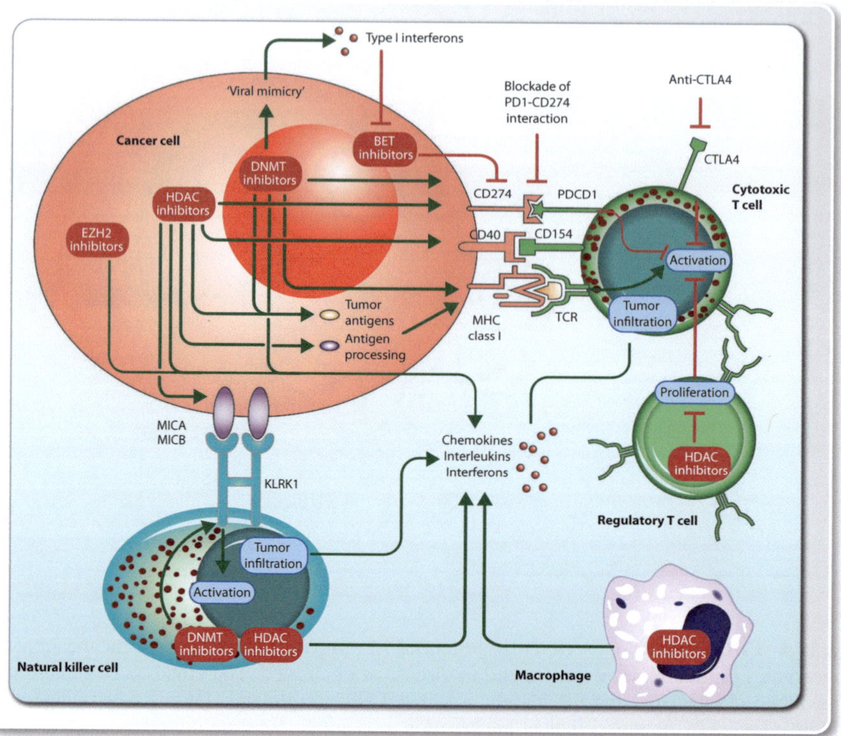

Fig. 14.5   Inhibitors of chromatin modifying enzymes in immunotherapy. Epigenetic inhibitors also may play an important role in immuno-oncology. HDAC inhibitors also modulate the expression of MHC proteins, costimulatory CD40 molecules and tumor antigens. Moreover, they affect the antigen-processing machinery and change in chemokine expression in both cancer and immune cells. Furthermore, they suppress T_H cells and induce of the NK cell receptor ligands MICA and MICB. Inhibitors of the PRC2 component EZH2 increase the expression of chemokines CXCL9 and CXCL10 attracting T cells and improving tumor clearance. BET = bromodomain and extraterminal, CXCL = chemokine C-X-C motif ligand, CTLA4 = cytotoxic T-lymphocyte associated protein 4, KLRK1 = killer cell lectin like receptor K1, MIC = MHC class I polypeptide-related sequence

encoding for MHCs or tumor antigens. This increases the visibility of the cancer cell to cytotoxic T cells and their subsequent elimination. In addition, decitabine increases the sensitivity of cancer cells to growth inhibition by type 1 INFs. This results in a "viral mimicry", in which DNA demethylation activates the transcription of endogenous retroviral elements in cancer cells leading to a double-stranded RNA-mediated immune response. The combination with an epigenetic mediator, such as chromatin modifying enzyme inhibitors, gives immune checkpoint blockage of T cells a wider approach for treating both cancer and chronic infections. For example, Hodgkin's lymphoma patients, who received a DNMT inhibitor before a treatment with immune checkpoint inhibitors showed a higher rate of complete remission in their cancer.

The contribution of altered epigenetic states to the phenotype of cancer cells suggests that **epigenetic cancer therapies** have a clinical impact. For example, genes encoding for epigenetic modifiers, such as KMTs and KDMs (Fig. 14.6), are frequent drivers in a larger range of cancer types. Since histone methylation marks have a far more selective function than histone acetylation marks, **KMT and KDM inhibitors promise to be more specific and may be less toxic than HDAC inhibitors or even DNMT inhibitors**. Several KMT inhibitors have been developed and the H3K27-KMT EZH2 inhibitor tazemetostat has already been approved for the treatment of epithelioid sarcoma and follicular lymphoma. Since EZH2 is the catalytic core of the PRC2 complex that also recruits DNMTs, EZH2 inhibitors may link both epigenetic repression mechanisms. Moreover, the inhibition of EZH2 results in reduced levels of H3K27me3 marks, upregulation of silenced genes and inhibition of the growth of cancer cells with EZH2 gain-of-function mutations or overexpression. KDMs use FAD, α-ketoglutarate or Fe(II) as cofactors and offer in this way a number of options for their inhibition. However, the catalytic domain of most KDMs is structurally highly conserved, which is a challenge for the design of specific KDM inhibitors.

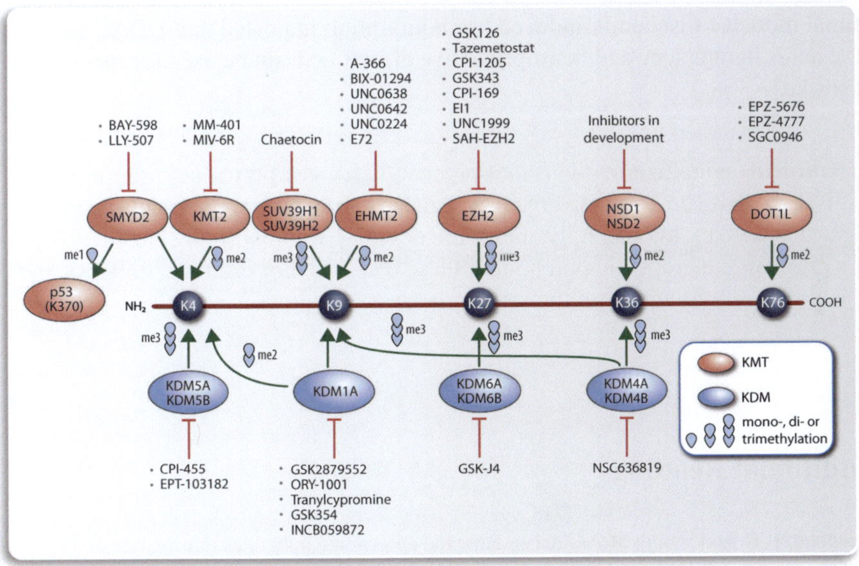

Fig. 14.6 Targeting KMT and KDM mutations. Inhibitors of KMTs (red) and KDMs (blue) are indicated that affect histone 3 lysines K4, K9, K27, K36 and K79 (dark blue). KMT inhibitors bind either within the SAM pocket, within the substrate pocket or at allosteric sites of the KMT protein. Respective inhibitors of DOT1L and EZH2 are already in clinical trials. Chaetocin is a non-selective inhibitor of KMTs, such as SUV39H (suppressor of variegation 3–9 homolog) 1 and SUV39H2. KDM1A is a FAD-dependent KDM, which can be inhibited by molecules blocking its cofactor binding site. The KDM1A inhibitors tranylcypromine, GSK2879552, INCB059872 and ORY-1001 are in clinical trials. Most KDMs carry a catalytic domain, which can be inhibited by iron-chelating molecules

Modulators of chromatin modifying enzymes are not only effective in the therapy of cancer but also in its prevention. Interestingly, a number of natural, food-derived compounds, such as epicatechin from green tea, resveratrol from grapes or curcumin from curcuma, are known to modulate the activity of chromatin modifying enzymes (Sect. 16.2). Thus, **ingredients of healthy diet have the potential to prevent cancer by keeping the activity of chromatin modifying enzymes under control**.

Targeting histone acetylation via the application of HDAC inhibitors is not only an option the therapy of cancer. Moreover, these compounds may also provide benefit for the treatment of complex neuronal diseases, such as Alzheimer's disease and Parkinson's disease as well as depression, schizophrenia, drug addiction and anxiety disorders. For example, in mouse models the HDAC inhibitor sodium butyrate showed antidepressant effects. Moreover, HDAC inhibitor-treated animals showed induced sprouting of dendrites, an increased number of synapses, re-established learning behavior and access to long-term memories. This suggests **a possible wide application for HDAC inhibitors in the therapy of cognitive disorders**. For example, valproic acid is approved as mood stabilizer and antiepileptic. HDAC inhibitors also widely affect gene expression in the immune system and showed to be effective in the treatment of inflammation and neuronal apoptosis. Furthermore, animal models of ischemia-induced brain infarction indicated that HDAC inhibitors have antiinflammatory and neuroprotective effects and can be used for the treatment of stroke.

> **(Clinical) conclusion**: Aberrant epigenetic changes play a key role in many diseases that can be as divergent as cancer and neurodevelopment disorders, such as autism. Interestingly, inhibitors of chromatin modifying enzymes have a promising therapeutic potential in these diseases. For example, epilepsy and leukemia are treated with the same type of compounds.

Additional Reading

Campbell, R.R. and Wood, M.A. (2019). How the epigenome integrates information and reshapes the synapse. Nat Rev Neurosci *20*, 133–147.

John, R.M., Higgs, M.J. and Isles, A.R. (2023). Imprinted genes and the manipulation of parenting in mammals. Nat Rev Genet *24*, 783–796.

Xiong, X., James, B.T., Boix, C.A., Park, Y.P., Galani, K., Victor, M.B., Sun, N., Hou, L., Ho, L.L., Mantero, J., et al. (2023). Epigenomic dissection of Alzheimer's disease pinpoints causal variants and reveals epigenome erosion. Cell *186*, 4422–4437.

Chapter 15
Cancer Epigenetics

Abstract In this chapter, we will learn that compared to normal cells, cancer cells show epigenetic drifts. These are genome-wide changes in DNA methylation, histone modifications and 3D chromatin structure representing epimutations. Moreover, many malignant tumors reactivate programs of fetal development indicating that tumorigenesis is associated with epigenetic reprogramming. Thus, **cancer is not only a genetic but also an epigenetic disease**. The mechanistic bases of cancer epigenomics are specific genetic, environmental and metabolic stimuli that disrupt the homeostatic balance of chromatin, which then either becomes very restrictive or permissive. Very most types of cancers display a clear epigenetic contribution to the different hallmarks of cancer suggesting that epigenetic modulators, modifiers and mediators form an additional classification system for cancer genes.

Keywords Chromatin · Chromatin modifying enzymes · Epimutation · Epigenetic reprogramming · DNA methylation · Histone modifications · Cell state transitions · Epigenetic mediators · Epigenetic modulators

15.1 Epigenetic Mechanisms of Cancer

Dysregulated epigenetic processes can act as drivers of early disruption of cellular homeostasis in precancerous and cancerous cells in the tumorigenesis process. An important epigenetic change in cancer is the deregulation of CpG methylation patterns, i.e., of the DNA methylome (Sect. 7.2). Like in aging (Sect. 12.3), tumorigenesis correlates with genome-wide DNA hypomethylation (Fig. 15.1, top right), which leads to genome instability via the reactivation of pluripotency transcription factors (acting as oncogenes, Sect. 15.4) and retrotransposons within repetitive genomic DNA (Box 1.3). In contrast, CpG islands at promoter regions of tumor suppressor genes get hypermethylated (Fig. 15.1, top left, Sect. 15.3), which causes

C. Carlberg, *Gene Regulation and Epigenetics*,
https://doi.org/10.1007/978-3-031-68730-3_15

the inactivation of the respective genes and their tumor-protective function. This is an example of an **epigenetic drift**. For example, silencing of the tumor suppressor gene *MLH1* (MutL homolog 1) via DNA hypermethylation harms the MMR (mismatch repair) DNA repair process and increases the risk of accumulating DNA mutations throughout the whole genome. Thus, **one epimutation can initiate a multitude of genetic changes**.

Mutations in genes encoding for chromatin modifying enzymes are either gain-of-function or loss-of-function. Abnormal histone methylation may be caused by mutations in genes encoding for KMTs and KDMs as well as for the histone variant H3.3, which reduces the genome-wide methylation of H3K27 and H3K36 (Fig. 15.1, center). Examples are gain-of-function and overexpression of EZH2 and loss-of-function of the H3K36-specific KMT SETD2 (SET domain containing 2). More-over, there are translocations of KMT2A (H3K4-specific) as well as translocations and overexpression of the H3K36-specific KMT NSD1 (nuclear receptor binding SET domain protein 1) and the H3K27-specific KMT NSD2. In addition, the amplification or overexpression of genes encoding for H3K4-, H3K9- and H3K36-specific KDMs have been described in the context of different types of cancer. Furthermore,

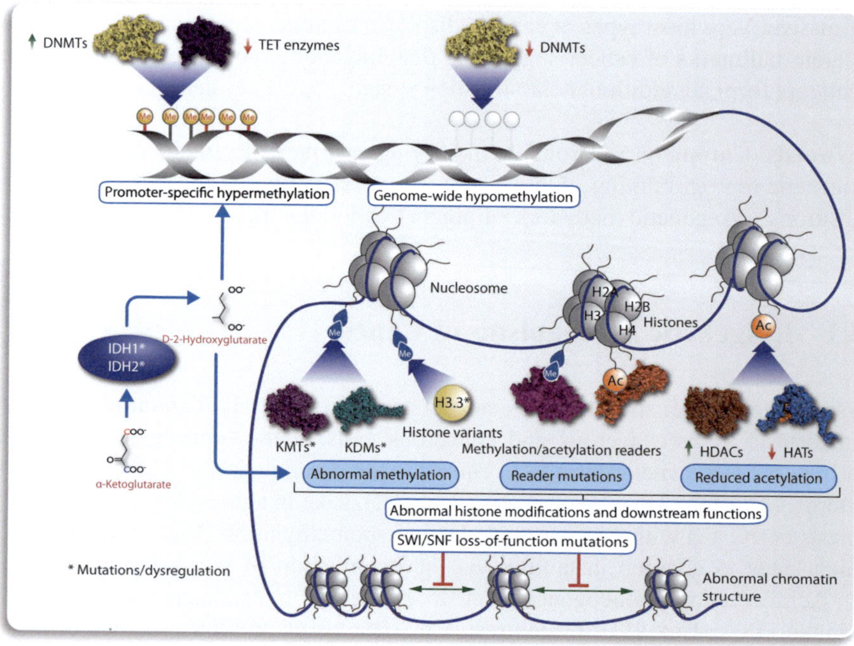

Fig. 15.1 Epimutations in cancer. There are four major types of epimutations affecting cancer: DNA hypermethylation at promoters (**top left**), genome-wide DNA hypomethylation (**top right**), abnormal modification of histones and/or their recognition (**center**) and abnormal chromatin structures caused by mal-functional chromatin remodelers (**bottom**). More details are provided in the text

also histone acetylation is reduced in cancer through the loss of the HATs EP300 and CREBBP and the overexpression of HDACs. Finally, **not only the writer and eraser function of chromatin modifying enzymes can be affected by mutations but also their reader task**. Examples are the overexpression or gain-of-function translocations of BRD4, which binds acetylated histones, or the overexpression of TRIM24, which recognizes H3K23ac. In many cancers, histone proteins or their variants, such as H3.3, are mutated (Fig. 15.1, center). For example, 90% of chondroblastomas and 20% of pediatric glioblastomas have a K36M mutation in histone H3.3, which inhibits H3K36-specific KMTs, reduces H3K36 methylation and alters gene expression.

Loss-of-function mutations in genes encoding for DNA demethylases (*TET1*, *TET2* and *TET3*) or increased expression of genes encoding for DNMTs (*DNMT1*, *DNMT3A* and *DNMT3B*) can cause promoter hypermethylation in some cancers (Fig. 15.1, top). In contrast, genome-wide hypomethylation is often based on loss-of-function mutations in the *DNMT3A* gene. Mutations in the genes encoding for the metabolic enzymes IDH (isocitrate dehydrogenase) 1 and 2 cause that the citric acid cycle intermediate α-ketoglutarate becomes transformed into the oncometabolite 2-hydroxyglutarate, which inhibits TETs and KDMs (Fig. 15.1, center left). This leads to the increased methylation of both DNA and histones. Taken together, **epimutations affect a large variety of changes in cellular homeostasis that result in the acceleration of tumorigenesis**.

The concept of the **hallmarks of cancer** initially highlighted the processes "sustained proliferative signaling", "evading growth suppressors", "resisting cell death", "enabling replicative immortality", "inducing angiogenesis" and "activating invasion and metastasis" as common during tumorigenesis of basically all types of cancer. Later on, the concept was extended to up to 14 hallmarks (Fig. 15.2, bottom), including "genomic instability and mutation" and "**epigenomic disruption**". This acknowledges the results of large-scale cancer genomics projects, such as *TCGA*, on both genetic and epigenetic drivers of different types of cancer. For example, the cancer projects indicated high frequency of mutations in genes encoding for epigenetic mediators, some of which form hot spots.

The cancer genome and epigenome influence each other in a multitude of ways and can work mutually (Fig. 15.2, top). Both genetics and epigenetics offer complementary mechanisms to achieve similar results. For example, a gain-of-function activation of the oncogene *PDGFRA* (platelet-derived growth factor receptor α) that is important for achieving the hallmark "sustained proliferative signaling" may be based either on a genetic mutation within the coding region of the gene or by an epimutation that disrupts the 3D chromatin strip carrying the gene. Moreover, the inactivation of tumor suppressor genes may occur either by a deletion within their coding region or epigenetic silencing of their promoter regions.

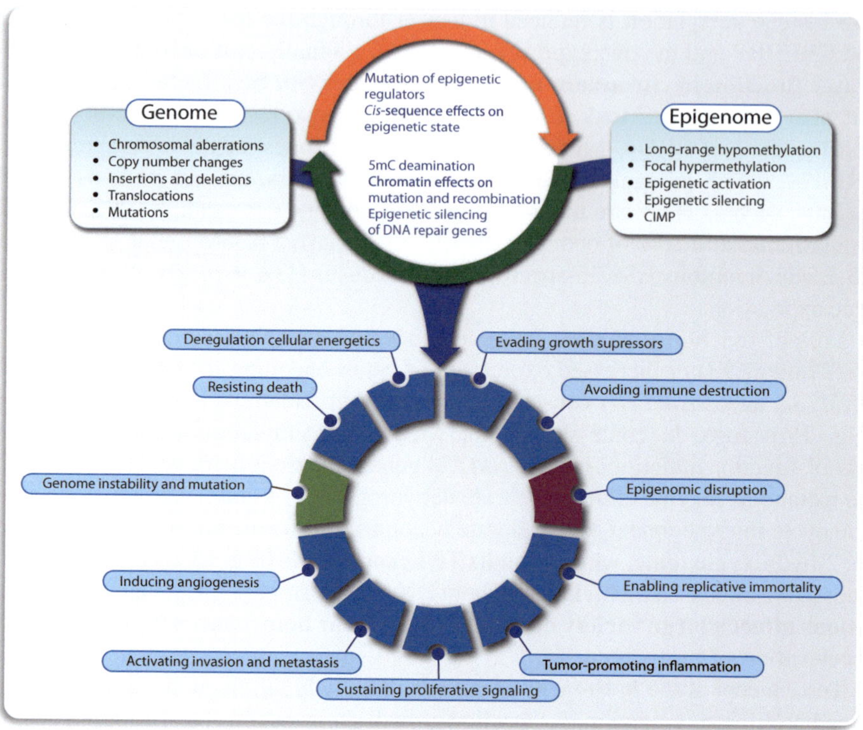

Fig. 15.2 Interplay between genome and epigenome in cancer. Changes in the genome can influence the epigenome and vice versa (**top**). This forms a network that produces genetically or epigenetically encoded variations in the phenotype that are subject to Darwinian selection for growth advantage and thus eventually achieving the hallmarks of cancer (**bottom**). CIMP = CpG island methylator phenotype

15.2   Epigenetic Reprogramming in the Context of Cancer

Changes in the self-renewal of adult stem cells can either result in premature aging, if it is impaired, or in predisposition to malignant transformation, if it is enhanced. Both epigenetic and cellular (re)programing as well as tumorigenesis show similar genome-wide changes in chromatin structure and DNA methylation. Compared with terminally differentiated cells, both iPS and cancer cells have reduced levels of H3K9 methylation and aberrant hypermethylation or hypomethylation. For example, reduced methylation levels, such as a result of low *DNMT1* expression, can cause T cell lymphomas but also promote iPS cell formation. Similarly, *DNMT3A* mutations are found in leukemia and knockdown of the gene facilitates human iPS cell formation. Thus, **cancer cells need to overcome some of the same epigenetic barriers as iPS cells in order to alter their cellular states** (Fig. 11.3).

The interpretation of cellular reprograming and transformation as biochemical reactions illustrates that both processes have to overcome a comparable epigenetic barrier that stabilized the starting cells. Both are based on multistep processes that involve proliferation, change of cell identity and finally lead to the formation of immortal cells with tumorigenic potential. Moreover, compared to terminally differentiated cells, adult stem cells or progenitor cells far more likely form either iPS cells or tumor cells. This suggests that **the epigenetic state of stem and progenitor cells is more sensitive to both cellular reprograming and tumorigenesis**. In addition, pluripotency transcription factors induce a metabolic switch to create ATP rather from glycolysis than from oxidative phosphorylation, which for cancer cells is known as the "Warburg effect". However, iPS cells keep the normal intact diploid genome of the starting cells, while cancer cells accumulate mutations and often get aneuploidy, i.e., they have an abnormal number of chromosomes.

Permissive chromatin has a high rate of plasticity. This allows cancer cells to acquire easily a number of different transcriptional states, e.g., shifting to alternative developmental programs, some of which are pro-oncogenic. When such an adaptive chromatin state propagates through mitosis, a new cell clone is created that overgrows other cells due to increased fitness (Fig. 15.3). This plasticity model can be considered as the epigenetic counterpart to the genetic model of genome instability being induced by carcinogen exposure or DNA repair defects. Interestingly, **in both models there are driver events, such as the activation of an oncogene and passenger events that do not alter the fitness of the cells**.

Chromatin homeostasis is closely linked to metabolic conditions. Many chromatin proteins, such as DNA- and histone-modifying enzymes, use metabolites as donors and cofactors, like α-ketoglutarate, the methyl donor SAM and acetyl-CoA (Sects. 7.1 and 8.3). Therefore, they are sensitive to shifts in the concentration of these metabolites. For example, mutations of IDH enzymes lead to the accumulation of 2-hydroxyglutarate inhibiting the demethylation of DNA via TET enzymes instead of producing α-ketoglutarate. The resulting DNA hypermethylation disrupts the binding of the methylation-sensitive transcription factor CTCF to insulator regions (Sect. 7.3). Accordingly, the partition of the genome into discrete functional domains, in which enhancers regulate their appropriate gene targets, is disturbed. For example, reduced CTCF binding in gliomas (a type of brain tumors) with *IDH* mutants leads to a transcriptome profile that indicates insulator dysfunction. Thus, **the oncogenicity of IDH mutants is primarily based on a loss of gene insulation**, i.e., on compromised

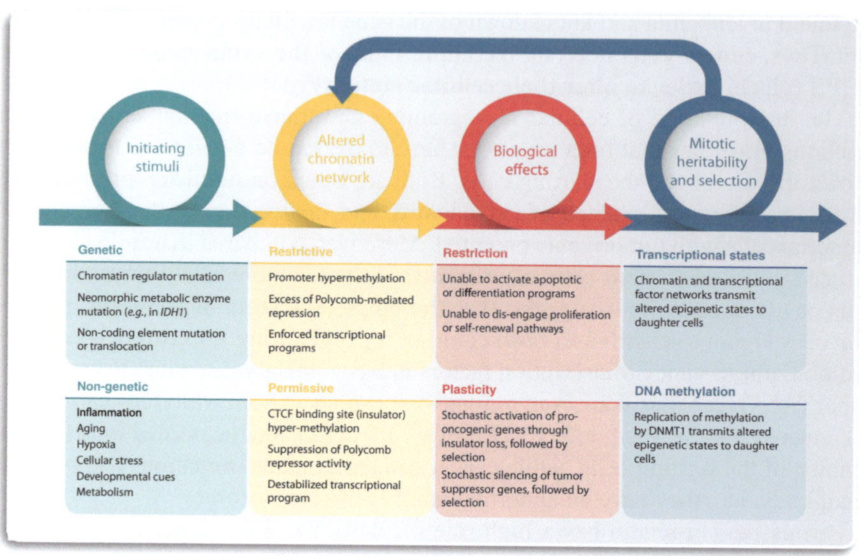

Fig. 15.3 Disruption of chromatin homeostasis in cancer. Chromatin homeostasis can be disrupted by genetic factors, such as mutations in chromatin proteins or translocation of regulatory elements, or by non-genetic factors, such as inflammation, stress or hypoxia. This can result either in very permissive or in restrictive chromatin networks. In permissive states oncogenic epigenetic changes may occur, such as silencing of tumor suppressor genes. Mitotically heritable, adaptive epigenetic changes will be selected and contribute to the hallmarks of cancer

CTCF-mediated genome topology, which is an effect of epigenetic dysregulation (Fig. 15.3).

The idea that cancer is fundamentally an epigenetic disease is also reflected by the relationship between cancer and the epigenetic landscape. Since epigenetic modifiers, such as genes encoding for chromatin modifying enzymes, are highly mutated in cancer, these mutations largely affect the stability of the epigenetic landscape (Fig. 15.3). While permissive chromatin states allow oncogene activation or non-physiologic cell fate transitions, restrictive states prevent the induction of tumor suppressors or block differentiation. For example, the cancer hallmark "evasion of growth suppressors" can be based on either a loss-of-function mutation of the tumor suppressor gene *CDKN2A* or by hypermethylation of its promoter. The relative contribution of genetic and epigenetic mechanisms to the hallmarks of cancer differs between cancer types. Interestingly, the example of the adult brain tumor glioblastoma in comparison to the childhood brain tumor ependymoma suggests that long-term tumorigenesis in adults may rather be based on genetic events, while short-term tumorigenesis has majorly an epigenetic origin (Fig. 15.4). General differences between adult and pediatric cancers are summarized in Box 15.1.

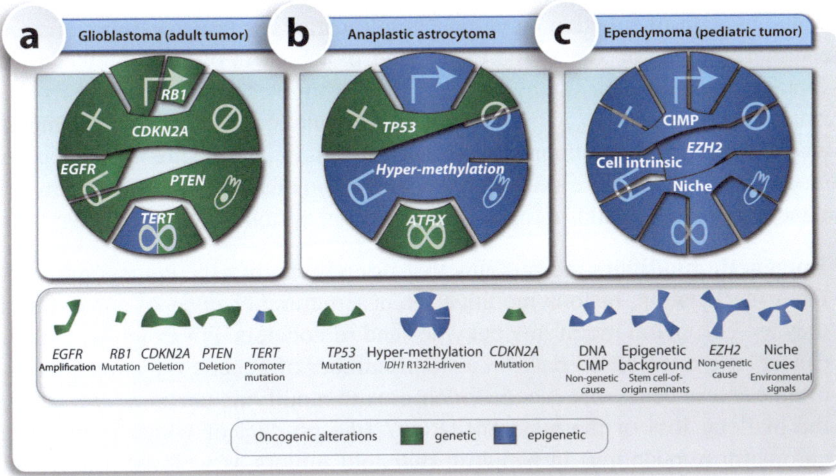

Fig. 15.4 Genetic and epigenetic mechanisms underlying the hallmarks of cancer. Both genetic (green) and epigenetic (blue) mechanisms are important factors in tumorigenesis, but their relative contribution to the hallmarks of cancer depends on the type of cancer. In the adult brain tumor glioblastoma (**a**), most hallmarks relate to genetic drivers, while in ependymoma (**c**), a childhood cancer, primarily epigenetic effects dominate. Anaplastic astrocytoma (**b**) represents an example where both genetic and epigenetic factors contribute to the hallmarks

Box 15.1: Differences Between Adult and Pediatric Malignant Tumors Cancers in adults and children cannot be compared, even if they carry the same name, such as ALL (acute lymphoblastic leukemia). Carcinomas, i.e., rather slowly growing tumors arising out of epithelial cells, are in adults the predominant form of cancer, while they are rarely found in children. In general, for children cancer is a rare disease. For them malignant tumors often arise out of the hematopoietic system, such as leukemias in younger children and lymphomas in adolescence, or the brain. These cancers develop from non-epithelial cells and show a very aggressive nature combined with a high growth rate. Moreover, biological processes underlying cancer development in children differ clearly from that of adults. Tumorigenesis of adult cancers is based on the accumulation of mutations in the genome, while malignant tumors in children most often develop due to an epigenetic dysregulation during embryogenesis. Accordingly, the treatment of pediatric cancers differs significantly from that of adult patients. Moreover, children tolerate far higher doses (per kg bodyweight) of chemotherapy than adults, since they have a higher stem cell division rate. This allows their healthy cells to regenerate faster from side effects of the therapy. Thus, **the cancer cure rate, i.e., the 5-year disease-free survival, is much higher in children than in adults.**

The known cancer genes are classified into dominant oncogenes, which can be activated by gain-of-function mutations, amplifications or translocations, and recessive tumor suppressor genes, the expression of which is often lost, like by loss-of-function mutations or methylation of their promoter regions (Table 15.1). An alternative classification is to divide all somatic mutations into drivers, the mutation of which directly affects tumorigenesis, and passengers, which are mutated as a side product but would not have a functional contribution on oncogenesis on its own. The epigenetic perspective adds a further classification option for cancer genes:

- **Epigenetic modifiers** are proteins that directly modify the epigenome through DNA methylation, histone modification or structural changes of chromatin, i.e., they are chromatin modifying enzymes and remodelers. For example, childhood cancers are often based only on a small number of genetic mutations, but these often occur in genes that encode for chromatin modifying enzymes. Interestingly, the biallelic loss of the gene *SMARCB1* (the product of which is involved in chromatin remodeling) in pediatric rhabdoid tumors as well as in lung cancer and Burkitt's lymphoma, was a first indication that the **disruption of epigenetic control can serve as a driver for cancer**.

Table 15.1 Three classification systems for cancer genes

Genetic classification		
Oncogene	A gene, when activated, gives advantages to cancer cells	*MYC, KRAS, PIK3CA, BRAF*
Tumor suppressor gene	A gene, when inactivated, gives advantages to cancer cells	*TP53, RB1, APC, CDKN2A*
Selection classification		
Driver gene	A gene, when mutated or showing aberrant gives advantages to cancer cells	*MYC, TP53, KRAS, RB1, PIK3CA*
Passenger gene	A gene, when mutated does not affect, the growth of cancer cells, i.e., not a driver	> 99% of all genes mutated in cancer
Epigenetic classification		
Epigenetic modifier	A gene that modifies the methylation of DNA, histone modification or 3D chromatin structure	*TET2, DNMT3A, ARID2, BRD4, EZH2, NSD1*
Epigenetic mediator	A gene regulated by an epigenetic modifier that gives advantages to cancer cells	*OCT4, NANOG, SOX2, KLF4*
Epigenetic modulator	A gene that activates or represses the epigenetic machineries	*IDH1, CTCF*

Example genes for each classification are provided

- **Epigenetic mediators** are targets of epigenetic modification, i.e., they are downstream of epigenetic modifiers. For example, due to the overactivity of an epigenetic modifier, such as a chromatin modifying enzyme of the HAT family, genes encoding for pluripotency transcription factors, such as NANOG, SOX2 or OCT4, get activated in a somatic cell. Thus, in this situation the pluripotency transcription factors are epigenetic mediators and may transform a terminally differentiated cell back to a pluripotency stage. In this way, the cell forms a cancer stem cell and initiates the tumorigenesis process.
- **Epigenetic modulators** are gene products that are located upstream of epigenetic modifiers and mediators in signal transduction pathways. Epigenetic modulators influence the activity or localization of epigenetic modifiers in order to destabilize differentiation-specific epigenetic states. They represent a bridge between environment and epigenome. Inflammatory responses mediated by the transcription factor NFκB (Sect. 13.4) are an example of an epigenetic modulator. They trigger an epigenetic switch to a positive feedback loop with the cytokine IL6 and the transcription factor STAT3 in the transformation of mammary epithelia. Thus, **the actions of epigenetic modulators are often the first steps in tumorigenesis resulting in changing epigenome patterns**.

15.3 DNA Methylation and Cancer

Large-scale cancer genome projects, such as *TCGA*, revealed that nearly all human cancers carry mutations in key chromatin-associated proteins. It is important to realize that the epigenetic signature of a cell allows more variation than its primary genetic status. The error rate in inheritance of DNA methylation is some 4% for a given CpG per cell division, while the mutation rate of the genome during DNA replication is far lower. Therefore, changes in DNA methylation patterns are key epigenetic dysregulations occurring during tumorigenesis. Compared with normal cells of the same individual, the epigenome of tumor cells shows a massive overall loss of DNA methylation, while for certain genes also hypermethylation at CpGs is observed (Fig. 15.5). Thus, **epigenetic variability leads in much shorter time to an epimutation and phenotypic selection, such as possible onset of cancer, than genetic mutations in a traditional view**. Interestingly, a child is too young to acquire severe genetic mutations leading to cancer but epigenetic mutations alter much faster the cellular homeostasis. Accordingly, the majority of mutations found in cancers arising in adults are traditional genetic mutations, while many childhood cancers are based on epimutations (Sect. 15.1).

Global **DNA hypomethylation** during tumorigenesis generates chromosomal instability, reactivates transposons and causes loss of imprinting. The resulting low DNA methylation favors mitotic recombination leading to deletions and promotes chromosomal rearrangements, such as translocations. The disruption of genomic imprinting, such as the loss of imprinting of the *IGF2* gene, is a risk factor for different types of cancer, such as colorectal cancer or Wilms' tumor. Furthermore,

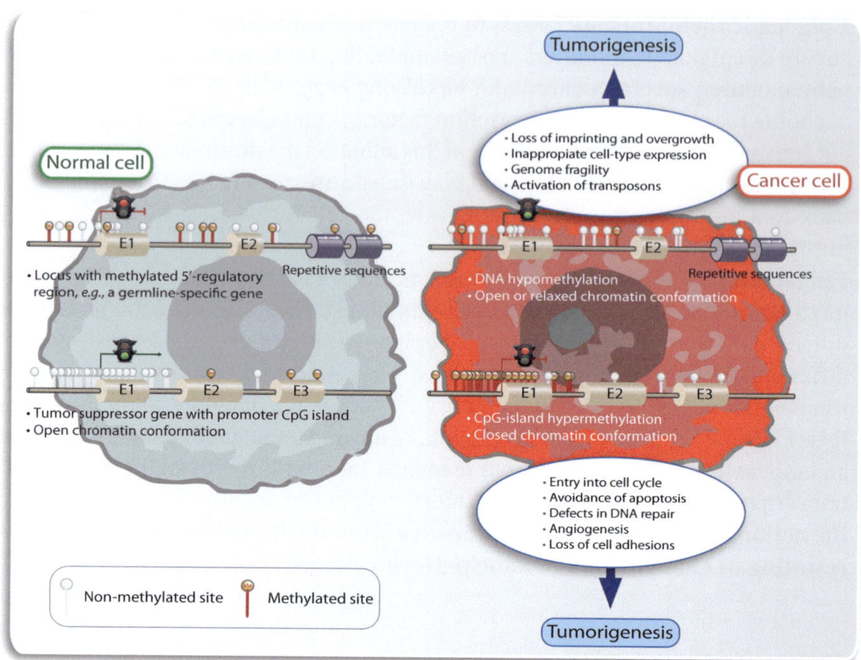

Fig. 15.5 Changes of DNA methylation patterns during tumorigenesis. Compared to normal cells (**left**), cancer cells are hypomethylated on the genome-wide scale (**top right**), in particular at repetitive sequences, such as transposons. In addition, imprinted and tissue-specific genes often get demethylated. Hypomethylation causes changes in the epigenetic landscape, such as the loss of imprinting and increases the genome instability that characterizes cancer cells. Another common alteration in cancer cells is the hypermethylation of CpGs within regulatory regions of tumor suppressor genes (**bottom right**). These genes are then transcriptionally silenced so that cancer cells lack functions, such as inhibition of the cell cycle

hypermethylated promoter regions of tumor suppressor genes, such as *TP53*, *RB1* and *MGMT* (O-6-methylguanine-DNA methyltransferase), can serve as biomarkers that provide significant diagnostic potential in the clinic, in particular in early detection screenings of individuals with a high familial risk of developing cancer. Many CpGs can become methylated already early in tumorigenesis, in particular in CIMP of colorectal cancer, glioma and neuroblastoma. The profiles of CpG hypermethylation vary with cancer types. Each type of cancer can be characterized by its specific DNA hypermethylome, i.e., these epigenetic marks are comparable to traditional genetic and cytogenetic markers. From about 200 genes that are regularly mutated in various forms of human breast and colorectal cancers, on average 11 carry a mutation in a single malignant tumor type. For comparison, 100–400 CpGs close to TSS regions are found to be hypermethylated in a given malignant tumor, i.e., **epigenetics is able to provide 10-times more information than genetics.**

Based on the model of an **epigenetic landscape** (Sect. 11.2), the epigenetic status of a cell, such as its methylation level, can be represented by a ball trapped in a valley. In case of normal differentiated cells, the borders of the valley are high and gene regulatory networks keep the cells in stable epigenetic homeostasis (Fig. 15.6a). This prevents the epigenetic state from moving too far from its equilibrium point in normal tissue. In contrast, a dysregulation of the epigenome during tumorigenesis, such as overexpression of an epigenetic modulator or an inflammatory insult, flattens the valley (Fig. 15.6b). Under these conditions of reduced regulation, the epigenetic status is more relaxed and influenced by stochastic variations. Thus, during tumorigenesis DNA methylation levels diffuse away from the initial state in normal cells (Fig. 15.6c). A plot of CpG methylation levels during the transformation of normal cells into adenoma and carcinoma cells shows a tight distribution in normal tissue, but a progression from adenoma to carcinoma. This explains the substantial level of epigenetic variation for a given cancer type across individuals or between metastatic cells originating from the same primary malignant tumor.

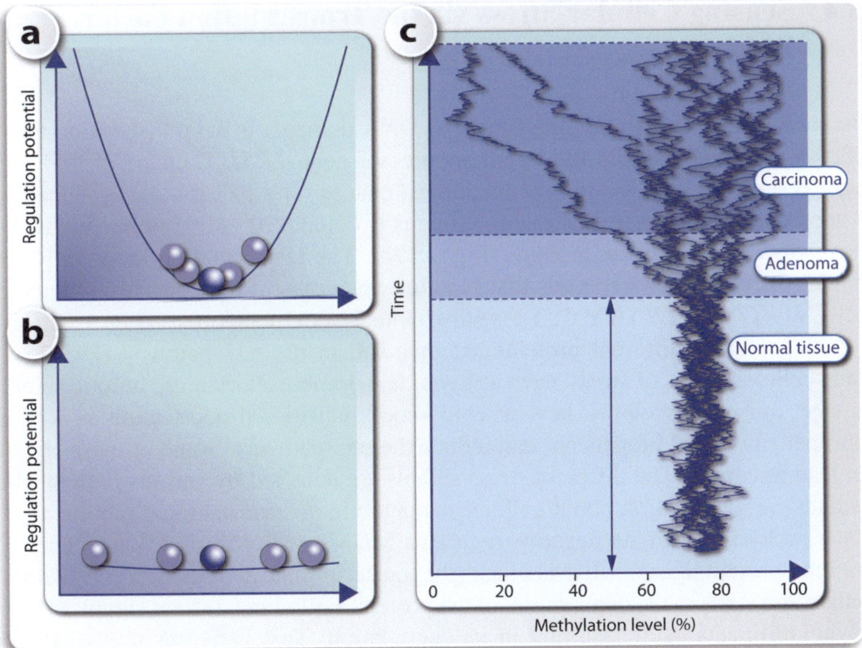

Fig. 15.6 Exemplary model of epigenetic dysregulation. DNA methylation is used here as an example for epigenetic dysregulation. The methylation level of a normal cell is illustrated as a ball at the bottom of a bin (**a**), where regulatory forces, such as gene regulatory networks, allow only minor changes in the epigenetic status. In contrast, during tumorigenesis (**b**) the landscape flattens and the methylation levels can be far more variable. When the methylation level is modeled over time (bottom to top) for 10 examples (**c**), the variations in the transition from normal tissue to adenoma and carcinoma become obvious as wider methylation ranges

Aberrant DNA methylation is not only a well-established marker of cancer and disturbed genomic imprinting, but it also can lead to general instabilities of the genome through reduced heterochromatin formation on repetitive sequences. One of the most frequent mutations found in human diseases is a C to T transition at methylated CpG islands, i.e., the epigenetic mark cytosine methylation induces a genetic point mutation by reducing the efficiency of DNA repair at these sites. DNA methylome profiles, e.g., of PBMCs that can be obtained with minimal invasion, may serve as biomarkers for evaluating the individual risk of cancer. Moreover, the DNA methylome indicates the progress of aging showing significant interindividual differences (Sect. 12.3). Although biomarkers often do not explain the causality of a disease, they can monitor the disease state and may suggest appropriate therapy. Thus, **epigenomic profiles, such as DNA methylation patterns, in combination with genetic predisposition and environmental exposure, may be prognostic for personal risk of disease onset**.

15.4 Sensing Cellular Stress via the Transcription Factor P53

The main sensor of cellular stress, such as DNA damage, is the transcription factor p53, which is encoded by the tumor suppressor gene *TP53*. Damage of this gene leads to severely reduced protection against cancer. The p53 protein is named by its apparent molecular weight and in humans is composed of 393 amino acids that are subdivided into seven domains (Fig. 15.7a). The DNA binding mode of p53 is unique, since it forms tetramers that bind to two copies of the consensus sequence RRRCWWGYYYYYYYYYYYYY with 10 intervening nucleotides (Fig. 15.7b, c).

A large set of different proteins are involved in the p53 pathway (Fig. 15.8). Many cellular forms of stress, such as hypoxia, telomere shortening, mitotic spindle damage, unfolded proteins, heat or cold shock, nutritional deprivation as well as improper ribosomal biogenesis, can induce the p53 pathway. Some of these signals can lead to cancer. The different stress signals are detected by various proteins that mediate the information about cellular damage via posttranslational modifications of the p53 protein or its negative regulator MDM2 (MDM2 proto-oncogene, E3 ubiquitin protein ligase). MDM2 blocks the transcriptional activity of p53 by a direct contact and leads to the degradation of the protein. Following a stress signal, MDM2 polyubiquitinates itself resulting in its degradation. This increases p53's half-life from minutes to hours. Depending on the interaction with other signal transduction pathways, the activation of p53 can lead either to cell cycle arrest, senescence or apoptosis. The cell cycle arrest permits cellular repair, reverse of damage and cell survival, while the two other processes lead to cellular death. p53 mediates activation as well as repression of its target genes, mostly via direct sequence-specific binding of p53 to their regulatory genomic regions. Through direct protein–protein contacts p53 interacts with GTFs, such as TBP, TAF6 and TAF9, as well as with HATs like

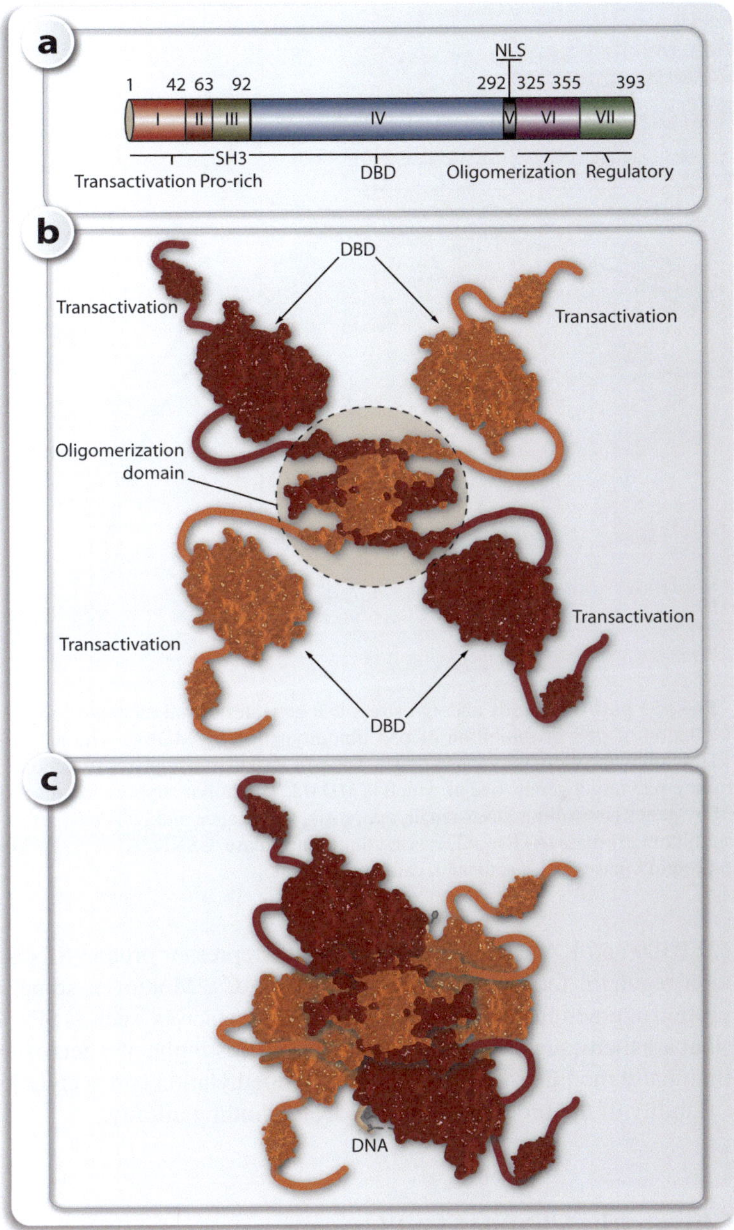

Fig. 15.7 Structure of p53. The principal structure of the human p53 protein with its seven subdomains is schematically depicted (**a**). Model of the p53 tetramer (**b**). The two different shades of gray refer to interacting p53 homodimers. The model is based on the folded, stable human oligomerization domain (1OLG, highlighted by a gray circle), the p53 DBD (2AC0) and the *X. laevis* transactivation domain (1YCQ). The disordered domains are represented by lines connecting the domains. DNA-bound p53 tetramer (**c**)

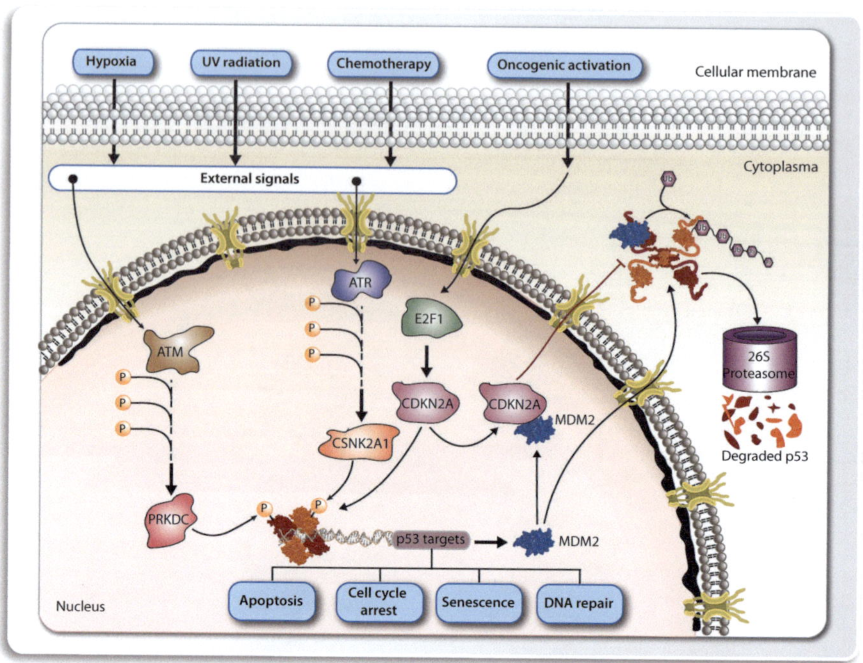

Fig. 15.8 The p53 pathway. Cells undergo stress that activates signal mediator proteins leading to phosphorylation of p53 or inhibition of p53 ubiquitynation by MDM2. The half-life of p53 increases in the following from minutes to hours. The p53 tetramer recognizes its genomic binding sites controlling p53 target genes, one of which is *MDM2*. The tumor suppression function of p53 is mediated by genes controlling DNA repair, apoptosis, senescence and cell cycle arrest. ATM = ATM serine/threonine kinase, ATR = ATR serine/threonine kinase, CSNK2A1 = casein kinase 2α 1, PRKDC = protein kinase, DNA-activated, catalytic subunit

CREBBP, EP300 and KAT2B (Sect. 8.2) or via the repressor protein SIN3A (SIN3 transcription regulator family member A) with HDACs. Moreover, some of these protein–protein interactions involve other transcriptions factors, such as SP1, CEBPα and AP1 that are then squelched, i.e., inactivated. Interestingly, p53 gets excessively posttranslationally modified via phosphorylation, methylation and acetylation. This alters the stability of the protein and thus its DNA-binding affinity.

15.5   Chromatin Changes and Cancer

In general, DNA methylation and histone modifications have different roles in gene silencing. **While most DNA methylation loci represent very stable silencing marks that are seldom reversed, histone modifications mostly lead to labile and reversible transcriptional repression**. For example, genes for pluripotency transcription factors in embryogenesis, such as *POU5F1* (encoding for OCT4) and *NANOG*, need to be permanently inactivated in later developmental stages, in order to prevent possible tumorigenesis. This happens via H3K9 methylation at unmethylated CpGs on the TSS regions of these genes, the attraction of HP1, de novo DNA methylation via DNMT3A and DNMT3B and finally transcriptional silencing for the rest of the life of the individual. In contrast, when in differentiated cells pluripotency genes are silenced only by histone modification, the cells can be rather easily converted to iPS cells.

Changes in DNA methylation during tumorigenesis are always combined with other epigenetic dysregulations, such as aberrant patterns of histone modifications and overall changes in the nuclear architecture, i.e., the overall epigenetic landscape of cancer cells is significantly distorted compared to somatic stem cells or differentiated cells (Fig. 15.9). These alterations in the 3D organization of chromatin exemplify epigenome changes during tumorigenesis. In differentiated cells developmentally repressed genes, i.e., genes that are not needed in a given cell type, are often found within LADs that constitutively localize close to the nuclear periphery. A significant fraction of these LADs represent so-called large organized chromatin K9-modifications (LOCKs), which are genomic regions that are enriched in repressive H3K9me2 and H3K9me3 histone marks (Fig. 15.9, right). This is further promoted by the recruitment of KDMs and HDACs to the repressive environment of the nuclear envelope and DNA hypermethylation at these regions. Thus, 3D chromatin compaction mediates gene repression during lineage specification and represents a form of **epigenetic memory**. This results in reduced transcriptional noise and provides barriers for de-differentiation. Although adult stem cells are in a less differentiated state than terminally differentiated cells, also in them specific LAD/LOCK structures are found (Fig. 15.9, left).

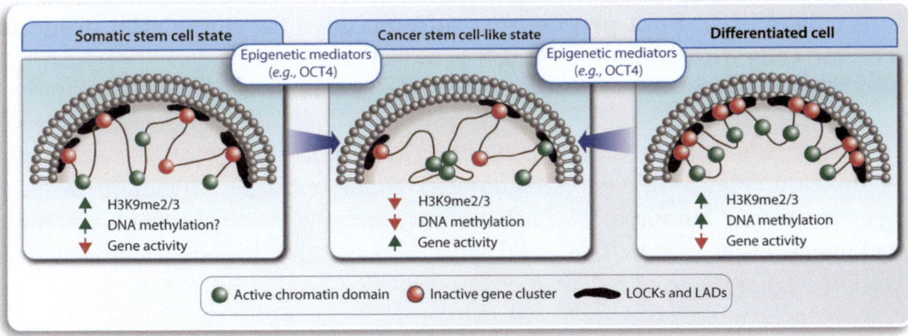

Fig. 15.9 Reprogramming of the nuclear architecture in cancer cells. Epigenetic mediators, such as the transcription factor OCT4, can reprogram the epigenome of somatic stem cells (**left**) or differentiated cells (**right**) into cancer stem cells (**center**). Normal cells are characterized by high levels of H3K9 di- and trimethylation as well as DNA methylation in LOCKs being a part of LADs. The latter are located close to the nuclear membrane and contain only a low number of active genes. In contrast, in cancer stem cells LOCKs and LADs are largely absent and a larger variety of genes are active. This leads to phenotypic heterogeneity

Proteins that regulate the interaction of chromatin with the lamina and recruit chromatin modifying enzymes to the nuclear periphery act as epigenetic mediators. For example, the reactivated pluripotency transcription factor OCT4 can reprogram the epigenome of both differentiated cells and adult stem cells into cancer stem cells (Fig. 15.9, center). The activation of the epigenetic mediator dissolves most of the LADs/LOCK structures, as a consequence of which a large number of genes are reactivated. This provides cancer cells with phenotypic heterogeneity, such as increased variability in gene expression, in order to switch between different cellular states within the cancer. The loss of LOCKs also affects enhancer-promoter region communication within and between TADs, so that oncogenic super-enhancers are able to cluster. A similar process happens during epithelial-mesenchymal transition (EMT), which is a key process in normal wound healing, but also the first step toward metastasis. In EMT, the activation of an H3K9-specific KDM, such as KDM1A, often is the initiating epigenetic event. When cancer cells have destabilized the epigenetic memory of the cells they originate from and form EMT-related chromatin structures, they gain phenotypic plasticity. Thus, the overall result of **the change in chromatin architecture might be an oncogenic transformation of the cell**.

The reprogramming of a somatic cell into an iPS cell or its transformation to a cancer cell (Sect. 11.2) are related events. In both cases, an epigenetic barrier has to be overcome in a multistep process that primarily involves epigenetic mediators, such as the transcription factors SOX2, KLF4, NANOG, OCT4 and MYC or the RNA-binding protein LIN28A (Lin-28 homolog A). Interestingly, all five transcription factors are encoded by oncogenes. Moreover, also the chromatin modifying enzymes SUV39H1 (KMT1A), EHMT2 (KMT1C), SETDB1 (KMT1E), KMT2A, KMT2D, KMT2C, DOT1L (DOT1 like histone lysine methyltransferase, also called KMT4), EZH2 (KMT6A), LSD1 (KDM1A), KDM2B, KDM6A, the SWI/SNF complex member ARID1A (AT-rich interaction domain 1A) as well as DNMT3A and DNMT3B have comparable roles both in cellular reprogramming and in tumorigenesis (Fig. 15.10). **Both processes acquire de novo developmental programs and create cells with an unlimited self-renewal potential**.

(Clinical) conclusion: Cancer is not only based on changes of the genome but also on epigenome-wide alterations. We cannot change mutations of the human genome, but via lifestyle changes we are able to affect our epigenome and may prevent cancer. This implies the responsibility for our own health as well as a potential for a holistic treatment of disease(s).

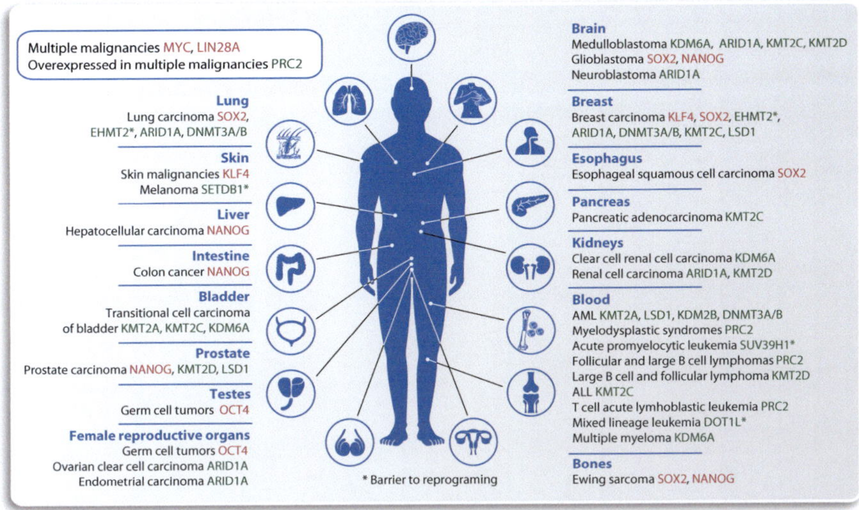

Fig. 15.10 Overlap of iPS nuclear reprogramming and cancer. The same transcription factors (red) and chromatin modifying enzymes (green) play a central role both in iPS cellular reprogramming and in different types of cancer. Some of them are encoded by oncogenes and tumor suppressor genes

Additional Reading

Bates, S.E. (2020). Epigenetic therapies for cancer. N Engl J Med *383*, 650–663.

Carlberg, C. and Velleuer, E. (2021). Cancer biology: how science works. Springer Textbook.

Corces, M.R., Granja, J.M., Shams, S., Louie, B.H., Seoane, J.A., Zhou, W., Silva, T.C., Groeneveld, C., Wong, C.K., Cho, S.W., et al. (2018). The chromatin accessibility landscape of primary human cancers. Science *362*, eaav1898.

Dubois, F., Sidiropoulos, N., Weischenfeldt, J. and Beroukhim, R. (2022). Structural variations in cancer and the 3D genome. Nature Reviews Cancer *22*, 533–546.

Feinberg, A.P. and Levchenko, A. (2023). Epigenetics as a mediator of plasticity in cancer. Science *379*, eaaw3835.

Filbin, M. and Monje, M. (2019). Developmental origins and emerging therapeutic opportunities for childhood cancer. Nat Med *25*, 367–376.

Heltberg, M.S., Lucchetti, A., Hsieh, F.S., Minh Nguyen, D.P., Chen, S.H. and Jensen, M.H. (2022). Enhanced DNA repair through droplet formation and p53 oscillations. Cell *185*, 4394–4408 e4310.

Mohammad, H.P., Barbash, O. and Creasy, C.L. (2019). Targeting epigenetic modifications in cancer therapy: erasing the roadmap to cancer. Nat Med *25*, 403–418.

Zhao, S., Allis, C.D. and Wang, G.G. (2021). The language of chromatin modification in human cancers. Nature Reviews Cancer *21*, 413–430.

Chapter 16
Nutritional Epigenetics

Abstract In this chapter, we will present how nutritional molecules can influence the activity and biological function of both transcription factors like nuclear receptors as well as chromatin modifying enzymes. The nuclear receptors PPARs, LXRs and FXR play a central role in sensing fatty acids, oxysterols and bile acids and regulating lipid metabolism. In this way, dietary compounds affect gene regulation and epigenetics. Different epigenetic mechanisms, such as posttranslational histone modifications and DNA methylation, process information provided by dietary molecules. Accordingly, many chromatin modifying enzymes use intermediary metabolites, such as acetyl-CoA, α-ketoglutarate, NAD^+ or ATP, as co-substrates and/or cofactors. Finally, we will discuss that rather epigenetic misregulation due to unfavorable lifestyle decision will affect the risk for developing the metabolic syndrome than genetic variations.

Keywords Nuclear receptors · PPAR · LXR · FXR · Lipid metabolism · DNA methylation · Histone modifications · Chromatin modifying enzymes · Intermediary metabolism · Acetyl-CoA · NAD^+ · Folate · Methylenetetrahydrofolate reductase · Metabolic syndrome · GWAS · DOHaD concept

16.1 Integration of Lipid Metabolism by Nuclear Receptors

The different steps in handling fatty acids, such as their resorption in the intestine, metabolism in the liver, oxidation in active tissues and collection of their excess for long-term storage in adipose tissue, is coordinated by the three members of the PPAR family (Fig. 5.2). **PPARs are activated by various fatty acids and their derivatives, such as PUFAs, eicosanoids and oxidized phospholipids.** Each PPAR subtype has unique functions, which is based on its distinct tissue distribution. PPARα is expressed predominantly in the liver, heart and kidney. PPARδ is ubiquitously expressed but has most important functions in skeletal muscle, liver and heart. PPARγ is highly expressed in adipose tissue and acts there both as a key regulator of adipogenesis and a potent modulator of lipid metabolism and insulin sensitivity. Due to

C. Carlberg, *Gene Regulation and Epigenetics*,
https://doi.org/10.1007/978-3-031-68730-3_16

alternative splicing and differential promoter usage, there are two PPARγ isoforms, of which PPARγ1 is expressed in many tissues, while the expression of PPARγ2 is restricted to adipose tissue.

During fasting or starvation, PPARα is the primary regulator of the adaptive response in the liver (Fig. 16.1a). This receptor senses the reversed flux of fatty acids and activates a gene network to oxidize fatty acids, in order to generate energy in liver and muscle and to convert fatty acids into a usable energy source, such as ketone bodies during starvation. PPARα is also the molecular target of fibrates, which are widely used drugs that reduce serum triglyceride levels through increased fatty acid β-oxidation. In addition, PPARα stimulates the production and secretion of fibroblast growth factor (FGF) 21. The hepatokine acts as a stress signal to other tissues, in order to adapt to an energy-deprived state in case of fasting. After a meal, PPARα and PPARδ are sensing increasing fatty acid efflux from the liver and start to manage lipid metabolism via the promotion of fatty acid β-oxidation and ATP production in mitochondria of skeletal muscles and the heart. Thus, **fatty acids stimulate their own breakdown**.

PPARα, PPARδ and PPARγ interfere with the transcription factors NFκB and AP1 in macrophages, endothelial cells, epithelial cells and other tissues, causing the attenuation of proinflammatory signaling by decreasing the expression of proinflammatory cytokines, chemokines and cell adhesion molecules (Fig. 16.1b). For example, PPARα and PPARδ neutralize the p65 subunit of NFκB and inhibit in this way the expression of NFκB-controlled cytokines (Sect. 13.4), such as TNF, IL1β and IL6. Thus, **PPARs are involved in the control of inflammation**.

PPARδ decreases serum triglyceride levels, prevents high-fat diet-induced obesity and increases insulin sensitivity through the regulation of genes encoding fatty acid metabolizing enzymes in skeletal muscle and genes encoding for lipogenic proteins in the liver. Moreover, PPARδ increases serum HDL (high-density lipoprotein)-cholesterol levels via stimulating the expression of the reverse cholesterol transporter ABC (ATP-binding cassette) A1 and APO (apolipoprotein) A1, which are specific for cholesterol efflux. Moreover, the activation of PPARα and PPARδ promotes osteoblast activity in bone, whereas the activation of PPARγ leads to osteoclast stimulation (Fig. 16.1c).

In adipose tissue, PPARγ controls glucose uptake via regulating the expression of the *SLC2A4* gene (encoding for the glucose transporter GLUT4). Furthermore, PPARγ together with FGF1 mediates adipose tissue remodeling for maintaining metabolic homeostasis during famine. Since high concentrations of circulating fatty acids can cause insulin resistance, enhanced uptake of fatty acids in adipose tissue, the stimulation of the secretion of adiponectin and the inhibition of resistin production via PPARγ improve the condition of the disease. Thus, **activating PPARs by specific synthetic ligands (glitazones) can inhibit obesity-related insulin resistance**. However, the increased uptake of fatty acids as well as the enhanced adipogenic capacity in white adipose tissue after PPARγ activation is responsible for

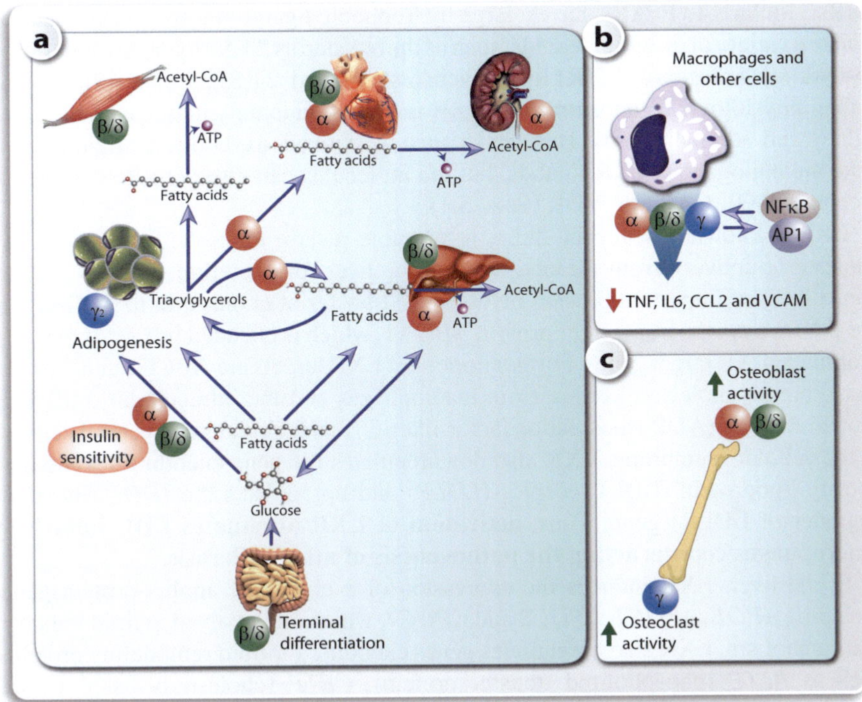

Fig. 16.1 Physiological roles of PPARs. PPARα regulates the expression of enzymes that lead to the mobilization of stored fatty acids in adipose tissue and of fatty acid catabolizing enzymes in the liver, heart and kidney (**a**). PPARδ (also called PPARβ in rodents) is expressed at high levels in the intestine where it mediates the induction of terminal differentiation of epithelium. Activating PPARδ or PPARγ can increase insulin sensitivity. PPARδ regulates the expression of fatty acid catabolizing enzymes in skeletal muscle where released fatty acids are oxidized to generate ATP. PPARγ promotes the differentiation of adipocytes. PPARα, PPARδ and PPARγ can interfere with the transcription factors NFκB and AP1 causing the attenuation of proinflammatory signaling (**b**). All tree PPAR subtypes also act in bone (**c**)

thiazolidinedione-associated weight gain. Moreover, the thiazolidinedione rosiglitazone was found to increase the risk of heart failure, myocardial infarction and CVD, leading to restricted access in the United States and a market withdrawal in Europe. Nevertheless, the activation PPARγ via its natural ligands, such as PUFAs taken up by healthy diet, may be sufficient.

LXRs and FXR are sensors for the cholesterol derivatives oxysterols and bile acids, respectively. These nuclear receptors do not only regulate cholesterol and bile acid metabolism but also have a central role in the integration of sterol, fatty acid and glucose metabolism. LXRα is expressed in tissues with a high metabolic activity, such as liver, white adipose tissue and macrophages, whereas LXRβ is found ubiquitously.

LXRs, similarly to PPARs, have a large hydrophobic ligand-binding pocket that can bind to a variety of different ligands, such as the oxysterols 24(S)-hydroxycholesterol, 25-hydroxycholesterol, 22(R)-hydroxycholesterol and 24(S),25-epoxycholesterol, at their physiological concentrations. FXR is expressed mainly in the liver, intestine, kidney and adrenal glands. Bile acids, such as chenodeoxycholic acid and cholic acid, are endogenous FXR ligands, but the molecules can also activate the nuclear receptors PXR, CAR and VDR (Sect. 5.1).

LXR is known best for its ability to promote reverse cholesterol transport, i.e., cholesterol delivery from the periphery to the liver for excretion (Fig. 16.2). This clinically very important process involves the transfer of cholesterol to APOA1 and pre-β HDLs via the transporter protein ABCA1, which is encoded by one of the most prominent LXR target genes. Further important LXR targets are ABCG1 and ABCA1 that promote cholesterol efflux from macrophages and the intracellular trafficking protein ARL4C (ADP-ribosylation factor-like 4C) that facilitates cholesterol delivery to the cellular membrane. LXR also downregulates the gene encoding for the low-density lipoprotein (LDL) receptor (*LDLR*) and upregulates the *IDOL* (inducible degrader of LDLR) gene. Thus, **activation of LXR attenuates LDL uptake by macrophages counteracting the pathogenesis of atherosclerosis**.

In the liver, LXR induces the expression of a cluster of apolipoprotein genes including *APOE*, *APOC1*, *APOC2* and *APOC4*, which are involved in lipid transport and catabolism. LXR also upregulates genes encoding for lipid remodeling proteins, such as *PLTP* (phospholipid transfer protein), *CETP* (cholesterol ester transfer protein) and *LPL* (lipoprotein lipase) (Fig. 16.2). Furthermore, LXR induces the expression of several genes that mediate the elongation and the desaturation of fatty acids, which leads to synthesis of long-chain PUFAs, such as ω-3 fatty acids. This increase in long-chain PUFAs results in decreased expression of NFκB target genes via epigenetic silencing of their regulatory regions. Long-chain PUFAs also function as substrates for the enzymes that synthesize eicosanoids and other specialized inflammation-resolving lipid mediators, such as resolvins and protectins. Furthermore, LXR induces the gene for the enzyme *LPCAT3* (lysophospholipid acyltransferase 3) mediating the synthesis of phospholipids containing long-chain PUFAs. Thus, **LXR activation decreases ER stress and inflammatory responses**.

A major function of LXR in the liver is the promotion of de novo biosynthesis of fatty acids through the stimulation of the transcription factor SREBF1 and the enzymes FASN (fatty acid synthase), ACC (acetyl-CoA carboxylase) and SCD1 (steroyl-CoA desaturase 1). Some of these fatty acids are esterified with cholesterol, in order to avoid toxic levels of free cholesterol. In the intestine, LXR induces the expression of genes encoding the transporters ABCG5 and ABCG8 that mediate the apical efflux of cholesterol from enterocytes. In adipose tissue, LXR also affects glucose metabolism via the stimulation of GLUT4 expression. In this tissue, LXR regulates the expression of lipid-binding and metabolic proteins, such as APOD and THRSP (thyroid hormone responsive) and increases fatty acid β-oxidation.

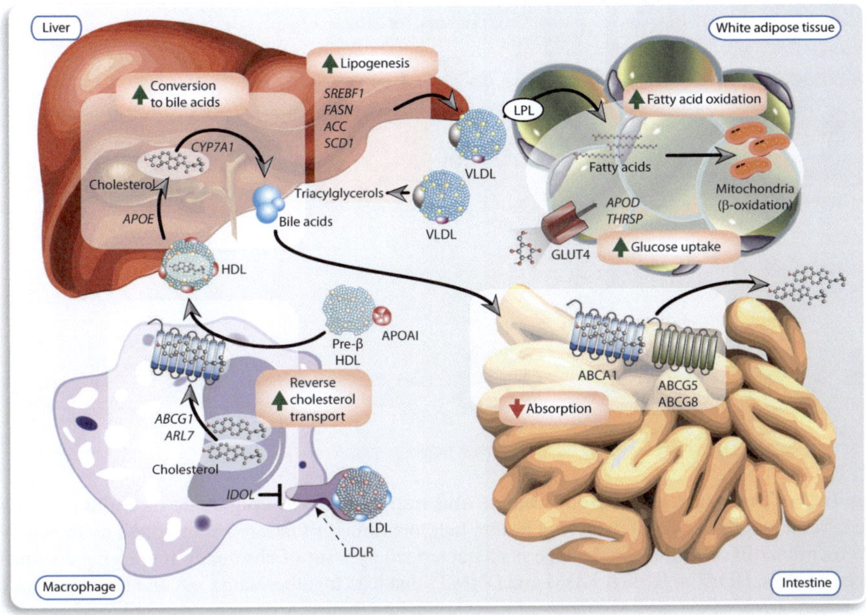

Fig. 16.2 Effects of LXR on metabolism. LXR has effects on multiple metabolic pathways. In macrophages, LXR induces the expression of the genes *IDOL*, *ARL7* and *ABCG1*. In the liver, LXR promotes fatty acid synthesis via induction of the transcription factor SREBF1 and its target genes *FASN*, *ACC* and *SCD1*. Triglyceride-rich very low-density lipoproteins (VLDLs) in the liver serve as transporters for lipids to peripheral tissues, including adipose tissue, where the action of LPL liberates fatty acids from VLDLs. In adipose tissue, LXR regulates the expression of *APOD* and *THRSP* and promotes fatty acid β-oxidation and glucose uptake via induction of *SLC2A4*. Finally, in the intestine, LXR inhibits cholesterol absorption by inducing the expression of the ABCG5-ABCG8 complex

FXR often acts in a complementary or reciprocal way to LXR in the control of lipid metabolism. Since high bile acid concentrations are toxic to cells, a central function for FXR is to control these levels. Bile acids are cholesterol derivatives that facilitate the efficient digestion and absorption of lipids and lipid-soluble vitamins after a meal, but they represent also the major way to eliminate cholesterol from the body. Nevertheless, most bile acids are recycled via enterohepatic circulation, i.e., they pass from the intestine back to the liver. FXR controls this bile acid flux via modulating their synthesis, modification, absorption and uptake. In the liver, FXR inhibits bile acid synthesis by repressing the expression of the genes *CYP7A1* and *CYP8B1*. FXR stimulates the secretion of FGF19 from the intestine, which then activates FGFR4 (FGF receptor 4) in the liver and in this way provides a complementary mechanism for the feedback inhibition of bile acid synthesis.

In the gall bladder, FXR upregulates the expression of the enzymes SLC27A7 (bile acid-CoA synthase) and BAAT (bile acid-CoA-amino acid N-acetyltransferase), which catalyze the conjugation of bile acid with the amino acids taurine or glycine

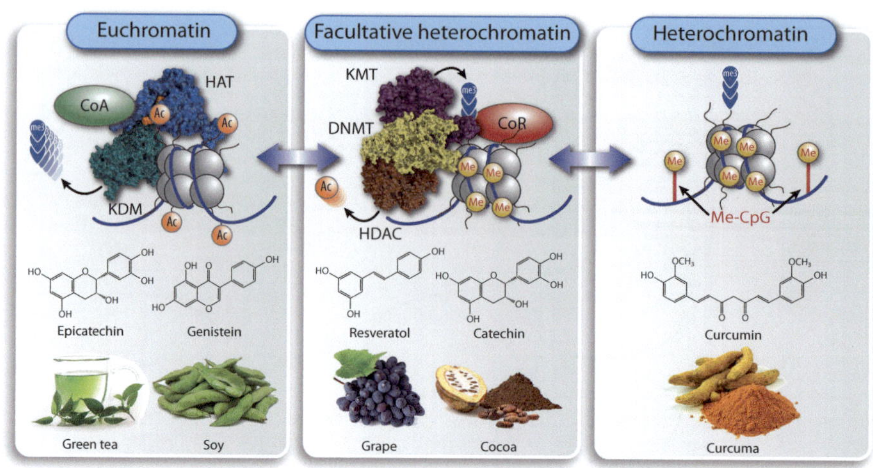

Fig. 16.3 Chromatin, associated proteins and nutrition-based epigenetic modulators. There are several stages between euchromatin and heterochromatin that are summarized as facultative heterochromatin (**middle**). Each stage is characterized by a set of chromatin modifying enzymes, such as HATs, HDACs, KMTs, KDM and DNMTs that lead together with CoA and CoR proteins to the schematically indicated scenarios of acetylation and methylation of histone tails and genomic DNA. The indicated plant-origin natural compounds have been shown to modulate the activity of chromatin modifying enzymes and to affect in this way the epigenetic status of cells and tissues

and that of the bile salt export pumps ABCB11 and ABCB4. In the intestine, FXR inhibits the absorption of bile salts via the downregulation of the apical sodium-dependent bile salt transporter SLC10A2 and promotes the movement of bile salts from the apical to the basolateral membrane of enterocytes via the upregulation of the *FABP6* (ileal fatty acid-binding protein) gene. FXR limits hepatic bile salt levels by the downregulation of the bile acid transporters solute carrier organic anion transporters (SLCOs) A1 and A2. Furthermore, in order to protect the liver from toxicity, FXR induces the expression of the enzymes CYP3A4 and CYP3A11, which hydroxylate bile acids, as well as SULT2A1 (sulfotransferase family 2A, member 1) and UGT2B4 (UDP glucuronosyltransferase 2 family, polypeptide B4), which sulphate and glucuronidate bile acids, respectively. In addition, in the liver FXR reduces lipogenesis via the repression of the *SREBF1* and *FASN* gene expression, i.e., the receptor decreases triglyceride levels. Finally, FXR also influences hepatic carbohydrate metabolism and inhibits gluconeogenesis via the downregulation of G6PC (glucose-6-phosphatase) and PCK (phosphoenolpyruvate carboxykinase) 2. Thus, **FXR is an important regulator of lipid metabolism**.

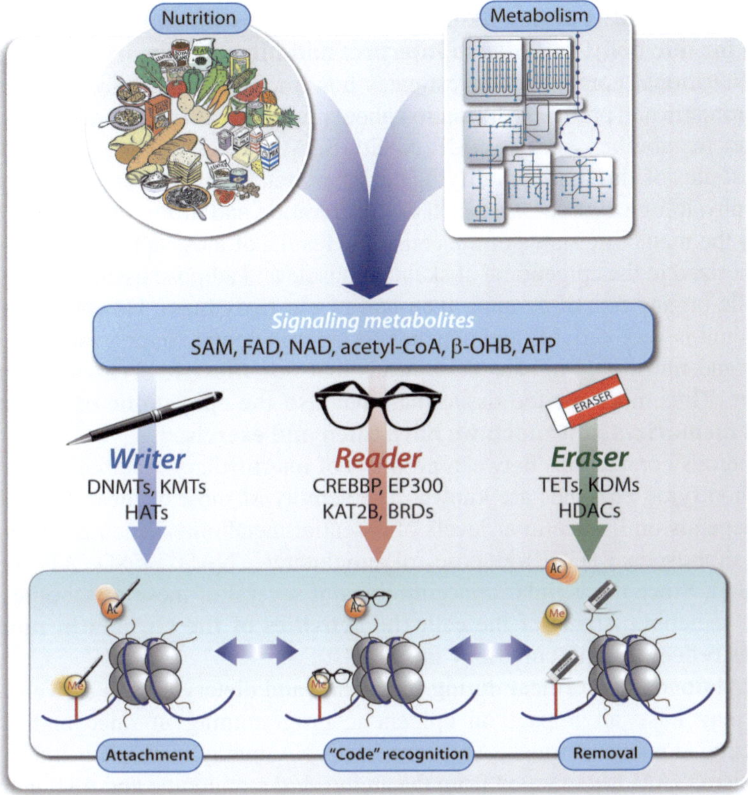

Fig. 16.4 Epigenetic mechanisms link metabolites and transcription. Changes in nutrition or fluctuations in metabolism affect the transcriptional responses of metabolic tissues. Several intermediary metabolites change the activity of chromatin modifying enzymes in a dose-dependent manner. These proteins use some of these metabolites as co-substrates and/or cofactors and act in this way as metabolic sensors. "Writer" enzymes create covalent chromatin marks, "reader" enzymes recognize these marks and "eraser" enzymes remove them. These histone tail modifications create changes in the local chromatin structure, which has consequences for the activity and regulation of the neighboring genes

16.2 Intermediary Metabolism and Epigenetic Signaling

Within a cell, signal transduction pathways result in the activation of gene expression programs that integrate signals originating from the environment. **A wide spectrum of secondary metabolites from fruits, vegetables, teas, spices and traditional medicinal herbs, such as genistein, resveratrol, curcumin and polyphenols from green tea, coffee and cocoa, respectively, are able to modulate the activity of transcription factors and chromatin modifying enzymes** (Fig. 16.3). NGS technologies allow the genome- and transcriptome-wide assessment of the specificity and efficacy of these compounds, e.g., for preventing and/or treating cancer.

The energy status of our tissues and cell types is the most important information for our body, in order to interpret and integrate environmental conditions. Nutritional epigenetics investigates how metabolic pathways communicate with chromatin and provide information about nutrient availability and energy status. Since key metabolites, such as AMP, NAD^+, SAM and acetyl-CoA, act as cofactors and substrates of chromatin modifying enzymes, gene expression programs of many central physiological processes, such as proliferation and differentiation, are modulated by the metabolic status of the cells. The results of these epigenetic events may be memorized in the epigenome of skeletal muscle and adipose tissue. The latter two metabolic organs constitute more than half of our body mass. However, their relative amount is very variable and depends on environmental factors, such as physical activity and nutritional intake. This means that **our lifestyle creates a metabolic memory**. Thus, not only the tissue mass but also **the epigenome of our muscles and fat memorizes how much we have eaten and exercised**.

Numerous connections between products of intermediary metabolism and chromatin modifying enzymes are known. The activity of most of these enzymes critically depends on intracellular levels of essential metabolites, such as acetyl-CoA, uridine diphosphate (UDP)-glucose, α-ketoglutarate, NAD^+, FAD, ATP or SAM (Fig. 16.4). Since the cellular concentrations of several of these metabolites represent the metabolic status of the cell, **the activities of the chromatin modifying enzymes reflect the intermediary metabolism**.

Methyl donors are critical during pregnancy and dietary excess as well as deficiency may have an impact on epigenetic programming in mice and humans. The methyl donor substrate SAM connects DNA methylation with intermediary metabolism. SAM is generated from the amino acid methionine and ATP in the one-carbon metabolism pathway (Box 16.1). When a methyl group of SAM is transferred to DNA or a histone, the product S-adenosylhomocysteine (SAH) is recycled back to SAM. Interestingly, SAH acts as a negative feedback regulator for KMTs, suggesting that **the SAM/SAH ratio, referred to as the "methylation index", is critical for histone and DNA methylation**. A derivative of the B vitamin folate, tetrahydrofolate, serves as a methyl group donor for feeding of the cyclic one-carbon pathway. The dependence of the pathway on folate and other micronutrients is another example of the direct connection between nutrition and epigenetics. The function of folate in normal neural tube closure in early gestation (in human 21–28 days after conception) is well known and maternal supplementation with folate is recommended for prevention of neural tube defects.

Box 16.1 Folate Metabolism and Methylation A derivative of the B vitamin folate, tetrahydrofolate, feeds the cyclic one-carbon pathway by serving as a methyl group donor. This demonstrates a direct connection between nutrition and epigenetics. **Methyl group donors are critical for epigenetic programing during embryogenesis.** A high homocysteine level is an established biomarker for the disturbance of the one-carbon metabolism and related to low concentrations of folate, vitamins B_6 and B_{12}, choline and betaine. This may cause an elevated risk of premature delivery, low birth weight and neural tube defects. Moreover, a low dietary intake of folate or methionine increases the risk of colon adenomas, while **in utero exposure to higher folate is associated with a reduced risk of childhood ALL, brain tumors and neuroblastoma.** The enzyme MTHFR methylenetetrahydrofolate reductase), catalyzes the conversion of 5,10-methylenetetrahydrofolate to 5-methyltetrahydrofolate. 10–15% of Europeans carry the missense SNP rs1801133 on both alleles of the *MTHFR* gene, which reduces the activity of the enzyme by more than 50%. Accordingly, individuals with a TT genotype are affected more by a low folate intake than those with the CC or CT allele. Accordingly, raised serum homocysteine concentrations cause an elevated risk of premature delivery, low birth weight and neural tube defects.

The ratio of the oxidized (NAD^+) and reduced (NADH) form of the cofactor NAD reflects the cellular redox state and is inversely proportional to the energy state of a cell. During fasting, i.e., at low levels of nutritional metabolites, the intracellular concentration of NAD^+ raises. This leads to an increase in the activity of HDACs of the SIRT family (which use NAD^+ as a cofactor) and the deacetylation of their target proteins (Fig. 16.5, left). The targets are often histones, but also transcription factors or their cofactors, such as p53 and PPARGC1A, are affected in their acetylation status. Since the NAD^+ concentrations fluctuate in a circadian manner, SIRT-mediated gene regulation is linked to the epigenetic clock (Sect. 12.4). Accordingly, time-restricted feeding can restore daily rhythms and improves number of metabolic responses, such as reducing insulin resistance and increasing glucose tolerance. Short-chain fatty acids, such as the ketone body β-hydroxybutyrate (β-OHB), are potent inhibitors of several HDACs and are produced from dietary fibers in the lumen of the colon. Fiber-rich diet can prevent colitis (i.e., a chronic inflammation of the colon) and colon cancer in human, which could be, at least in part, explained via butyrate-mediated inhibition of HDAC-dependent transcriptional programs of colonocyte proliferation.

In contrast, nutrients ingested in the feeding state enter the catabolic pathways of intermediary metabolism and acetyl-CoA is produced. Augmented acetyl-CoA concentrations stimulate HAT activity, so that their target proteins get acetylated (Fig. 16.5, right). **When the target proteins are histones, the acetylation of chromatin leads to open chromatin.** This stimulates the expression of genes involved in metabolic processes, such as lipogenesis and adipocyte differentiation. Moreover, a

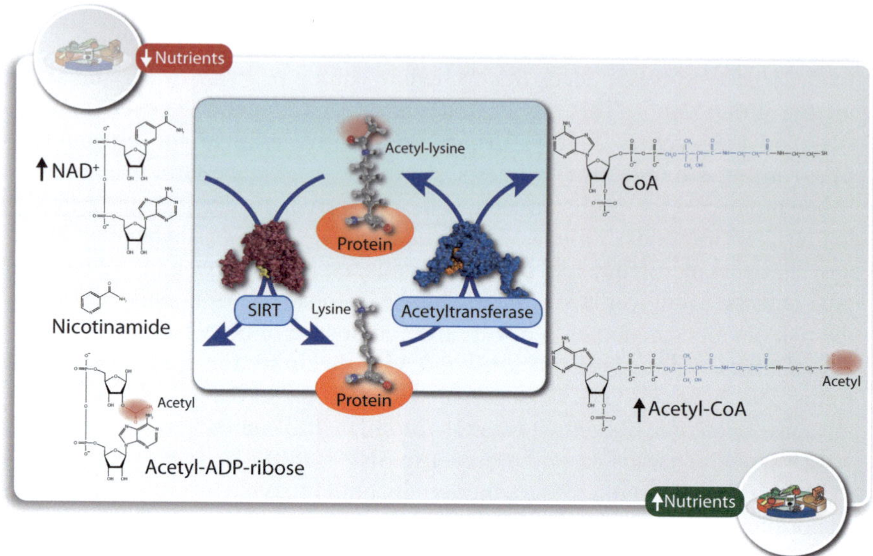

Fig. 16.5 The relation of protein acetylation and cellular metabolism. NAD$^+$ (**left**) acts as a cofactor for HDACs of the SIRT family that deacetylate proteins, which had been acetylated by HATs using acetyl-CoA (**right**). Thus, the acetylation status of key regulatory proteins reflects the cellular concentration of NAD$^+$ and acetyl-CoA, i.e., of low (**left**) or high (**right**) nutritional status, respectively

large proportion of acetyl-CoA-responsive genes are involved in cell cycle progression, i.e., an increase in histone acetylation is associated with cellular proliferation. However, upon induction of cellular differentiation the acetyl-CoA level decreases significantly. Accordingly, loss of pluripotency is associated with decreased glycolysis and lower levels of acetyl-CoA and histone deacetylation. Moreover, the acetyl-CoA level also affects cell survival and death decisions. For example, a low acetyl-CoA level induces the catabolic process of autophagy, which is crucial for organelle quality control and cell survival during metabolic stress. Thus, **the acetyl-CoA/CoA ratio is an important regulator of major cellular decisions**.

Another example of metabolite sensing is that of the enzyme AMPK, which is controlled in its activity by the AMP/ATP ratio. When cells consume more ATP than they are producing, i.e., at conditions of low nutrient availability, AMP concentrations raise as a signal of energetic stress. AMP binds to the γ-subunit of the AMPK heterotrimer and activates the kinase. Since histones are AMPK substrates, a low energy status of the cell is marked via histone phosphorylation. Thus, **insults to the energy status of a cell are memorized on the level of histone modifications and can be translated into functional outputs via adaptive gene regulation**. In contrast, a high nutritional level results in low AMP levels, no AMPK activity, a modified histone phosphorylation pattern and the activity of a different set of genes. Thus, the metabolic state of a cell can be expressed by the ATP/AMP ratio, the

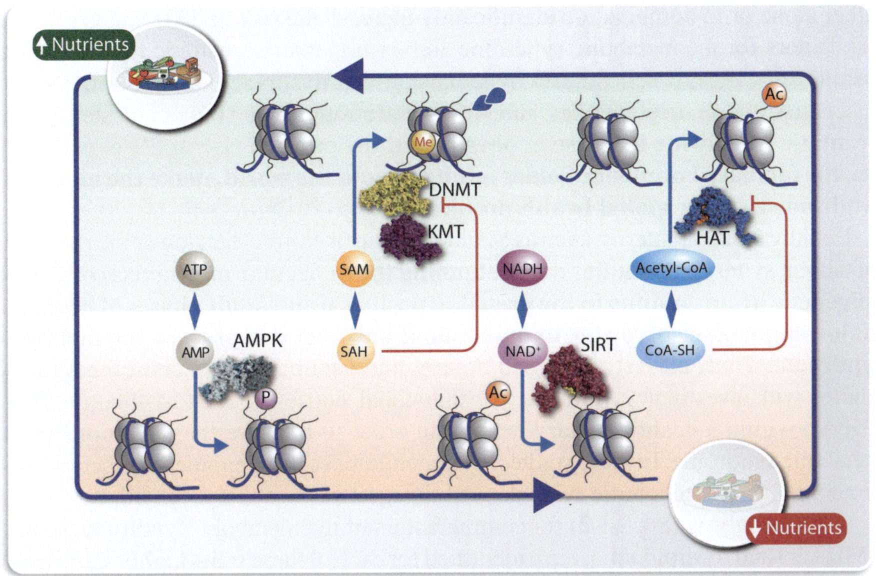

Fig. 16.6 Epigenetic sensing of the nutritional state. A high nutritional state of a cell (**top**) is represented by the abundance of the metabolites ATP, SAM, NADH and acetyl-CoA, while in the case of low nutrient levels (**bottom**) the metabolites AMP, SAH, NAD$^+$ and CoA are predominant. Accordingly, at high nutrient concentrations, KMTs and HATs are stimulated, while at low concentrations AMPK and HDACs of the SIRT family are activated and DNMTs and HATs are repressed. This results in histone methylation and acetylation or histone phosphorylation and deacetylation, respectively

SAM/SAH ratio, the NADH/NAD$^+$ ratio and the acetyl-CoA/CoA ratio (Fig. 16.6). Under high nutrient concentrations, such as abundant availability of methionine and glucose, SAM activates KMTs and acetyl-CoA stimulates HATs, thus leading to histone methylation and acetylation, respectively. In contrast, at low nutrient levels, such as during fasting, AMP activates AMPK and NAD$^+$ stimulates SIRTs resulting in histone phosphorylation and deacetylation. Moreover, in parallel SAH inhibits DNMTs and CoA blocks HATs.

16.3 Epigenetics of the Metabolic Syndrome

The metabolic syndrome is nowadays a very common age-related condition that occurs primarily as a result of overweight and obesity caused by a sedentary lifestyle, i.e., physical inactivity and the consumption of diet containing excess of calories. Organizations like the WHO and others use slightly different thresholds to define the metabolic syndrome based on rate of obesity, hyperglycemia, dyslipidemias and hypertension (Table 16.1). The syndrome is composed of different factors that either alone or in combination significantly increase the risk of T2D and CVD. The risk factors for the metabolic syndrome are visceral obesity, ectopic lipid overload, insulin resistance, β cell failure, hypertension and dyslipidemias with high serum concentrations of triglycerides and low concentrations of HDL cholesterol. **The dramatic worldwide increase in obesity and the parallel rise in life expectancy, i.e., the increased number of older adults around the world, make the metabolic syndrome a major global health problem.**

Healthy dietary patterns, such as Mediterranean or Nordic diet, lower the risk of the metabolic syndrome. **Studies understanding the molecular mechanism of diet on epigenetic programming in the prenatal, postnatal and adult phases of life are of major importance, in order to understand how diet can prevent the metabolic syndrome.** Thus, cleverly designed dietary intervention studies and observational studies will investigate the impact of individual nutrients, such as vitamin D_3 or PUFAs, within a healthy dietary pattern, in order to improve the conditions of the metabolic syndrome. In these studies, a larger number of epigenome-, transcriptome-, proteome- and metabolome-wide data are integrated.

GWAS analysis (Sect. 1.2) for central factors of the metabolic syndrome, such as BMI, T2D and dyslipidemia, have identified for each of these traits highly statistically significant associations with 40–100 genetic variants. The key genes of these lists, such as *LPL*, *APOE*, *MC4R* (melanocortin 4 receptor), *FTO* (fat mass and obesity

Table 16.1 Definitions of the metabolic syndrome

	NCEP (2005)	WHO (1998)	EGIR (1999)	IDF (2005)
Absolutely required	None	Insulin resistance	Hyperinsulinemia	Central obesity
Criteria	Any of the 5 below	Insulin resistance or T2D plus 2 of 5 below	Hyperinsulinemia plus 2 of the 4 below	Obesity plus 2 of the 4 below
Obesity	Waist circumference: Males > 101.6 cm Females > 88.9 cm	Waist/hip ratio: Males > 0.90 Females > 0.85 or BMI > 30 kg/cm^2	Waist circumference: Males > 94 cm Females > 80 cm	Central obesity
Hyperglycemia	Fasting glucose > 5.6 mM	Insulin resistance	Insulin resistance	Fasting glucose > 5.6 mM
Dyslipidemia I	Triacylglycerols > 1.7 mM	Triacylglycerols > 1.7 mM or HDL < 0.9 mM	Triacylglycerols > 2.0 mM or HDL < 1.0 mM	Triacylglycerols > 1.7 mM
Dyslipidemia II	HDL: Males < 1.0 mM Females < 1.25 mM			HDL: Males < 1.0 mM Females < 1.25 mM
Hypertension	> 130 mm Hg systolic or > 85 mm Hg diastolic	> 140/90 mm Hg	> 140/90 mm Hg	> 130 mm Hg systolic or > 85 mm Hg diastolic
Other criteria		Microalbuminuria		

associated) and *TCF7L2* (Table 16.2), are also the central determinants for the genetic risk for the metabolic syndrome. However, the dilemma with all these common SNPs remains that their individual ORs are clearly below 2, i.e., they contribute considerably less than 100% increased risk for the disease. This implies that the common SNPs can explain only a small fraction of the cases of the metabolic syndrome. This suggests that **socio-environmental factors and epigenetic mechanisms rather than variations of the core genome play a role in the obesity epidemic and its associated metabolic abnormalities**.

Integrative genomic approaches can combine large-scale genome- and transcriptome-wide data, in order to construct gene networks that underlie metabolic traits, such as serum lipoproteins levels. For example, serum HDL-cholesterol levels were linked to variants in the regulatory region of the vanin 1 (*VNN1*) gene using RNA expression profiles in lymphocytes. This type of analysis indicates that metabolic traits are the products of molecular networks being modulated by sets of complex genetic loci and environmental factors. This re-emphasizes that **the genetic predisposition for a metabolic disease, such as atherosclerosis, is comprised of multiple common genetic variants that each have a small to moderate effect on the trait, either alone or in combination with modifier genes or environmental factors**.

There are more and more epidemiological and clinical evidences that the DOHaD concept (Sect. 12.1), i.e., a prenatal epigenetic programming in utero, may be the key cause for the metabolic syndrome. So far, there is no comprehensive analysis of the epigenome of persons suffering from the metabolic syndrome, i.e., no concrete genomic regions of elevated risks have been identified. However, it can be assumed that due to the complexity of insulin signaling and its interference with multiple pathways a high number of regions will be affected in an individual-specific way. Nevertheless, since epigenetic modifications respond dynamically to environmental conditions, there is potential for intervention and reversibility.

In general, common SNPs are characterized by low ORs, while rare monogenic forms of T2D have high ORs. However, both extremes do not explain all genetic basis of T2D. Like for many other common diseases and traits, also for T2D risk, there is a large number of low frequency SNPs with intermediate ORs. Some of these genetic variants are expected to be identified by the use of whole genome sequencing (Sect. 2.5), but will not be able to explain all heritability of T2D. Thus, **prenatal and postnatal epigenetic programming will demonstrate its contribution to the disease risk**.

Table 16.2 Central genes in the development of metabolic syndrome

Gene locus	Gene function	Disease context	Affected parameter
LPL	Hydrolyzes triacylglycerols	Cardiovascular	HDL concentration
APOE	Removing lipoproteins from circulation	Cardiovascular	HDL concentration
MC4R	Membrane receptor on neurons binding α-MSH	Obesity	Waist circumference
FTO	Function mediated by IRX3 and IRX5	Obesity	Waist circumference
TCF7L2	Transcription factor in β cells	T2D	Glucose concentration

The experience from the already discussed *Dutch Hunger Winter* (Sect. 12.1) provides a molecular explanation for an increased obesity and T2D risk. Similar observations for a rise in T2D prevalence have been made for survivors of the *Ukraine Famine* (1932–33) and the *Chinese Famine* (1959–61). The in utero environment has a strong impact on fetal epigenetic programming, as individuals exposed to either famine or maternal gestational diabetes during fetal development develop more likely obesity and/or T2D later in life. **Food-deprived conditions of the mother change the epigenome of the fetus, so that genes involved in energy homeostasis are more sensitive to food intake**. The DOHaD concept (Sect. 12.1) indicates that in times of continued famine, this epigenetic sensitizing is a survival advantage, while in times of plenty food it may drive the individual into obesity and T2D (Fig. 12.2). DNA methylation is particularly sensitive to events during development in utero, because genomic DNA gets almost fully demethylated in the days after zygote formation and specific methylation is reestablished throughout embryogenesis. For example, patterns of DNA methylation at candidate genes, such as the *LEP* gene, associate with in utero exposure to famine and with maternal impaired glucose tolerance during pregnancy. Thus, epigenetic markers will help to detect children or families, for whom **intensive lifestyle interventions are likely to prevent the onset of metabolic disease**.

> **(Clinical) conclusion**: The molecular sensing of nutrition via ligands of the nuclear receptors PPAR, LXR and FXR or through substrates of energy metabolism on the level of chromatin modifying enzymes are the key mechanisms of gene regulation and epigenetics of dietary compounds. Up to 90% of elderly people develop the metabolic syndrome although this collection of diseases, such as obesity, hypertension, T2D and dyslipidemias is largely preventable. Moreover, there is also the chance to positively modulate nutritional signaling, e.g., by caloric restriction. The metabolic syndrome is a whole body's disease and affects many organs including the immune system. In most people the metabolic syndrome leads to fatal CVD, i.e., the metabolic syndrome is the major reason for death.

Additional Reading

Afshin, A., Sur, P.J., Fay, K.A., Cornaby, L., Ferrara, G., Salama, J.S., Mullany, E.C., Abate, K.H., Abbafati, C., Abebe, Z., et al. (2019). Health effects of dietary risks in 195 countries, 1990–2017: a systematic analysis for the Global Burden of Disease Study 2017. Lancet *393*, 1958–1972.

Campisi, J., Kapahi, P., Lithgow, G.J., Melov, S., Newman, J.C. and Verdin, E. (2019). From discoveries in ageing research to therapeutics for healthy ageing. Nature *571*, 183–192.

Carlberg, C., Ulven, S.M. and Molnár, F. (2020). Nutrigenomics: how science works. Springer textbook.

Di Francesco, A., Di Germanio, C., Bernier, M. and de Cabo, R. (2018). A time to fast. Science *362*, 770–775.

Haeusler, R.A., McGraw, T.E. and Accili, D. (2018). Biochemical and cellular properties of insulin receptor signalling. Nat Rev Mol Cell Biol *19*, 31–44.

Kupkova, K., Shetty, S.J., Haque, R., Petri, W.A., Jr. and Auble, D.T. (2021). Histone H3 lysine 27 acetylation profile undergoes two global shifts in undernourished children and suggests altered one-carbon metabolism. Clin Epigenetics *13*, 182.

Ling, C. and Ronn, T. (2019). Epigenetics in human obesity and type 2 diabetes. Cell Metab *29*, 1028–1044.

Zimmet, P., Shi, Z., El-Osta, A. and Ji, L. (2018). Epidemic T2DM, early development and epigenetics: implications of the Chinese Famine. Nat Rev Endocrinol *14*, 738–746.

Glossary

Acute inflammation is a short-term immunological process occurring in response to tissue injury or microbe infection usually appearing within minutes or hours. It is characterized by pain, redness, swelling and heat.

Anatomically modern humans individuals classified as *Homo sapiens* on the basis of a set of morphological characteristics that distinguish them from other, now extinct, archaic humans. According to the fossil record they emerged approximately 300,000 years ago.

Assay for transposase accessible chromatin using sequencing (ATAC-seq) is a method similar to DNase I hypersensitivity and FAIRE-seq mapping, which is used to identify active regulatory sites characterized by lower density of nucleosomes. ATAC-seq uses the Tn5 transposase, which can insert sequencing adaptor sequences only into regions free of nucleosomes.

Autistic spectrum disorders is a group of neurodevelopmental diseases characterized by deficits in social and communicative interaction and stereotypic behaviors.

Basal transcriptional machinery (also called preinitiation complex) is composed of a large number of GTFs (many of which are summarized as the TFIID complex) located at the TSS region and using Pol II as its core. The basal transcriptional machinery is connected with activating and repressing cell- and site-specific transcription factors binding to enhancer regions via another multiprotein complex of coactivators termed the MED complex.

Beckwith-Wiedemann syndrome is a predominantly maternally transmitted epigenetic disorder, which involves fetal and postnatal overgrowth and a predisposition to embryonic tumors. The disease locus includes several imprinted genes, including *IGF2*, *H19* and *KCNQ1* and loss of imprinting at *IGF2* is seen in ~ 20% of cases.

C. Carlberg, *Gene Regulation and Epigenetics*,
https://doi.org/10.1007/978-3-031-68730-3

Bisulfite sequencing is a method to study 5mC DNA methylation. Native DNA is exposed to sodium bisulfite, as a result of which non-methylated cytosines undergo deamination and are converted to uracils (which are read as thymines), whereas methylated cytosines remain unconverted. Sequencing libraries are generated from the converted template and they allow the study of methylation at single-base resolution.

Bivalent chromatin is composed of chromatin regions that harbor active and repressive histone modifications. Bivalent chromatin domains mark genes that are expressed at low levels only but are poised for activation upon an intra- or extracellular signal.

Blastocysts are early stage embryos that have undergone the first cell lineage specification, which results in two primary cell types: cells of the inner cell mass and trophoblasts.

Broad promoter is, in contrast to a sharp promoter, typical of ubiquitously expressed genes, has a more dispersed pattern of transcriptional initiation and does not contain a TATA box.

Bromodomain is a protein module of ~ 110 amino acids that mediates interaction with acetylated lysines and is often found in HATs and ATP-dependent chromatin remodeling proteins.

Cancer is a group of diseases involving abnormal cell growth with the potential to invade or spread to other parts of the body.

Cardiovascular disease (CVD) is a class of diseases that involve the heart or blood vessels, such as coronary heart disease, stroke, heart failure and more.

Cellular reprograming is the conversion of a differentiated cell to an embryonic state.

CG dinucleotides (CpGs) are the only dinucleotides that can be methylated symmetrically, i.e., DNA methylation can be inherited only via CpGs to both daughter cells. The "p" indicates the phosphate linking the two nucleosides.

Chromatin conformation capture (3C) is a method for studying chromosomal 3D structure by proximity ligation. The assay relies on cross-linking chromatin with a fixing agent (usually formaldehyde), digestion of the DNA with a restriction enzyme and ligation of the fixed chromatin. In the resulting chimeric DNA template, regions that were close spatially are now closed linearly.

Chromatin immunoprecipitation followed by sequencing (ChIP-seq) is a method for genome-wide mapping of the distribution of histone modifications and chromatin associated proteins, such as transcription factors, that relies on immunoprecipitation with antibodies to modified histones or other chromatin proteins. The enriched DNA is sequenced to create genome-wide profiles.

Chromatin modifying enzyme is an enzyme either recognizing (reading) chromatin (i.e., posttranslationally modified histones or methylated genomic DNA), adding (writing) marks or removing (erasing) them.

Chromatin is the molecular substance of chromosomes being a complex of genomic DNA and histone proteins.

Chromodomain is a modular methyl-binding domain of 40–50 amino acids that is commonly found in chromatin remodeling proteins.

Chromosome territories are nuclear volumes that are occupied by each specific chromosome within the interphase nucleus.

Chronic inflammation represents long-term inflammation lasting for prolonged periods of several months to years. Chronic inflammation plays a central role in most common non-communicable diseases, such as cancer, T2D, CVD and Alzheimer's.

Circadian clock is a biochemical oscillator that cycles with a stable phase and is synchronized with solar time.

Cis **regulatory elements** are stretches of genomic DNA, *e.g.*, TFBSs in enhancers, that regulate a target gene by a mechanism that depends on their residing on the same TAD, i.e., mostly not more than 1 MB in distance.

Cistrome is the set of *cis* regulatory elements of a *trans*-acting factor on a genome-wide scale, i.e., in most cases the complete set of experimentally verified binding sites, *e.g.*, by ChIP-seq, of a transcription factor.

Coactivator is a nuclear protein that binds to an activator (mostly a transcription factor), in order to increase the rate of transcription of a gene. Most coactivators cannot bind DNA but some chromatin modifying enzymatic activity.

Corepressor is a nuclear protein that like a coactivator binds to a transcription factor, but results in its repression, so that the rate of gene expression decreases. There are different mechanisms of repression and often corepressors and coactivators compete for the same binding sites. Some corepressors have chromatin modifying enzymatic activity.

Comparative genomics a subdiscipline of genomics, in which DNA sequence, genes, gene order, regulatory sequences and other genomic structural landmarks of different organisms are compared, in order to understand basic biological similarities and differences as well as evolutionary relationships between species.

Constitutive heterochromatin is a subtype of heterochromatin that is present at the highly repetitive DNA sequences found at the centromeres and telomeres of chromosomes, where it hinders transposable elements from becoming activated and thereby ensures genome stability and integrity.

CpG island is a genomic region of at least 200 base pairs showing a CG percentage of higher than 55%. However, typically CpG islands are 300–3000 base pairs long.

CTCF is a transcription factor with an 11 zinc finger DBD that is involved in many cellular processes, such as transcriptional regulation, insulator activity and regulation of chromatin architecture.

Damage-associated molecular patterns (DAMPs) are also known as alarmins, are molecules often released by stressed cells undergoing necrosis that act as endogenous danger signals to promote and exacerbate inflammatory responses. Examples of non-protein DAMPs include cholesterol crystals and saturated fatty acids. DAMPs are associated with many inflammatory diseases, including arthritis, atherosclerosis, Crohn's disease and cancer.

Developmental Origins of Health and Disease (DOHaD) is an approach to investigate the role of prenatal and perinatal exposure to environmental factors, such as undernutrition, in determining the development of human diseases in adulthood.

Diploid is an organism or cell with a double set of chromosomes, so that each position is represented by two genes or alleles.

DNA methylation is the covalent addition of a methyl group to the C5 position of cytosine.

DNA methyltransferases (DNMTs) are a family of enzymes catalyzing the transfer of a methyl group to cytosines of genomic DNA.

Driver gene mutation is a mutation within an oncogene or tumor suppressor gene that directly or indirectly confers a selective growth advantage to the cell in which it occurs.

Ectoderm is the outermost layer of the three embryonic germ layers that gives rise to the epidermis, like skin, hair and eyes and the nervous system.

Embryogenesis is also called embryonic development, i.e., the process by which the embryo forms and develops. In mammals, the term is use exclusively to the early stages of prenatal development, whereas the terms fetus and fetal development describe later stages.

Embryonic stem (ES) cell is a pluripotent stem cell that is derived from the inner cell mass of the early embryo. Pluripotent cells are capable of generating virtually all cell types of the organism.

Endoderm is the innermost layer of the three embryonic germ layers that gives rise to the epithelia of the digestive and respiratory systems, such as liver, pancreas and lungs.

Enhancer RNAs (eRNAs) are a class of short (<2 kb) unstable ncRNAs that are usually not spliced or polyadenylated. They are transcribed from enhancers and are rapidly degraded by the exosome.

Enhancer is a stretch of genomic sequence that (like a promoter) contains clusters of TFBSs that regulate a gene within the same TAD.

Epidrugs are small molecule inhibitors that target chromatin modifying enzymes, such as DNMTs, HATs, HDACs, KMTs or KDMs.

Epigenetic clock is a term used to measure age on DNA methylation levels.

Epigenetic drift is a divergence of the epigenome as a function of age due to stochastic changes in DNA methylation or stable histone modifications.

Epigenetic epidemiology is the study of the relationship between epigenetic variants and disease phenotype in the population.

Epigenetic landscape is a metaphor of cellular development, in which a landscape valleys and ridges illustrate how a pluripotent cell is guided to a well-defined differentiated state, represented by a ball rolling down the landscape.

Epigenetic mediators are genes whose products are the targets of the epigenetic modifiers.

Epigenetic memory is a heritable change in gene expression that is induced by a previous developmental or environmental stimulus. It requires chromatin-based changes, such as DNA methylation, histone modifications or incorporation of variant histones.

Epigenetic modifier is mostly identical to a chromatin modifying enzyme or the genes encoding for them.

Epigenetic modulators are genes upstream of the epigenetic modifiers and mediators. The products of these genes mediate injury, inflammation and other forms of cellular stress.

Epigenetic programing is the process leading to stable and long-lasting alterations of the epigenome based on specific covalent modifications of the DNA and histones.

Epigenetics the study of heritable changes in gene function that do not involve changes in the DNA sequence. Epigenetic mechanisms include the covalent modifications of DNA and histones.

Epigenome is the complete set of epigenetic modifications across an individual's genome.

Epigenomics are studies of the epigenome.

Epimutation is a heritable change in the chromatin state at a given position or region. In the context of cytosine methylation, epimutations are defined as changes in the methylation status of a single cytosine or of a region or cluster of cytosines. Epimutations do not necessarily imply changes in gene expression.

Erasers are chromatin modifying enzymes that remove histone modifications from chromatin, such as HDACs or KDMs.

Euchromatin is light-staining, decondensed and transcriptionally accessible regions of the genome.

Evolution is represented by changes in the heritable characteristics of biological populations over successive generations.

Evolutionary pressure is any cause that reduces reproductive success in a portion of a population driving natural selection.

Exome is the collection of exons in the human genome. Exome sequencing generally refers to the collection of exons that encode proteins.

Facultative heterochromatin is a dynamic form of heterochromatin that can change its density and activity in response to intra and extracellular signals.

Gene body is the DNA sequence of a gene from the TSS to the end of the mRNA transcript.

Gene expression is the process by which information from a gene is used in the synthesis of a functional gene product. These products are often proteins, but can also be ncRNAs.

Gene regulatory networks represent units of interacting proteins that are functionally constrained by defined regulatory relationships. These interactions provide a structure and determine an output in the form of a pattern of gene expression. The networks are usually visualized by nodes (proteins) and edges (their interactions).

General transcription factors (GTFs) are proteins forming together with Pol II the basal transcriptional machinery (preinitiation complex).

Genetic architecture is the landscape of genetic contributions to a given phenotype. It comprises the number of genetic variants that influence a phenotype, the size of their effects on the phenotype, the frequency of those variants in the population and their interactions with each other and the environment.

Genetic drifts are changes in genetic variation over time that are due to random (by chance) processes, i.e., different from natural selection in evolution.

Genome-wide association study (GWAS) is a study that aims to identify genetic loci (mostly SNPs) associated with an observable trait, disease or condition.

Genome is the complete haploid DNA sequence of an organism comprising all coding genes and far larger non-coding regions. The genome of all 400 tissues and cell types of an individual is identical and constant over time (with the exception of cancer cells).

Genomic imprinting is an epigenetic phenomenon in which expression of a gene is restricted to a single allele based on parental origin.

Genotype is the complete heritable genetic identity.

Germline variants are variations in sequences observed in different individuals. Two randomly chosen individuals differ by ~ 20,000 genetic variations distributed throughout the exome and some 4 million SNPs in total.

Haploid is a single set of chromosomes.

Haploinsufficiency occurs when a single copy of the wildtype allele in heterozygous combination with a variant allele is insufficient to produce the wildtype phenotype.

Haplotype block is a genomic region with no evidence of a history of genetic recombination.

Hematopoietic stem cells (HSCs) are stem cells located in the bone marrow that can develop into all types of blood cells.

Heritability is the proportion of total variation between individuals within a population that is due to genetic factors.

Heterochromatin are dark-staining, condensed and gene-poor regions of the genome.

High-throughput chromosome capture (Hi-C) is a 3C-based method for genome-wide analysis of chromosome conformation. Hi-C involves massive parallel sequencing of chimeric 3C DNA templates and subsequent statistical analysis of the distribution of ligation junctions over two-dimensional contact matrices.

Histone acetyltransferases (HATs) are enzymes that acetylate lysine amino acids on histone proteins by transferring an acetyl group from acetyl-CoA to form ε-N-acetyl-lysine. When they show specificity for lysines they are also abbreviated KATs.

Histone code is an epigenetic code that is based on posttranslational modifications of histone proteins. The histone modifications serve to recruit other proteins by specific recognition of the modified histone via specialized protein domains. The code comprises more than 130 posttranslational modifications serving as an "alphabet" for the instructions, how the epigenome directs transcriptional regulation and stores information.

Histone deacetylases (HDACs) are enzymes that remove acetyl groups from an ε-N-acetyl lysine amino acid on a histone.

Histone modification is a covalent posttranslational modification to histone proteins, which includes methylation, phosphorylation, acetylation, ubiquitylation and sumoylation. They can impact gene expression by altering chromatin structure or recruiting histone modifiers.

Histone proteins are lysine- and arginine-rich, positively charged proteins forming octamer cores on which genomic DNA is wrapped. They are the key protein components of chromatin.

Homeostasis is the state of steady internal physical and chemical conditions maintained by living systems.

Homodimer is a complex formed by two identical proteins.

Housekeeping genes are constitutively expressed genes that are required for the maintenance of basic cellular function, such as glucose metabolism.

Hypertension is a long-term medical condition, in which the arterial blood pressure is persistently elevated.

Immune checkpoints are regulators of the immune system being crucial for self-tolerance. Immune checkpoint blocking molecules, such as CTLA4 and PDCD1, are targets for cancer immunotherapy.

Immunity refers historically to the protection of politicians from legal prosecution in the old Rome (*immunitas* in Latin). Applied to science, immunity means protection from disease, in particular from infectious disease.

Implantation is an early developmental stage at which the embryo adheres to the endometrium.

Inducible pluripotent stem (iPS) cells are pluripotent stem cells that can be generated directly from terminally differentiated adult cells.

Inner cell mass is a group of cells inside a mammalian blastocyst that gives rise to the embryo.

Insertions or deletions (indels) are mutations due to small insertion or deletion of one or a few nucleotides.

Insulator is a chromatin element that acts as a barrier against the influence of positive signals from enhancers or negative signals from silencers, i.e., binding sites for repressive transcription factors and heterochromatin. They are often bound by the protein CTCF.

Interphase is the resting phase between successive mitotic divisions of a cell. Most cells of the human body are terminally differentiated and in the interphase.

Kinase is a protein that catalyzes the addition of phosphate groups to other molecules, such as proteins or lipids. These proteins are essential to nearly all signal transduction pathways.

Lamin-associated domains (LADs) is a DNA loop of heterochromatin interacting with lamina at the nuclear membrane. They are a subgroup of TADs.

Lineage is a descent in a line from a common progenitor cell.

Linkage disequilibrium is an association between two alleles that are located so close to each other on the genome that they are inherited together more frequently than expected by chance.

Long-term potentiation is a persistent strengthening of synapses of neurons based on recent patterns of activity.

Lysine methyltransferases (KMTs) are chromatin modifying enzymes that catalyze the transfer of one, two or three methyl groups to lysine residues of histone proteins.

Macrophage is a leukocyte that engulfs and digests cellular debris, foreign substances, microbes and cancer cells in a process called phagocytosis.

Massive parallel sequencing is a high-throughput approach to DNA sequencing using the concept of massively parallel processing. It is presently the most used sequencing approach of various NGS methods.

Mediator (MED) is a multiprotein complex (some 30 components) that functions as a transcriptional coactivator in all eukaryotes. The main function of the MED complex is the transmission of signals from the transcription factors to the basal transcriptional machinery.

Mesoderm is the middle layer of the three embryonic germ layers that gives rise to the muscle, cartilage, bone, blood, connective tissue etc.

Metabolic syndrome is a cluster of conditions, such as increased blood pressure, high blood sugar, excess body fat around the waist and abnormal cholesterol or triglyceride levels, that occur together, increasing the risk of heart disease, stroke and T2D.

Metastable epiallele are alleles that are variably expressed in genetically identical individuals due to epigenetic modifications established during early development and are particularly vulnerable to environmental influences.

Missense mutation is a single-nucleotide substitution (*e.g.*, C to T) that results in an amino acid substitution (*e.g.*, histidine to arginine).

Missing heritability is the fact that genetic variations cannot account for all of the heritability of diseases, behaviors and other phenotypes.

Mitosis is cell division that results in two daughter cells each having the same number and kind of chromosomes as the parent nucleus.

Monocytes are leukocytes of the innate immune system that can differentiate into macrophages and myeloid lineage DCs.

Multigenic disease is a disorder that, in contrast to a monogenic disease, is caused by changes in multiple genes. Most common diseases, such as T2D, atherosclerosis and Alzheimer's, belong to this category.

Multipotent the ability of a cell to differentiate into multiple but a limited range of cell types. For example, cells of the embryonic germ layers and adult stem cells are multipotent.

Next-generation sequencing (NGS) also known as high-throughput or "deep" sequencing, summarizes a number of modern technologies, such as ChIP-seq and RNA-seq, allowing for sequencing of DNA and RNA much more quickly and cheaply than the previously used Sanger sequencing. It has revolutionized the study of genomics and molecular biology.

Non-coding RNA (ncRNA) is an RNA molecule that is not translated into a protein.

Non-communicable disease is a disease that is not transmissible directly from one person to another, such as autoimmune diseases, CDVs, most cancers, T2D and Alzheimer's disease.

Non-synonymous mutation is a mutation that alters the encoded amino acid sequence of a protein. These include missense, nonsense, splice site, translation start, translation stop and indel mutations.

Nonsense mutation is a single-nucleotide substitution (*e.g.*, C to T) that results in the production of a stop codon.

Nuclear receptor is a transcription factor that can be activated by a small lipophilic ligand in the size of cholesterol.

Nucleosome-depleted region is a genomic region depleted of nucleosomes being associated with active regulatory elements, such as promoters and enhancers.

Nucleosome is a basic unit of DNA packaging in eukaryotes, consisting of 147 base pairs of genomic DNA wound around a histone octamer.

Odds ratio (OR) is the mathematical expression of the relation between the presence or absence of a variant, *e.g.*, a SNP and the presence or absence of a trait, *e.g.*, a disease, in the population.

Pathogen-associated molecular patterns (PAMPs) are small molecular motifs derived from microbes, such a LPS. They are recognized by TLRs and other PRRs on the surface of cells of the innate immune system.

Pattern recognition receptors (PRRs) are evolutionarily conserved receptors that elicit inflammation and innate immunity upon recognition of conserved microbial products (PAMPs) or endogenous danger signals (DAMPs).

Phenotype is the set of observable characteristics of an individual resulting from the interaction of its genotype with the environment.

Pioneer factors are transcription factors that can directly bind to heterochromatin. They can have positive and negative effects on transcription and are important in recruiting other transcription factors and histone modification enzymes as well as controlling DNA methylation.

Plasticity is the reversibility of epigenetic marks on DNA and proteins.

Pleiotropic are genetic or epigenetic changes that affect multiple unrelated phenotypic traits.

Pluripotency is the ability of a cell to differentiate into all three germ layers and to give rise to all fetal or adult cell types. For example, cells of the inner cell mass of blastocysts are pluripotent.

Poised promoter or enhancer is an inactive status of a genomic region carrying histone markers, such as H3K4me1, that enable, after appropriate stimulation, the rapid reactivation of the region. Often "poised" and "paused" are used in the same meaning.

Polycomb repressive complexes (PRCs) are large protein complexes limiting access of chromatin to transcription factors and therefore limit gene expression.

Polygenic risk scores is a weighted sum of the number of risk alleles carried by an individual, where the risk alleles and their weights are defined by the loci and their measured effects as detected by GWAS.

Posttranslational modifications are covalent modifications by which most proteins reach their full functional profile. Due to posttranslational modifications the proteome is far more complex than the transcriptome and also varies a lot in response to extra and intracellular signals.

Primordial germ cells (PGCs) are the common origins of oocytes and spermatozoa, i.e., they represent the ancestors of the germline. They occur in primary ectoderm already in the second week of embryogenesis.

Promoter is a stretch of genomic DNA for productive transcription initiation encompassing at least one TSS.

Proteome is in analogy to the transcriptome, the complete set of all expressed proteins in a given tissue of cell type. The proteome depends on the transcriptome, but is not its 1:1 translation, i.e., transcriptome analyses provide only a very rough description of the resulting proteome.

Pseudogenes are non-functional segments of genomic DNA that resemble functional genes. They are often superfluous copies of functional genes, either directly by DNA duplication or indirectly by reverse transcription of an mRNA transcript.

Readers are nuclear proteins that recognize and bind chromatin through histone modification recognition domains.

RNA sequencing (RNA-seq) is a method using massive parallel sequencing to reveal the presence and quantity of RNA in a biological sample at a given moment.

Senescence (also called biological aging) is the gradual deterioration of functional characteristics. It can refer either to cellular senescence or to senescence of the whole organism.

Sharp promoter is a type of core promoter (also referred to a narrow promoter) being typical of genes with restricted tissue-specific expression. It is often associated with positioned motifs, such as the TATA box or Inr and has a single predominant TSS.

Signal transduction cascade is the process by which a chemical or physical signal is transmitted through a cell membrane as a series of molecular events, such as protein phosphorylation catalyzed by protein kinases. Mostly, signal transduction cascades end in the activation of a transcription factor or a chromatin modifying enzyme.

Silencer is a genomic region that causes reduced expression of their target gene(s).

Single nucleotide polymorphism (SNP) is a substitution of a single nucleotide at a specific position in the genome, which is present to some appreciable degree within a population (*e.g.*, more than 1%).

Sirtuins (SIRTs) is a family of seven NAD^+-dependent HDACs that are structurally and mechanistically distinct from Zn^{2+}-dependent HDACs. SIRTs influence a wide range of cellular processes such as aging, transcription, apoptosis, inflammation and stress resistance.

Somatic mutations are mutations that occur in any non-germ cell of the body after conception, such as those that initiate tumorigenesis.

Super-enhancer is a genomic region comprising multiple enhancers that is collectively bound by multiple transcription factor proteins driving transcription of genes involved in cell identity.

SWI/SNF complex is protein complex that uses the energy of ATP hydrolysis to mobilize nucleosomes and remodel chromatin.

TET family proteins are α-ketoglutarate-dependent dioxygenases that catalyze the oxidation of 5mC to 5hmC and further products. Genes encoding these enzymes are frequently mutated in human cancers.

Tn5 transposase is a member of the RNase superfamily of proteins, which includes retroviral integrases. Tn5 is utilized for fragmentation of the DNA in next generation sequencing methods, such as ATAC-seq and ChIPmentation.

Toll-like receptor (TLR) is a class of patter-recognition receptors that play a key role in the innate immune system.

Topologically associated domain (TAD) is a large genomic region promoting regulatory interactions by forming higher-order chromatin structures separated by boundary regions.

Totipotent is the ability of a cell to give rise to differentiated cells of all tissues, including embryonic and extraembryonic tissues, in an organism. For example, a zygote is totipotent.

Trained immunity is a memory system of innate immunity which is based on epigenetic programing.

Trait is a distinguishing quality or characteristic belonging to a person.

Transactivation is the increased rate of gene expression triggered by an activating transcription factor.

Transcription factor binding site (TFBS) is a short (4–12 base pairs) DNA sequence pattern that summarizes the DNA sequence binding preference of a transcription factor. These motifs are usually represented as sequence logos based on position weight matrices.

Transcription factories are discrete sites within the nucleus where transcription occurs. The factories contain Pol II, transcription factors, chromatin modifying enzymes and chromatin modelers. In recent literature they are also described as "condensates".

Transcription factors are proteins that sequence-specifically bind to genomic DNA. The human genome encodes approximately 1600 transcription factors, referred to as *trans*-acting factors, since they are not encoded by the same genomic regions, which they are controlling. Accordingly, the process of transcriptional regulation by transcription factors is often called *trans*-activation.

Transcription start site (TSS) is a nucleotide within a promoter that is the first to be transcribed by Pol II into a particular RNA.

Transcriptome is the complete set of all transcribed RNA molecules of a tissue or cell type. It significantly differs between tissues and depends on extra and intracellular signals.

Transgenerational epigenetic inheritance is the transmission of epigenetic information that is passed on to gametes without alteration of the DNA sequence. If only the F1 generation is concerned, the correct term is **intergenerational epigenetic inheritance**.

Translocation breakpoints are locations where two fragments of chromosome(s) are joined subsequent to chromosomal translocation.

Translocation is a specific type of rearrangement where regions from two non-homologous chromosomes are joined.

Transrepression is a process whereby one protein represses (i.e., inhibits) the activity of a second protein through a protein–protein interaction.

Trithorax group proteins are large protein complexes maintaining the stable and heritable expression of certain genes throughout development.

Trophoblast is the outer layer of the mammalian blastocyst that eventually develops to form part of the placenta.

Tumor suppressor gene is a gene that, when inactivated by mutation, increases the selective growth advantage of the cell in which it resides.

Type 2 diabetes (T2D) is a form of diabetes being characterized by high serum glucose levels, insulin resistance and relative lack of insulin.

Western diet is a dietary pattern characterized by high intakes of red meat, processed meat, prepackaged foods, butter, fried foods, high-fat dairy products, eggs, refined grains, potatoes, corn and high-sugar drinks.

Writers are chromatin modifying enzymes that add histone modifications to chromatin, such as DNMTs, HATs and KTMs.

X chromosome inactivation (XCI) is a process in which one of the two X chromosomes is randomly inactivated in female mammalian cells early in development.

Zygote is fertilized egg before cleavage occurs, i.e., the one-cell stage embryo.